INTERNATIONAL SERIES OF MONOGRAPHS IN

NATURAL PHILOSOPHY

GENERAL EDITOR: D. TER HAAR

VOLUME 19

FOUNDATIONS OF CLASSICAL AND QUANTUM STATISTICAL MECHANICS

OTHER TITLES IN THE SERIES
IN NATURAL PHILOSOPHY

Vol. 1. DAVYDOV—Quantum Mechanics
Vol. 2. FOKKER—Time and Space, Weight and Inertia
Vol. 3. KAPLAN—Interstellar Gas Dynamics
Vol. 4. ABRIKOSOV, GOR'KOV and DZYALOSHINSKII—Quantum Field Theoretical Methods in Statistical Physics
Vol. 5. OKUN'—Weak Interaction of Elementary Particles
Vol. 6. SHKLOVSKII—Physics of the Solar Corona
Vol. 7. AKHIEZER *et al.*—Collective Oscillations in a Plasma
Vol. 8. KIRZHNITS—Field Theoretical Methods in Many-body Systems
Vol. 9. KLIMONTOVICH—The Statistical Theory of Non-equilibrium Processes in a Plasma
Vol. 10. KURTH—Introduction to Stellar Statistics
Vol. 11. CHALMERS—Atmospheric Electricity (2nd edition)
Vol. 12. RENNER—Current Algebras and their Applications
Vol. 13. FAIN and KHANIN—Quantum Electronics
Vol. 1. Basic Theory
Vol. 14. FAIN and KHANIN—Quantum Electronics
Vol. 2. Maser Amplifiers and Oscillators
Vol. 15. MARCH—Liquid Metals
Vol. 16. HORI—Spectral Properties of Disordered Chains and Lattices
Vol. 17. SAINT JAMES, THOMAS and SARMA—Type II Superconductivitity
Vol. 18. MARGENAU and KESTNER—Theory of Intermolecular Forces
Vol. 20. TAKAHASHI—An Introduction to Field Quantization
Vol. 21. YVON—Correlations and Entropy in Classical Statistical Mechanics

FOUNDATIONS OF CLASSICAL AND QUANTUM STATISTICAL MECHANICS

BY

R. JANCEL

WITH A PREFACE BY

L. DE BROGLIE

TRANSLATED BY

W. E. JONES

EDITED BY

D. TER HAAR

THE QUEEN'S AWARD TO INDUSTRY 1966

PERGAMON PRESS

OXFORD · LONDON · EDINBURGH · NEW YORK
TORONTO · SYDNEY · PARIS · BRAUNSCHWEIG

Pergamon Press Ltd., Headington Hill Hall, Oxford
4 & 5 Fitzroy Square, London W.1

Pergamon Press (Scotland) Ltd., 2 & 3 Teviot Place, Edinburgh 1

Pergamon Press Inc., Maxwell House, Fairview Park, Elmsford, New York 10523

Pergamon of Canada Ltd., 207 Queen's Quay West, Toronto 1

Pergamon Press (Aust.) Pty. Ltd., 19a Boundary Street, Rushcutters Bay, N. S. W. 2011, Australia

Pergamon Press S.A.R.L., 24 rue des Écoles, Paris 5e

Vieweg & Sohn GmbH, Burgplatz 1, Braunschweig

First English Edition 1969

Library of Congress Catalog Card No. 68-30729

This is a translation of the original French book *Les Fondements de la Mécanique Statistique Classique et Quantique*, published by Gauthier-Villars, Paris, 1963
Printed in Germany

08 012823 8

Contents

Contents

Contents

Preface

M. RAYMOND JANCEL has written, during the last decade, numerous very exhaustive papers on the foundations of statistical mechanics, both in its classical and quantum form. He has followed closely all research which has been undertaken in recent times on these difficult subjects and the researches which he has pursued, in collaboration with M. Kahan on the theory of plasmas, have also led him to study the evolution of non-equilibrium systems and the range of validity of Boltzmann's equation.

M. Jancel was thus particularly well qualified for presenting to us a general study on the foundations of classical and quantum statistical mechanics. The difficulties of the subject are well known and the author has reviewed them in the very interesting introduction to his book. Although nobody is in doubt today of the validity of the remarkable interpretation of Thermodynamics with which Statistical Mechanics, following the efforts of Boltzmann and Gibbs, has recently provided us, it still remains extremely difficult to give a completely accurate justification for it. The method, which consists in beginning with Liouville's theorem, can only be developed in a satisfactory manner by introducing hypotheses of the ergodic type; however, despite the numerous efforts which have been made in this direction and the remarkable progress which it has been possible to make, these hypotheses could not be justified in a completely satisfactory manner, even in classical statistical mechanics. It had been hoped that in quantum statistical mechanics the situation would have been more favourable, because of the quantum uncertainties, but M. Jancel, who had already studied this question thoroughly in his doctoral thesis in 1957, shows us that it is nothing of the kind and that the difficulties are even greater here than in classical mechanics.

It is true, there exists another method of justifying the principles of statistical mechanics, which starts from the hypothesis of "molecular chaos" of Boltzmann and which relies on the more or less generalised H-theorems. But it likewise comes up against the difficulties, which M. Jancel has painstakingly studied and very shrewdly analysed, as much in classical mechanics as in quantum mechanics. The discussion becomes even more tricky when we consider non-equilibrium states, the conservation of "molecular chaos", the validity of Boltzmann's equation and the whole range of problems of this type which the author has dealt with in the last part of his book.

We shall not try to analyse further an account where each chapter would require an exhaustive study. M. Jancel's book, where all the problems are studied in a very detailed manner, reveals that the author has a very profound knowledge of the difficult questions which he explains clearly after having thought about them for a long time. He has looked at every aspect of these questions and his conclusions rely on very extensive information, as proved by the copious bibliography at the end of the book. It is certain that this book does its author the greatest credit and that it will constitute, henceforth, a source book which is indispensable for all researchers who are interested in the difficult task of examining in detail the theoretical foundations of statistical mechanics as they are at this moment in the present state of the problem.

LOUIS DE BROGLIE

Preface to the English edition

In recent years, research into statistical mechanics has continued to develop in directions which had already become evident at the moment of appearance of the French edition of this book. They have given rise to numerous and important works which have been based in particular on the study of the evolution of non-equilibrium systems and on the derivation of kinetic equations and generalised "Master Equations" starting from Liouville's equation. Whilst these researches have, up to now, not involved a widespread evolution of the problem related to the interpretation of Statistical Mechanics, nevertheless they have contributed to making more precise the conditions of irreversibility and they have enabled us to analyse, in certain cases, certain detailed mechanisms. Thus, recent works, in particular, have improved the theory of irreversible processes and they have enabled us to study the return to equilibrium (relaxation times, etc.), but they have not contributed any significant change in the ergodic theory. Another important contribution of recent years has been to emphasise the major role played by certain limiting processes, such as the limit of weak coupling ($\lambda \rightarrow 0$), the well-known limit as $T \rightarrow \infty$ and, especially, the limit of "large systems" ($N \rightarrow \infty$) which appears to be an essential element of irreversibility (infinite duration of Poincaré cycles). In this connection we note that another concept has been evolved which is capable of opening new vistas of research: it concerns the non-commutation of the limiting processes $N \rightarrow \infty$ and $T \rightarrow \infty$, the first limit being taken *before* any time-average is evaluated whenever one wishes to obtain a macroscopic or thermodynamic description of the observed system.

This is why we have believed it necessary to proceed with publication of this book, by trying to take into account the principal views of research actually used in these domains. As it was unfortunately not possible for us to take into account all the works relating to these questions, we have been obliged to choose from amongst the various theories those which appeared to us at the time to be the most representative of a certain flow of ideas and the richest in new ideas about the problem of the foundations of Statistical Mechanics.

First of all it appeared essential to us to give a detailed account of perturbation methods based on the development of the resolvant operator associated with Liouville's equation. We have presented this study at the end of Chapter V by choosing the framework of the theory of Prigogine and his co-workers. These authors, by the systematic use of the Fourier coefficients of the distribution functions, write down the equations for a "correlation dynamics" which makes possible a deep insight into the mechanics of irreversibility.

In parallel with these techniques, it seemed to be equally useful for us to explain the major steps of Zwanzig's method which has the great advantage of avoiding, at least in its early developments, the complex formalism of perturbation theory. By relying on the necessarily incomplete nature of any macroscopic description, this method introduces suitable projection operators which have the effect of retaining only part of the information contained in the density-matrix (or in the distribution function for N particles). Thus, we end in a very simple way with so-called "pre-Master" equations, which are still equivalent to Liouville's equation and therefore reversible, but which constitute an important intermediate stage for the derivation of the truly irreversible kinetic equations. Although we have explained this theory in its quantum aspects at the end of Chapter VI, it is important to note that it can be applied equally well to the classical case provided Liouville's quantum operator is replaced by the corresponding classical operator.

Finally, it seemed to us to be equally necessary to come back to certain aspects of the ergodic theory and particularly to the important role played by the primary integrals other than that of the energy. These problems, which we study at the end of Chapter I, have led us to explain Lewis's theorems and to discuss the difficulties encountered by the "probability" ergodic theorems because of the possible existence of these supplementary integrals. At the end of Chapter III, we have also given a brief account of Ludwig's ideas concerning macroscopic observation and of an ergodic theorem of Golden and Longuet-Higgins, which stresses once more the importance of the limit process as $N \to \infty$.

In Appendix IV we have given a brief account of some recent Russian work in classical ergodic theory.

I would like to take advantage of expressing my most sincere thanks on this occasion to D. ter Haar, who has been kind enough to interest himself in the publication of this book in English and who has devoted much time to the critical work of translation and editing.

General Introduction

1. Recent years have seen the achievement of important advances in the domain of applications of classical and quantum statistical mechanics. By way of illustration, it is sufficient to recall the multiplicity of projects instigated by the study of classical and quantum plasmas, as well as the large extent of research on low-temperature physics, where the problems dealt with relate exclusively to quantum statistical mechanics (analysis of the effect of the condensation of bosons in gases, problems of the superfluid state, spin temperature, etc.). These various advances have been accomplished as much by the introduction into statistical mechanics of new methods of calculation—this is the case, for example, of perturbation methods borrowed from the quantum theory of fields [see, for example, Abrikosov, Gor'kov and Dzyaloshinskii, 1965; Kirzhnits, 1967]—as by the impetus given to research by the exploration of relatively new domains such as plasma physics.

Even though the field of application of statistical mechanics is extending itself to the point that this discipline now plays a fundamental role in modern physics, the often discussed theoretical problems raised by the introduction of statistical methods in physics remain meanwhile without a definitive answer. Thus, it is necessary to return to a study of the foundations of statistical mechanics and to study the actual state of this question, the more so because certain new results have been established recently in this field, particularly in quantum ergodic theory. The interest of such an improvement, moreover, is not only theoretical, because the various applications which come to our mind are often concerned with fundamental problems, sometimes by throwing new light on them: such is the case, for example, with the problem of describing non-equilibrium systems and that of the domain of validity of Boltzmann's equation, completely revived by research in plasma physics.

Thus, the present book has the objective of presenting as complete a critical account as possible of the foundations of statistical mechanics and of the various problems set by the theoretical justification of its methods. In doing this, I have laid down the following general rules: on the one hand to compare the classical and quantum aspects of the questions dealt with at each stage of reasoning, whilst highlighting the characteristics belonging to each of the two theories; on the other hand to analyse, independently of all possible applications, the nature and physical contents of the methods employed for justifying the fundamental postulates of statistical mechanics. Thus defined, the point of view which I have put forward is appreciably different from that adopted by Khinchin (1949, 1960) in his two fundamental works, where greater importance is given to the mathematical problems set by the calculation of average values and the application of the asymptotic theorems of probability calculus.

2. The object of statistical mechanics is the atomistic interpretation of the macroscopic properties of matter and radiation; it has thus the task of bridging the gap between the microscopic description of elementary phenomena and the macroscopic description of systems at our scale. Because of the corpuscular structure of matter, statistical mechanics considers any physical system as a mechanical system composed of a large number of constituents (elementary particles, atoms, molecules); the object of this discipline is thus reduced to studying the macroscopic properties of mechanical systems with a large number of degrees of freedom. Consequently, the first problem which is posed, is that of defining the actually observed macroscopic quantities, starting from the microscopic mechanical quantities of the system.

We achieve this by analysing the nature and properties of macroscopic observations, which lead directly to the introduction of the concept of probability in physics. In fact, macroscopic observers such as we are, are under no circumstances capable of observing, let alone of measuring, the microscopic dynamic state of a system which involves the determination of an enormous number of parameters, of the order of 10^{23}. Against this very large number of microscopic parameters, it is necessary to set the very small number of variables used in the macroscopic description, such as the temperature and pressure of a gas in the equilibrium state. The result is thus that macroscopic observation is always—and here is its essential character—a highly incomplete observation of the underlying mechanical state of the system and that a whole *ensemble* of possible dynamic states corresponds to the *same* macroscopic state, compatible with our knowledge. Thus, the fundamental concept of statistical ensembles of systems, is introduced quite naturally and, with it, that of a probability density defining a statistical weight for the various systems in the ensemble. The method of statistical mechanics will thus consist in associating with each macroscopic observation a certain *representative ensemble* of virtual systems, which are independent of one another, and in adding to these ensembles a probability density in phase in the classical theory, or a statistical operator or density matrix in the quantum theory. The concept of a stationary ensemble will correspond to the concept of thermodynamic equilibrium and the macroscopic quantities observed will then be defined as the values of the microscopic observables associated with the measured quantities, averaged over all systems of the ensemble.

Given the essentially statistical nature of this method, one would not know how to escape from the necessity of defining the *a priori* probabilities for constructing the representative ensembles. The simplest postulate, compatible with the requirements of invariance under the equations of motion, is that of equal probability of equal volumes of phase space in classical theory and that of equal probability of states and uniformly distributed phases in quantum theory; this is the fundamental hypothesis of statistical mechanics. In this connection we emphasise as from now that the concept of probability comes into quantum statistical mechanics under two essentially different aspects: on the one hand in the quantum sense, because of the very nature of the definition of the state of a system in wave mechanics, and on the other hand in the classical sense, by virtue of the incomplete nature of any macroscopic observation. Thus, the method based on the concept of a representative ensemble is found to be completely defined in classical theory as well as in quantum theory; we shall make constant use of it in this book, whilst still reserving discussion of certain aspects of the method of the kinetic theory of gases during the study of the H-theorems.

3. It is a well-known fact that in the history of the sciences, the historical development of a theory rarely proceeds by following a logical order. Such, indeed, is the case of statistical mechanics; it is, in fact, in the kinetic theory of gases that the probability concept first appeared with the idea of the velocity distribution function, to which the names of Clausius, Maxwell, and Boltzmann remain attached. It was necessary to await the efforts of Gibbs in order that the properties of statistical ensembles of systems be brought to light, as well as their thermodynamic analogies, which are the foundation of the whole of statistical mechanics. Although the kinetic method assumes, to a certain extent, a more physical aspect, the role of the probability concept is not so clearly distinguished here, as the long and famous controversies about the *H*-theorem between Boltzmann, on the one hand, and Loschmidt and Zermelo, on the other hand, bear witness; moreover, it is in the course of these controversies and in reply to the arguments of his opponents that Boltzmann elaborated little by little the statistical interpretation of the kinetic theory of gases, and that he set in motion his ideas concerning ergodic theory.

Thus, the ground had already been largely cleared when Gibbs' works appeared; the concept of statistical ensembles of systems had just been born. It had to reveal itself subsequently most fruitfully as much on the theoretical plane where the probability concept is introduced naturally, as we have just seen, as well as in the field of applications where the results obtained are at once independent of the structure and the generally unknown interactions of the particles which constitute the observed system. If Gibbs was thus the first to show the thermodynamic analogies of the properties of ensembles, he did not go so far as to affirm that the mean values of the microscopic observations represented effectively the physically observed quantities; actually, he considered those ensembles of systems as a rational model, which presented interesting similarities with the principles of thermodynamics without, however, furnishing a true explanation. This latter step had to be surmounted by other authors, first of all by Einstein in his work on the Brownian movement, contemporary with but independent of Gibbs, and later by the founders of modern statistical mechanics (Ehrenfest, Uhlenbeck, Fowler, Tolman, etc). The reasons for these reservations of Gibbs lay in the difficulties which classical statistical mechanics encountered with the theory of the specific heat of diatomic gases and with radiation phenomena. Actually, the principles of statistical mechanics were not concerned, since the transition to the quantum description was to make possible the solution of this paradox; we have here an illustration of the close relations which are established between the development of statistical mechanics and that of the quantum concepts, of which another particularly striking example is furnished by the origin of Planck's quantum of action itself.

4. The method of statistical mechanics being thus well defined and the validity of its hypotheses being confirmed in addition by experimental results, it remains to justify from the theoretical point of view the fundamental assumptions of this discipline. In fact although the introduction of the Gibbs ensembles is a natural consequence of the incomplete nature of our macroscopic observations, all the same a single system of the ensemble, at least from the mechanical point of view, corresponds in effect to the observed physical system, and a single trajectory in Γ-space or a single vector Ψ in Hilbert space is associated with it. Thus, whilst noting that a precise and maximum determination of the state of the system is macroscopically impossible, it appears, however, to be difficult to state, without a more thorough examination, that the system is "by right" in an undetermined state, which would mean that we should question the physical reality of the state; we note that such a statement, however, would be more acceptable within the framework of the usual interpretation of wave mechanics, where the function Ψ represents only the state of our knowledge of the observed system. Apart from the introduction of such virtual ensembles of systems, it is necessary once more for us to justify the fundamental postulate of statistical mechanics relative to *a priori* probabilities, these being in fact completely foreign to the principles of mechanics.

We are thus in the presence of two possible descriptions of a physical system: (i) the microscopic description, where only a single system is considered whose evolution is reversible and in general quasiperiodic, and where the physical observables are either phase functions in classical mechanics or Hermitian operators in wave mechanics; (ii) the macroscopic description, involving a statistical ensemble of systems whose average behaviour, which is the only one measured by experiment, must in particular possess the necessary properties of irreversibility. This being so, the justification of the method of statistical mechanics then poses the following problems: A: For what reasons can we represent the macroscopic quantities observed by the averages of the microscopic observables in a representative ensemble? B: Is it possible to prove the tendency of these mean values to develop towards equilibrium states represented by stationary ensembles, as the irreversibility of macroscopic evolution demands? C: If the answer to B is affirmative, how is it possible to reconcile this irreversibility with the reversibility and the quasiperiodicity of mechanical systems? In order to answer these questions, various attitudes are possible and we follow ter Haar (1955) in distinguishing three of them:

The utilitarian point of view—for which the assumptions introduced are considered to be sufficiently justified by their consequences, that is to say by the agreement of the theoretical conjectures with the experimental facts; any method leading to correct results can then be held to be valid. If several methods of calculation lead to the same results, the choice between them is reduced to a simple matter of convenience: such is the case, for example, with microcanonical and grand-canonical ensembles which give the same average values for a system having a large number of degrees of freedom. It is hardly necessary to stress the fact that this attitude is unsatisfactory for the theoretician. Even within the field of applications, such a point of view lends itself to criticism: thus according to Khinchin, microcanonical and canonical ensembles only provide equivalent results for certain types of phase functions (sum-functions); for other phase functions, the utilitarian attitude does not allow a choice to be made, without recourse to experiment, of the ensemble which it is convenient to use. Be that as it may, the utilitarian point of view does not make it possible to avoid questions B and C about the evolution of the system and its tendency towards a state of equilibrium: the generalised H-theorem is needed to provide an answer to these questions.

The formalistic point of view—which considers mechanical systems as isolated systems whose energy and number of particles are fixed; in classical mechanics it corresponds to trajectories on a hypersurface of constant energy. The only stationary ensemble adapted to these conditions is the microcanonical ensemble which thus plays a privileged role; canonical and grand-canonical ensembles appear either as mathematical procedures suitable for simplifying the calculation of average values, or as representative ensembles of subsystems in thermal equilibrium in the midst of a vaster, isolated system. From this point of view, the main problem is to justify the use of microcanonical averages in analysing the properties of the dynamic trajectory of a system observed on a hypersurface of constant energy: this is the aim of the ergodic theory which can, moreover, be stated in formally parallel terms in statistical quantum mechanics, the trajectory of the system on a hypersurface of phase space being replaced by that of the vector Ψ on the unit hypersphere of Hilbert space.

The physical point of view—which consists in attributing a fundamental character to the incomplete nature of our macroscopic observations and which leads to the concept of an ensemble representative of the state of a system. This concept is thus considered to be the starting point and sufficiently justified by our absence of information on the observed system; as we have already mentioned, this point of view is defended more easily within the framework of quantum ideas, as there does not exist in the normal interpretation of wave mechanics an objective and intrinsic state of a system, independent of measurements. The different stationary ensembles are then introduced naturally by the analysis of the physical condition of the observed system and by looking carefully at the concept of an isolated system. According to the degree of isolation of the system, one arrives with Gibbs, and above all with Tolman, at the concepts of perfectly isolated systems, represented by a microcanonical ensemble and systems which are "essentially" isolated and which are described by canonical or grand-canonical ensembles. This being so, it remains to answer questions B and C, as in the utilitarian point of view, and to justify moreover the postulate of equal *a priori* probabilities; for this purpose, one must turn again to the generalised H-theorem.

5. Such are the various possible attitudes facing the problems which are raised by the justification of the fundamental postulates of statistical mechanics. In final analysis, they lead to the use of two distinct general methods: that of the ergodic theory and that of the generalised H-theorem. In order to analyse the contents of these methods, it is necessary for us to refer once again to two characteristic properties of macroscopic observation which we have as yet not considered: (a) any observation is necessarily extended over a finite interval of time, which is generally large on the microscopic scale; (b) a macroscopic measurement always involves a certain inaccuracy of the measurement result.

Let us consider, then, the ergodic theory in classical theory in order to fix the ideas. In this case we have to deal with an isolated and conservative mechanical system, whose evolution is described by the trajectory of the representative point P_t of a system on a hypersurface Σ of phase space; the microscopic quantities associated with the system are the instantaneous values of phase functions $f(P_t)$. Now, according to (a) experiment never measures these instantaneous values, but only the time-average of the phase functions over an interval T. If we are interested in the equilibrium properties, we must observe the system over a sufficiently long period in order that the equilibrium state be established. Thus, we are led to consider the limit of the time average of the phase functions for T tending to infinity, so that the ergodic theory reduces to the following problem: Can we prove the equality of the limit of the time averages and of the phase averages taken over the hypersurface Σ, which corresponds here to the microcanonical ensemble? It is clear that the complete solution of this problem would contribute a satisfactory answer to the questions raised by the justification of the principles of statistical mechanics. Actually, the equality of the time and of the phase averages justifies, first of all, the introduction of uniformly distributed statistical ensembles; it thus contributes an answer to question A and knowledge of the trajectory of the actual physical system is no longer necessary for calculating the time averages. Moreover, as the limits of these time averages are equal to the statistical averages at equilibrium, it follows that the system exists for most of the time in a state of equilibrium and that it is sufficient to observe it only over a period which is sufficiently large for seeing it tend towards such an equilibrium state. In this manner, it leads to an interpretation of macroscopic irreversibility which is compatible with microscopic reversibility and which enables us to anticipate the existence of fluctuations: such is the answer to questions B and C; it assigns an essential role to the limit $T \to \infty$. The justification for the methods of statistical mechanics then rests entirely on the dynamic properties of the observed systems. Despite the remarkable efforts of Birkhoff, von Neumann, Hopf, etc., we shall see that such a programme does not fail to raise some very great difficulties and that we are obliged to be satisfied, within the actual state of our knowledge, with the "probability" ergodic theorems. (I have denoted it thus, to underline the fact that statistical concepts play a major role here, contrary to what occurs in the ergodic theory proper.)

The method, based on the proof of a generalised H-theorem, is very much different in spirit: it accepts at the outset the concept of statistical ensembles of systems and the concept of a probability density in phase. In this point of view it is thus sufficient to show that these probability densities tend, in the course of evolution, towards one of the microcanonical, canonical or grand-canonical stationary distributions. However, such a result cannot be obtained with the fine-grained densities that satisfy Liouville's fundamental equation, which is reversible like all the equations of dynamics. We are thus obliged to turn to the second property of macroscopic observation mentioned above: in fact, because of the inaccuracy inherent in any observation, phase space can be divided up into finite cells of extension in phase (coarse graining), by which a *coarse-grained* probability density is defined, the importance of which has been particularly emphasised by P. and T. Ehrenfest. As this coarse-grained distribution does no longer satisfy Liouville's equation, one is now in a position to establish a generalised H-theorem which shows us the probable direction of evolution. However, this theorem has, above all, a qualitative value and it is quite inadequate for serving as the basis for a description of the irreversible evolution of non-equilibrium systems. Questions B and C thus receive a qualitative answer, which leaves alone a detailed analysis of the irreversible processes themselves. Let us emphasise, in addition, that the generalised H-theorem does not permit in any way the justification for the hypothesis of equal *a priori* probabilities—necessary for statistical mechanics—since these are used in the definition itself of coarse-grained densities; the justification for this hypothesis can thus be attempted only within the framework of the ergodic theory.

These general considerations highlight certain essential points: first of all, the special role played by the properties of macroscopic observation at each stage of reasoning; it is for this reason that the incomplete nature of observations on our scale is the origin of the basic idea of representative ensembles and that the ergodic theory and the generalised H-theorem depend on the finite duration and on the inherent inaccuracy in any macroscopic measurement. On the other hand, only the ergodic theory could be in a position to answer all the fundamental questions and especially question A; thus, the formalistic attitude will be important for the physical point of view. Because of the impossibility of verifying the conditions for ergodicity, we shall see that we are led, in fact, to adopt an intermediate attitude, resting on the demonstration of the "probability" ergodic theorem. Finally, as the generalised H-theorem is reduced to a qualitative interpretation of the property of irreversibility, a precise analysis of the evolution of non-equilibrium systems would have to rely on supplementary principles or hypotheses.

6. If we now place ourselves within the framework of quantum concepts, the foregoing conclusions remain valid. Actually, apart from any formal differences, the logical structure of quantum statistical mechanics is identical, in essence, to that of classical statistical mechanics. The concept of the representative ensemble is introduced for the same reasons as in classical theory and the probability distributions thus met with must be carefully distinguished from specifically quantum probabilities. Contrary to what one might expect, the indeterministic nature of wave mechanics does not facilitate the solution of problems set by the foundations of statistical mechanics. We shall even see new difficulties arising because of certain features which are inherent in quantum theory.

Thus, the concept of phase of classical mechanics cannot be transposed to wave mechanics all at once, because of the Heisenberg relations. The quantum theory imposes on us a functional representation of the state of a system described by a vector Ψ in Hilbert space. A set of vectors Ψ in the functional space must thus be associated with an ensemble of quantum systems and the role of the probability density in phase in classical mechanics is played by the statistical operator. Moreover, by virtue of the essentially statistical nature of the quantum theory, the probability concept appears in statistical mechanics under two radically different aspects: on the one hand, under the form of quantum probabilites which do not satisfy the normal rules of probability calculus and which are connected with the very nature of the quantum description of the state of a system; on the other hand, under the form of classical probabilities which have the same origin as in classical statistical mechanics. The distinction between these two types of probability, which corresponds moreover to the difference between mixtures and pure cases, turns out to be fundamental. However, as the statistical nature of the quantum theory is completely independent of the methods of statistical physics, it does not introduce any essential change into the principles of statistical mechanics, but only some modifications of its mathematical form, which we shall study in Chapter II.

Another characteristic property of wave mechanics without a classical parallel relates to the fact that any two quantities are not, in general, measurable simultaneously, unless their associated operators commute with one another. Because the quantities observed macroscopically are simultaneously measurable and as they correspond, moreover, from the point of view of classical physics to the intrinsic properties of the system, we might ask ourselves if we shall not encounter a new difficulty here. This obstacle is avoided by ascribing a system of operators to the observed microscopic quantities, so-called macroscopic operators, which satisfy the required properties of commutability. Because of their strong degeneracy and because of their non-commutation with the microscopic Hamiltonian, these macroscopic operators play an important role in the ergodic quantum theory as well as in the proof of the generalised H-theorem; we have devoted the second section of Chapter II to them.

We come now to another important consequence of the quantum theory, to which we shall return often, especially in the ergodic quantum theorem in Chapters III and IV: this is the absence of evolution of a quantum system when it is in a stationary state. This refers both to the properties of the integrals of motion and to the nature of the quantum description of the state of a system: in fact, only the probabilities relative to the various eigenstates of a quantity have a physical significance and the time-dependence of these probabilities alone determines the evolution of the system. If a system exists, for example, in a well-defined microscopic energy state the probabilities of any physical quantity associated with the system remain invariant with time so long as no external perturbation interferes. We note, moreover, that the microcanonical ensemble is then reduced to a single vector Ψ, associated with the energy state considered, unless this state is degenerate. This property of stationary states raises serious difficulties in the quantum ergodic theory: they can only be overcome by making use of the strong degeneracy of the macroscopic observables.

Finally, it still remains for us to evoke the specifically quantum relationship between spin and statistics, which is the origin of another essential difference between the classical and quantum aspects of statistical mechanics. This relationship imposes a restrictive condition on the choice of possible wave functions, to which corresponds a reduction of the "accessible" states for the system, the group of symmetrical states being associated with particles having integral spin, and that of antisymmetrical states being associated with particles having half-odd-integral spin. This condition thus leads us to the two types of quantum statistics, the Bose–Einstein and Fermi–Dirac statistics, which are applied with so much success, especially in low-temperature physics. Even though this separation into symmetrical and antisymmetrical states is the origin of a certain complexity of the mathematical equipment of the theory, in particular for calculating average values, it does not appear to play an important role in the problem of the foundations of statistical mechanics; in fact, it corresponds simply to the existence of a uniform integral of motion and it is necessary only to consider from the start one or the other of the two groups of states: this is what we shall do, in general, in this book.

7. The foregoing considerations are directed towards the plan of this book; quite naturally, it comprises two parts: the first part is devoted to the ergodic theory and the second part to the study of H-theorems and kinetic equations. In each part I have endeavoured to treat in parallel the classical and quantum aspects of the problems studied, stressing at the same time the similarities and differences between the classical and quantum theories. In addition, I have developed particularly the subjects where some recent advances have been achieved, confining myself simply to mentioning results acquired long ago and which can be considered as well known. Thus, only a single chapter—the first—is devoted to classical ergodic theory, where the modern ergodic theorems of Birkhoff, von Neumann and Hopf are reviewed, in addition to Khinchin's method for evaluating—for a system with a large number of degrees of freedom—the mean squares of phase functions of the "sum" type. In order to facilitate the explanation I have set out, in Appendix I, a detailed analysis of the concept of metrical transitivity and the study of the asymptotic properties of structure functions for systems with weakly coupled constituents; in this connection, I leave aside the calculation of averages of phase functions, which is outside the scope of our subject.

On the other hand, I have reserved three chapters for an account of the quantum ergodic theorem, because of the progress accomplished recently in the domain of the probability ergodic theory, thanks to the work of various authors and to the research undertaken in my doctoral thesis and developed subsequently. Chapter II deals with the definition of statistical ensembles of quantum systems and with the introduction of macroscopic operators, which play a principal role in all the arguments. In Chapter III, I then deal with the quantum ergodic theory proper, with an account of the methods of Ludwig, of von Neumann, and of Pauli and Fierz. After having analysed the method of averages over all possible macroscopic observers, I show how we are led to the concept of the quantum probability ergodic theorems, the proof of which occupies the whole of Chapter IV and which, together with Chapter VI, comprises the most original part of the book. As all the results of this chapter rest on the asymptotic properties of the unit hypersphere of *n*-dimensional space, real or complex, I have—as in classical theory—given in Appendix II a study of these hyperspheres and of the corresponding probability distributions.

In the second part, devoted to the study of irreversible processes, I have, on the other hand, developed at greater length certain classical problems, because of the recent papers in the field of classical kinetic equations which were stimulated by the progress of plasma physics. This is why I have devoted the whole of Section V of Chapter V to the detailed study of the various methods of integrating Liouville's classical equation, on the deduction of the "Master Equation" in quantum theory, and on the quantum parallel of the hypothesis of molecular chaos. Finally, in the last chapter, taking the place of the conclusion, I return to the properties of macroscopic observation by emphasising the differences which distinguish it from quantum measurements and which certain authors have considered, together with Klein's lemma, as the foundation of macroscopic irreversibility.

8. In conclusion, let us point out the essential themes of this book. In the first place, it is shown how the ergodic problem can be stated in terms which are formally similar in classical theory and in quantum theory, despite a few important differences which we have already emphasised. In both cases, the proof of the ergodic theorems encounters difficulties which are still unresolved; in fact, it would be necessary to satisfy certain *conditions of ergodicity*, which are unverifiable in practice: these are the hypotheses of metrical transitivity in classical theory, the absence of degeneracy and of resonance frequencies in quantum theory. This situation is

complicated again in the quantum ergodic theorem, where the properties of stationary states force us to make use of macroscopic observables and to take the averages over the ensemble of all the possible macroscopic observers (von Neumann, Pauli–Fierz). The existence of quantum ergodic systems could only thus be established relative to macroscopic observers; once again, such a result would be valid only as a "probability" result. It thus appears that the ergodic problem raises greater difficulties in quantum theory than in classical theory and that we are still far from its definitive solution in both cases. Moreover, the ergodic theorems always allow the possibility of exceptional trajectories whose set is of measure zero; we cannot eliminate them without accepting the following postulate, whose statistical nature is obvious: the states belonging to a set of zero measure are never observed in reality.

This is why the second theme of the book is aimed at getting rid of these various conditions, by relying on the following statement: the ergodic theorems are generally stated for any dynamic systems, independently of their number of degrees of freedom; these are, in fact, statements of general mechanics, bearing on ensembles of trajectories. However, macroscopic systems always have an enormous number of degrees of freedom, which give to the phase functions or to the quantum averages representing physical quantities, particular properties connected with the geometrical properties of hypersurfaces in a space with a large number of dimensions, n. We note that n, which is proportional in the classical case to the number N of particles of the system, is in the quantum case a very rapidly increasing function of N, because of the functional description of the system. In relying on the asymptotic properties of these hypersurfaces and using from the start the point of view of the theory of classical or quantum statistical ensembles, we can then prove without any particular hypothesis about the dynamic structure of the system the probability ergodic theorems, which establish the equality of the time and microcanonical averages, except on a set whose measure decreases very rapidly as N increases. In quantum theory, microscopic and macroscopic theorems can be proved, a comparison of which emphasises the fundamental importance of the strong degeneracy of the macroscopic operators. In all cases, these results rely exclusively on the limit $N \to \infty$, which thus plays a privileged role in place of the limit $T \to \infty$. By interchanging these two limit processes, it is shown in addition that the quantum ergodic theorems of von Neumann enter again into the framework of the probability ergodic theorems and that the ergodicity conditions, which were required, are finally superfluous. Finally, it is shown that the proof of the quantum probability ergodic theorems rest, in fact, on the asymptotic properties of the macroscopic observables themselves. These latter results have been obtained recently, as a consequence of work to which we have already made reference: they have thus enabled us to clarify the problems posed by the ergodic quantum theory and they are presented as the quantum extension of Khinchin's classical arguments.

In the second part, we approach another fundamental theme of the book, namely, the study of the irreversible evolution of non-equilibrium systems. We deal first of all with the generalised *H*-theorem which gives rise to developments entirely analogous in classical and quantum theory. The quantum concept of sub-spaces or cells in Hilbert space associated with a macroscopic operator corresponds to the classical concept of cells in extension in phase. These operators enable us, in addition, as we have already emphasised, to overcome the specific quantum difficulties associated with the existence of stationary states; because of their non-commutability with the microscopic Hamiltonian they lead, in fact, to a definition of non-stationary statistical ensembles which describe the evolution of the system. In classical, as in quantum, theory we arrive at the same conclusions: the generalised *H*-theorem gives only an indication to the direction of evolution of a system immediately after the moment of a macroscopic observation; its value is thus essentially qualitative. In order to obtain kinetic equations, we must analyse the process of evolution itself in more detail.

If we wish to draw conclusions from the generalised *H*-theorem which are valid for longer durations (of the order of the relaxation time) and which give rise to quantitative predictions, we must try to renew at each instant the hypotheses made for the initial moment of the macroscopic observation; we have here the problem of conservation with time of the property of molecular chaos which is expressed, in quantum theory, by the vanishing of the off-diagonal elements of the density matrix. By accepting the existence of such a property, we are led to a "Master Equation", which leads to representing the evolution of a gas by a Markovian process. However, one has substituted in a more or less explicit manner for the dynamic description of the evolution of the system, a stochastic description which is not in general compatible with the laws of mechanics; on the other hand, the same is true for the Boltzmann equation, whose stochastic significance is far from being completely clarified. In final analysis, we can only discuss the validity of such descriptions by reverting to Liouville's fundamental equation in its classical or quantum form. The various methods of approximate integration of this equation (the B.B.G.K.Y. equations, Bogolyubov's method, etc.) have thus as objective a study of which hypotheses and which approximations are necessary for deriving the kinetic equations such as the Master Equation, the Boltzmann equation, the Fokker–Planck equation, etc.; this analysis depends on the large number of degrees of freedom of the system (limit $N \to \infty$) and enables us, in certain cases, to remove the contradictions between microscopic reversibility and macroscopic irreversibility, by distinguishing between different scales of observation.

9. In the terms of this study, what can we say about the actual state of the problems raised by the foundations of statistical mechanics? In the light of the foregoing results can we make a choice between the formalistic point of view and the physical point of view? In order to attempt to answer these questions, we emphasise first of all that the programme of the ergodic theory is far from being completed in its entirety and that, despite the numerous remarkable efforts—especially in the classical domain—it is in fact illusory to attempt to found the methods of statistical mechanics on a purely dynamic basis. It would thus appear that the formalistic attitude must lose much of its interest, since we must always have recourse to statistical concepts which are foreign to mechanics, be it only to eliminate exceptional trajectories, which continue to exist in the ergodic theory. However, we have succeeded in proving the probability ergodic theorems, by heavily making use of the limit $N \to \infty$; the theorems are sufficient both for justifying the hypothesis of equal *a priori* probabilities of statistical mechanics and to establish the principles of statistical thermodynamics. On the other hand, as we have seen that the generalised H-theorem has mainly a qualitative value and that it is inadequate for serving as the basis for the principles of statistical mechanics, one might think that the physical attitude itself must be partially revised and that the truth is contained, probably, in a reconciliation of the two points of view, formalistic and physical.

Actually, because of the physical conditions which are particular to macroscopic systems, the concept of the statistical ensemble must necessarily be retained as it is an essential concept; it depends both on an analysis of the properties of macroscopic observation, which constitutes the physical aspect of the problem, and on the interpretation of the role played in physics by the concept of probability itself; this raises an important epistemological problem. Thus, statistical mechanics appears in final analysis as a theory of macroscopic observation of systems with a large number of degrees of freedom. This being accepted, the "probability" ergodic theory suffices to justify the postulate of equal *a priori* probabilities and the use of microcanonical ensembles for describing isolated systems; canonical and grand-canonical ensembles are then introduced as either representative ensembles of sub-systems in equilibrium in a vaster system, or as mathematical procedures which are convenient for calculating average values in systems with a large number of degrees of freedom. The fundamental questions A, B and C thus receive a satisfactory answer and it can be stated that a microscopic system exists most of the time in an equilibrium state represented by a stationary ensemble; if, as a result of fluctuations, it diverges significantly, we can expect to see a rapid return towards equilibrium. Moreover, as we have emphasised, the analysis of this return to equilibrium and more generally of irreversible processes, raises problems which are especially associated with the conservation with time of the property of molecular chaos.

Nevertheless, we cannot say that the problems raised by the foundations of statistical mechanics have thus received a definitive solution; quite the contrary, the numerous points which remain to be explained or defined more precisely can give birth to interesting projects. Within the framework of the ergodic theory, for example, two paths of research are open to us: the one, by extending the work of Birkhoff and von Neumann, would carry out the study of the dynamic properties of systems which have a large number of degrees of freedom, from the point of view of mechanics; the other, following the ideas of the probability ergodic theorem, would lead to a more precise statement of the statistical laws satisfied by ensembles of macroscopic systems. The problem which offers the most possibilities in this respect, however, is surely that of non-equilibrium systems; on the one hand, we encounter here, starting from the establishing of kinetic equations, various still unsolved problems, such as the conservation of the property of molecular chaos; on the other hand, we may think that the numerous researches carried out in the sphere of plasma physics would supply us with new methods, amongst others, the diagram method which appears to be destined to be extended considerably in the years to come.

Finally, to change the subject completely, the problem of the foundations of statistical mechanics would appear in a new light if mechanics itself were to undergo complete modifications. Such would be the case with the substitution of non-linear equations for the present linear equations, following the researches undertaken by M. Louis de Broglie and his school with a view to a causal interpretation of wave mechanics. The properties of such equations, in particular the existence of limit cycles, would be in the nature of contributing new elements into the principles of statistical mechanics and eliminating, perhaps, certain of the difficulties that we have encountered.

In concluding this introduction, I am pleased to express my sincere gratitude to M. Louis de Broglie for the great honour which he has afforded me in writing the Preface to this book and to express to him my profound gratitude for the interest which he has always shown towards my researches, and which owe so much to his teaching.

To M. Jean-Louis Destouches, who has shown such interest in this problem and who has always encouraged me and guided me in the achievement of this work, I express my sincere gratitude and my most hearty thanks for having been willing to accept this book into his series.

I also owe my thanks to M. Theo Kahan, who was willing to check the proofs of this book and I am pleased to express my sincere gratitude for the valuable and friendly support which he has always given to me during the interesting and fruitful discussions that we have held together.

PART I

Ergodic Theory

CHAPTER I

The Ergodic Theory in Classical Statistical Mechanics

I. Statistical Ensembles of Classical Systems

1. *Definition of ensembles of systems. Liouville's theorem*

In classical mechanics, the state of a system with n degrees of freedom is described by its n coordinates q_i and their n conjugate momenta p_i ($i = 1, 2, \ldots, n$); the set (q_i, p_i) defines a point P in a phase space (Γ-space) of $2n$ dimensions. The motion of this point is described by the Hamiltonian equations

$$\dot{q}_i = \frac{\partial H}{\partial p_i} = [q_i, H], \quad \dot{p}_i = -\frac{\partial H}{\partial q_i} = [p_i, H], \qquad \text{(I.1)}$$

where the [ ] are Poisson brackets and H is the Hamiltonian of the system. These equations determine the trajectory of the point P in phase space: from t_0 to t, the representative point of the system passes from P_0 to P_t. If the system is conservative (H independent of t) and if there are no singularities, equations (I.1) are the infinitesimal transformations of a continuous group of transformations ($P_0 \rightarrow P_t$) with parameter t which transforms the phase space into itself. If T_t denotes the transformations of this group, we have

$$P_t = T_t P_0,$$

with

$$T_{s+t} = T_s T_t \quad \text{and} \quad T_0 P_0 = P_0. \qquad \text{(I.1')}$$

Thus, to the motion described by (I.1') there corresponds an automorphism of Γ-space whose invariants are the integrals of motion of the system (I.1). The integral of motion of the energy,

$H = \text{const}$, defines a hypersurface Σ of $(2n-1)$ dimensions in Γ-space; it plays a privileged role for isolated systems. The group of automorphisms (I.1′) will be without singularities and T_t will be determined analytically over Σ, if the hypersurface Σ is located entirely in a region of Γ-space where the Hamiltonian is uniform and analytical.

The Hamiltonian of an "isolated" system includes internal interactions, interactions with external bodies being expressed by a conservative (time-independent) potential, depending only on the coordinates of the system considered, as well as interactions with the walls involving external parameters (fixing the position and shape of these walls) which do not depend explicitly on the time. The energy of such a system will be constant, so that the point P_t will remain on a hypersurface Σ; because of the walls, this hypersurface will be completely bounded and the measure $\mu(\Sigma)$ will be finite, which is essential for the arguments of the ergodic theory.

An ensemble of systems is defined by a set of points in Γ-space; such an ensemble is thus defined by a phase density $D(P, t)$ which represents the number of points per unit volume with $2n$ dimensions in the vicinity of the point P at time t. By definition, this density must satisfy the relation

$$\int_{\Gamma} D(P, t)\, d\Gamma = \text{const.} \tag{I.2}$$

The function $D(P, t)$ must also satisfy the hydrodynamic equation of continuity

$$\frac{\partial D}{\partial t} + \sum_{i=1}^{n} \left\{ \frac{\partial}{\partial q_i}(D\dot{q}_i) + \frac{\partial}{\partial p_i}(D\dot{p}_i) \right\} = 0 \tag{I.3}$$

which, because of (I.1), assumes the form

$$\frac{\partial D}{\partial t} = -\sum_{i=1}^{n} \left\{ \frac{\partial D}{\partial q_i}\frac{\partial H}{\partial p_i} - \frac{\partial D}{\partial p_i}\frac{\partial H}{\partial q_i} \right\} = -[D, H], \tag{I.4}$$

because we have according to (I.1)

$$\sum_i \left(\frac{\partial \dot{q}_i}{\partial q_i} + \frac{\partial \dot{p}_i}{\partial p_i} \right) = 0. \tag{I.5}$$

Since the total variation of D as a function of time is

$$\frac{dD}{dt} = \frac{\partial D}{\partial t} + [D, H], \tag{I.6}$$

we have, by comparison with (I.4)

$$\frac{dD}{dt} = 0. \tag{I.7}$$

This latter equation (I.7) is the differential expression of *Liouville's theorem*, because it states that $D(P, t)$ is an invariant integral of motion, i.e. that

$$\int_{V_0} D(P, t_0)\, d\Gamma = \int_{V_t} D(P, t)\, d\Gamma, \tag{I.8}$$

where V_t is the region occupied at t by the points which occupy V_0 at the initial time.

If we take D = const, expression (I.8) shows that the measure of a hypervolume of phase space is invariant with respect to the equations of motion: *the elementary measure $d\Gamma$ thus constitutes an invariant measure in Γ-space with respect to T_t*. We recall also, without proving it, that the measure $d\Gamma$ is invariant under canonical transformations of the coordinates and momenta of the system. In what follows, we shall always use the normalised density $\varrho(P, t)$, defined by

$$\varrho(P, t) = \frac{D(P, t)}{\int_{\Gamma} D(P, t)\, d\Gamma}. \tag{I.9}$$

Relation (I.2) can then be written as

$$\int_{\Gamma} \varrho(P, t)\, d\Gamma = 1 \tag{I.10}$$

and $\varrho(P, t)$, naturally, satisfies the same equations as $D(P, t)$.

2. *Stationary ensembles*

These correspond to the case of statistical equilibrium; in order that an ensemble be stationary, it is necessary and sufficient that the density ϱ be time-independent:

$$\frac{\partial \varrho}{\partial t} = 0 \tag{I.11}$$

or, according to (I.4)

$$[\varrho, H] = 0. \tag{I.11'}$$

It can be seen immediately that $\varrho = \text{const}$ is a solution of the equations (I.11) so that if a uniform distribution is realised at the initial time, it will be maintained in the course of evolution. Apart from this trivial solution, equations (I.11) allow as a solution any function $\varrho = \varrho(f)$, where f is an integral of motion of the system considered. In fact, we have by definition

$$\frac{df}{dt} = \sum_{i=1}^{n} \left(\frac{\partial f}{\partial q_i} \dot{q}_i + \frac{\partial f}{\partial p_i} \dot{p}_i \right) = 0, \tag{I.12}$$

whence, according to (I.4),

$$\frac{\partial \varrho}{\partial t} = -\frac{d\varrho}{df} \sum_{i=1}^{n} \left(\frac{\partial f}{\partial q_i} \dot{q}_i + \frac{\partial f}{\partial p_i} \dot{p}_i \right) = 0. \tag{I.13}$$

In the case of a conservative and isolated system it is natural to use the total energy of the system as the integral of motion; we are thus led to introduce the microcanonical ensemble for describing the statistical equilibrium of isolated systems. On the other hand, we also use canonical and grand-canonical ensembles in statistical mechanics; we have thus four fundamental stationary ensembles:

(a) *Uniform ensemble.* This is defined by $\varrho = \text{const}$. It corresponds to an ensemble whose representative points are distributed uniformly over phase space. It is of interest as it shows that there is no mechanical reason which can lead to assigning a privileged role to any region of Γ-space. This ensemble also has the property of being invariant under canonical transformations, so that if we have a uniform distribution for a system of variables (q_k, p_k), we will have the same distribution for any other system of canonically conjugate variables (Q_k, P_k).

(b) *Microcanonical ensemble.* For isolated conservative systems, the energy integral plays a privileged role; statistical equilibrium being assured by any density of the form $\varrho(H)$, the microcanonical ensemble is formed by taking the uniform distribution in an energy shell $(E, E + \delta E)$ and by taking $\varrho = 0$ over the rest of space: this ensemble describes the state of statistical equilibrium of an isolated system whose energy lies between E and $E + \delta E$.

In the case where δE tends to zero, the microcanonical ensemble is reduced to the constant energy hypersurface Σ: $H = E$. For such an ensemble the invariant density over the hypersurface is given by†

$$d\mu(\Sigma) = \frac{d\Sigma}{\text{grad}\, E}; \tag{I.14}$$

this enables us to define an invariant measure and, eventually, a probability law over the hypersurfaces Σ of Γ-space.

We note, moreover, that if we had m uniform integrals of motion in addition to the energy, a subset $\mathcal{M}$ with $(2n - m - 1)$ dimensions would correspond to it in Γ-space and that it would define, starting from $d\Gamma$, an invariant measure over $\mathcal{M}$. However, in an isolated system the only known uniform integrals are the momentum and the angular momentum, and it can be shown that they need not be taken into consideration because of the role of the walls: for example, if the walls are perfectly reflecting, the momentum of a molecule which rebounds at the wall is not conserved. This justifies the assigning of only a single uniform integral, the energy integral, to an isolated system and describing such a system by a microcanonical ensemble.‡

(c) *Canonical ensemble.* This is defined by

$$\varrho = e^{(\Psi - E)/\theta}, \tag{I.15}$$

where E is the energy of the system (a function of the coordinates and momenta) and θ is the modulus of the distribution; Ψ and θ are parameters with the dimensions of energy and are connected by the relation

$$e^{-\Psi/\theta} = \int_{\Gamma} e^{-E/\theta}\, d\Gamma. \tag{I.16}$$

The various possible values of the parameters Ψ and θ correspond to different macroscopic conditions of the observed system; θ is

† The proof of this result will be found in § 4 of Appendix I.

‡ For a more detailed discussion of the role of uniform integrals of motion, see also Appendix I, § 3.

proportional to the absolute temperature and Ψ is equal to the free energy apart from an additive constant. It can be shown that a canonical ensemble represents the distribution of the states of a system coupled with a thermostat (for the proof, see Chapter IV) and that it is thus particularly appropriate to the description of macroscopic systems in thermostatic equilibrium.

(d) *Grand canonical ensemble.* In everything up to now, we have considered systems having a specified number of particles; if we wish to describe a system in which chemical reactions are taking place, we must break away from this condition. We are thus led to introduce ensembles of systems having a variable number of particles: Gibbs (1902) has called these *grand ensembles* in contrast to the preceding ensembles (having a specified number of particles) called *petit ensembles.*

If we suppose that the system considered contains h kinds of particles and if n_i denotes the number of particles of the ith kind, the density of a grand ensemble will be a function of n_i and of the variables (q_i, p_i); let $\varrho(n_i, \ldots, n_h;\ P)$ be this density: then $\varrho\, d\Gamma$ represents the number of systems of the ensemble having n_i particles of the ith type and found in the element $d\Gamma$. (It should be mentioned that the element $d\Gamma$ depends on n_i, since the number of degrees of freedom, n, of the system is expressed as a function of n_i by the formula $n = \sum_i n_i r_i$, where r_i is the number of degrees of freedom of the particle of the ith type.)

In addition, we must discriminate between phases characterised by *specific* or *generic definitions*: if we consider two states for a system which differ only by the exchange of two particles of the same type, the specific definition considers that these states correspond to two different phases (specific phases) whilst the generic definition considers them as belonging to the same phase (generic phase). The number of specific phases contained in one generic phase is obviously $\prod_i n_i!$.

This being granted, a stationary grand ensemble, distributed canonically, or a *grand canonical* ensemble is defined by a specific density ϱ of the form

$$\varrho_s = \frac{\mathrm{I}}{\prod_i n_i!} \exp\left[\left(\Omega + \sum_{i=1}^{h} \mu_i n_i - E\right)\Big/\theta\right], \qquad \text{(I.17)}$$

where E is the total energy of the system and θ is the modulus; Ω, θ and μ_i are parameters to be determined from macroscopic conditions of the system.

For the generic density ϱ_g we have

$$\varrho_g = \sum \varrho_s = \exp\left[\left(\Omega + \sum_{i=1}^{h} \mu_i n_i - E\right)\Big/\theta\right] \tag{I.18}$$

since the energies of all the specific phases corresponding to the same generic phase are equal. According to equations (I.17) and (I.18), the grand canonical ensemble can be considered as a weighted collection of petit canonical ensembles: in fact, it suffices to fix the value of n_i in order to obtain the formulae of the preceding subsection (in particular, the modulus θ is also proportional to the absolute temperature). The quantity Ω is determined by the equation

$$\sum_{n_1 \ldots n_h} \int_{\Gamma} \frac{\exp\left[\left(\Omega + \sum_{i=1}^{h} \mu_i n_i - E\right)\Big/\theta\right]}{\prod_i n_i!} \, d\Gamma = 1 \tag{I.19}$$

which can be written as

$$e^{-\Omega/\theta} = \sum_{n_1 \ldots n_h} \frac{\exp\left[\left(\sum_i \mu_i n_i\right)\Big/\theta\right]}{\prod_i n_i!} \int_{\Gamma} e^{-E/\theta} \, d\Gamma. \tag{I.19$'$}$$

The integral in (I.19′) is equal to $e^{-\Psi/\theta}$ according to (I.16); we thus have the relation

$$e^{-\Omega/\theta} = \sum_{n_1 \ldots n_h} \frac{\exp\left[\left(\sum_i \mu_i n_i - \Psi\right)\Big/\theta\right]}{\prod n_i!} \tag{I.20}$$

These grand canonical ensembles are well suited for describing systems in equilibrium in contact with a thermostat and a "particle reservoir" (playing the same role as a thermostat). In Chapter IV we shall see the relations which exist between canonical and grand canonical ensembles on the one hand and the microcanonical ensembles, used exclusively in the ergodic theory, on the other hand.

3. *Fundamental hypothesis of classical statistical mechanics*

Statistical mechanics considers the correspondence of an ensemble of systems to a macroscopic physical system. This ensemble, called the *representative ensemble*, is made up from systems, which are independent of one another and with a structure similar to the system studied, with an appropriate statistical distribution. Such an ensemble thus enters into the framework of the ensembles of systems studied above and possesses all the properties which we have mentioned. This ensemble must be constructed, obviously, in such a way that the states of the constituent systems correspond to the partial knowledge of the state of the real system (*accessible states*), knowledge acquired by macroscopic observation; it will thus include all systems whose states are compatible with our macroscopic knowledge. In order to conclude the determination of this representative ensemble, we must define a probability density for it: thus, an essentially statistical hypothesis must be made concerning the *a priori* probabilities appropriate to the various states of the system compatible with our knowledge. We note that we can choose, for this purpose, any law of *a priori* probability over Γ-space (or over the hypersurface Σ), provided that it is an invariant measure with respect to T_t.†

The assumption which is usually made (and which is the most natural one) is that of *equal a priori probability for equal volumes of Γ-space*. The result is that the relative probabilities of finding the system in specified regions of the Γ-space are proportional to the measure of these regions; the probability thus defined is indeed an invariant measure with respect to T_t, because of the invariance of $d\Gamma$ demonstrated in § 1‡ (in addition, we note that as the measures of the volumes of the extension in phase are invariant under canonical transformations, our assumption of equal *a priori* probabilities is thus valid for any system of canonical variables).

This postulate is of a purely statistical nature and, even if it cannot be deduced from the principles of mechanics, it cannot contradict them; in fact, there would be incompatibility between the principles of mechanics and this statistical postulate if it

† Cf. Appendix I, § 4.

‡ In the case of a hypersurface Σ, this assumption leads us to adopt the density (I.14); see Appendix I, § 4.

limited the set of possible mechanical movements, or if the dynamic laws had led us to choosing another assumption than the one adopted: the previously-stated assumption only gives weights to all possible movements (without restricting them) and we have seen, on the other hand, in studying the uniform ensemble, that the Hamiltonian equations do not lead to a concentration of representative points of the systems in any particular region of phase space. This assumption is only finally justified by the agreement of the theoretical predictions with the experimental results; this is why certain authors, such as Tolman (1938), accept it as a basic assumption for founding statistical mechanics and, in adopting a utilitarian point of view, they consider it as sufficiently justified by its consequences.

With these definitions, macroscopic quantities are expressed by statistical averages of the corresponding functions of mechanics, or, for a quantity $f(P)$,

$$\bar{f} = \int_{\Gamma} \varrho(P, t) f(P)\, d\Gamma. \tag{I.21}$$

Thus, these averages are the only quantities which, from the point of view of statistical mechanics, are accessible to macroscopic observations: they offer the advantage that they can be calculated without integrating the equations of motion, at least in the stationary case.

If we wish to free ourselves from the utilitarian point of view, the fundamental problem of statistical mechanics thus is to justify the use of these statistical averages by considerations which bear on the dynamic behaviour of the actual system whose evolution is determined by the Hamiltonian equations: we shall see how the ergodic theory is attempting to resolve this problem.

II. Ergodic Theorems in Classical Mechanics

1. *The ergodic problem*

We are interested here in the case of an isolated system whose energy can be considered as fixed: the representative point P of the system moves thus on a hypersurface of constant energy $H(P) = E$ and the statistical ensemble which correspond to this situation is, as we have seen, the microcanonical ensemble. If we accept that macroscopic quantities are represented by the phase averages of the corresponding functions of mechanics, these will be given according to (I.21) by expressions of the type

$$\overline{f(P)}^{\Sigma} = \frac{1}{\mu(\Sigma)} \int_{\Sigma} f(P) \frac{d\Sigma}{\text{grad } E}, \qquad \text{(I.22)}$$

where $\mu(\Sigma)$ is the measure of the hypersurface $H(P) = E$ and where $d\Sigma/\text{grad } E$ is the invariant measure on Σ, according to (I.14).

On the other hand, however, it is known that the mechanical evolution of the actual system is described by the point P, which moves on a single trajectory defined by the equations (I.1′) (or by the $2n - 1$ integrals of motion of (I.1)). Thus, to each quantity $f(P)$ corresponds a function of P_t: $f(P_t) = f(T_t P)$, whose value would be defined at each instant if the trajectory and the motion of the point P_t along it are known. However, since this instantaneous value is not found by macroscopic observation, it is accepted that this observation only enables us to determine the time-average of these functions over an interval of time T:† this interval T corresponds to the duration of the observation, which is very large in relation to the microscopic evolution of the system. The quantities observed in actual fact will then be described by expressions of the form

$$\overline{f(P_t)}^{T} = \frac{1}{T} \int_{t_0}^{t_0+T} f(P_t)\, dt. \qquad \text{(I.23)}$$

The fundamental problem of statistical mechanics will be resolved if we can justify the replacement of the time-average $\overline{f(P_t)}^{T}$, taken over a trajectory, by the phase mean $\overline{f(P)}^{\Sigma}$, taken over the microcanonical ensemble: this is the more general form under which the ergodic problem is posed and its solution would enable statistical mechanics to be founded on a purely dynamic basis; we shall see, however, that the statistical element cannot be eliminated completely.

† This definition of observed macroscopic quantities depends on the fact that the recording devices always behave as integrators, whatever may be the apparatus used. We shall use another definition of macroscopic quantities in the second section, based on the inherent inaccuracy of any macroscopic observation (see Chapter V, § II.1).

The macroscopically small time T is generally assumed to be a very long one in relation to the process of the return to equilibrium† (the time of free flight of the particles, relaxation time, etc.), which leads us to consider the limit of $\overline{f(P_t)}^T$ for $T \to \infty$. The macroscopic quantities will then be represented by

$$\overline{f(P_t)}^{\infty} = \lim_{T \to \infty} \overline{f(P_t)}^T, \tag{I.23'}$$

if we can establish the convergence of $\overline{f(P_t)}^T$ towards $\overline{f(P_t)}^{\infty}$ and if the relaxation time of the system is small. We note, moreover, that the passage to the limit $T \to \infty$ cannot give rise to any serious objection against the ergodic theory, if we take account of the properties which we shall establish later on both for the "sum" phase functions in classical theory (see Section III of this chapter) and for the macroscopic observables in quantum theory (see Chapter IV).

Thus, in order to solve the ergodic problem, we are led to proving:

(a) the existence of the limit $\overline{f(P_t)}^{\infty}$;

(b) that this limit, if it exists, is the same for all trajectories.

According as we search for an almost certain convergence or convergence in the quadratic mean, the first part of the proof assumes either the form of Birkhoff's theorem or that of von Neumann's theorem; the second part can be proved only by making an assumption concerning the dynamic structure of the system—an assumption which is the same in the two types of convergence.

2. *Birkhoff's theorem* (1931 a, b)

If Ω is an invariant set in phase space, of finite volume, and if $f(P)$ is a function summable over Ω and determined at every point $P \in \Omega$, the limit $\lim_{T \to \infty} \frac{1}{T} \int_{t_0}^{t_0+T} f(P_t)\, dt$ *exists for all points P of the set Ω, except at the most for a certain set of zero measure.*

It is then easy to show that this limit is independent of the initial time t_0 (it is thus an integral of the system). Birkhoff's theorem,

† The order of magnitude of T is thus much larger than that of the duration τ introduced by Kirkwood (cf. Chapter V, § V.3).

which we are stating here without proof,† enables us to establish property (a) since the hypersurface Σ is an invariant set of Γ-space; it is applicable to very general cases, with the sole condition that the group of transformations T_t ("flux" T_t) be measurable.‡ Thus, for every $f(P) \in L_1(\Omega)$ we have an $f^*(P) \in L_1(\Omega)$, invariant over Ω, such that we have almost everywhere (a.e.)

$$\overline{f(P_t)}^{\infty} \equiv \lim_{T \to \infty} \overline{f(P_t)}^{T} \overset{\text{a.e.}}{=} f^*(P); \tag{I.24}$$

if, moreover, $m(\Omega)$ is finite (this is the case if the system is enclosed in a completely bounded region) we have also

$$\int_{\Omega} f^*(P)\, dP = \int_{\Omega} f(P)\, dP. \tag{I.24'}$$

Property (b) is only satisfied, on the contrary, by means of a fundamental assumption concerning the system: the set Ω (or, in our case the hypersurface Σ) must be metrically transitive, that is to say that this set (or this hypersurface) cannot be separated into two invariant subsets of non-zero measure.

In fact, it is then easy to show that $\overline{f(P_t)}^{\infty} = f^*(P)$ is constant almost everywhere over Ω; for, if it were not so, a number a would exist such that the conditions $\overline{f(P_t)}^{\infty} < a$ and $\overline{f(P_t)}^{\infty} > a$ would define a separation of the set into two invariant subsets of non-zero measure, which is contradictory to our hypothesis. It can be established without difficulty that this constant is equal to the microcanonical average of $\overline{f(P)}^{\Sigma}$.†† Conversely, since we can show that if every $f^*(P)$ is constant almost everywhere over Ω, the set Ω is metrically transitive,‡‡ the assumption of metrical transitivity of Ω is thus equivalent to the *ergodic hypothesis* stated in the following form: *all time limits $f^*(P)$ are constant almost everywhere over Ω.*

A system for which the hypersurface of constant energy is metrically transitive is called metrically transitive. Thus, we have

† For this proof, see Appendix I, § 2.

‡ Suppose that $\mathscr{T}$ is the straight line of the variable t and $A \subset \Omega$ is a measurable set; the flux T_t is measurable if the set produces the points $(P \times t)$ of the space $\Omega \times \mathscr{T}$, such that $P_t \in A$ is measurable relative to the measure produced in $\Omega \times \mathscr{T}$.

†† See Appendix I, § 3.

‡‡ See also Appendix I, § 3.

the following result: *if a system has the property of metrical transitivity, one has*

$$\overline{f(P_t)}^{\infty} = \frac{1}{\mu(\Sigma)} \int_{\Sigma} f(P) \frac{d\Sigma}{\text{grad } E}. \tag{I.25}$$

We note first of all that the replacement of the time averages by the phase averages is thus assured for all trajectories of the hypersurface, except perhaps on a set of zero measure. We have thus a stochastic convergence of the almost certain type, so that the set of exceptional trajectories—for which the theorem could not be used — can be neglected only on condition that the previous result is considered as a statistical statement: a zero *a priori* probability must be assigned to a set of initial conditions of zero measure.† Above all, we note that there is no criterion which permits us to recognise whether a physical system possesses the property of metric transitivity: the existence of this property must therefore be assumed in order that relation (I.25) be satisfied; this assumption concerning the structure of the system corresponds to the ergodic hypothesis introduced by Boltzmann‡ (which was unacceptable for topological reasons) or to the quasi-ergodic hypothesis (which was inadequate for establishing the equality of the time and phase-averages). We shall encounter the same difficulty with von Neumann's theorem.

3. *Von Neumann's theorem*

Von Neumann has proposed another solution to the ergodic problem by drawing attention to the convergence in quadratic mean which is entirely satisfactory for application to physical problems. From this point of view it is, in fact, sufficient to prove that there is a constant C such that the error incurred by replacing $\overline{f(P_t)}^T$ by C is always physically negligible (von Neumann, 1932b); for this, it is necessary that the statistical dispersion of $\overline{f(P_t)}^T$ around C is small, or

$$\int_{\Omega} \left| \frac{1}{T} \int_{t_0}^{t_0+T} f(P_t)\, dt - C \right|^2 d\Gamma < \varepsilon, \tag{I.26}$$

† We emphasise, however, that this hypothesis has a very different character from the statistical postulate adopted from the utilitarian point of view by certain authors such as Tolman.

‡ Cf. Appendix I, § 1.

where Ω is an invariant set of Γ-space; the constant C is independent of the point P and it can be shown easily that it can be replaced by the phase average of $f(P)$, with the result that (I.26) becomes

$$\int_{\Omega} \left| \overline{f(P_t)}^{T} - \overline{f(P)}^{\Omega} \right|^2 d\Gamma < \varepsilon. \tag{I.27}$$

In order to be able to establish this result, we shall state the ergodic theorem of von Neumann. For this purpose, Koopman's method is used (Koopman, 1931) which associates a Hilbert space $\mathscr{H}$ with the quadratically summable functions $f(P)$ in Ω; if $m(\Omega)$ is an invariant measure over Ω, the scalar product of $f(P) \in L_2(\Omega)$ and $g(P) \in L_2(\Omega)$ is defined by

$$(f, g) = \int_{\Omega} f(P)\, g(P)\, dm(\Omega).$$

Since T_t is an analytical group of analytical transformations of Ω into itself (or more generally a measurable flux), the scalar product $(f(P_t), g)$ is a measurable function of t and we define the time average $\overline{f(P_t)}^{T}$ by

$$\left(\overline{f(P_t)}^{T}, g\right) = \frac{1}{T}\int_0^T (f(P_t), g)\, dt, \tag{I.28}$$

with the property

$$\left\| \overline{f(P_t)}^{T} \right\| \leqq \|f\|. \tag{I.28$'$}$$

We then introduce the U_t transformations induced in this space $\mathscr{H}$ by the group T_t, by putting

$$U_t f(P) = f(T_t P) = f(P_t), \tag{I.29}$$

where U_t has the group properties (if H is conservative, uniform and analytic):

$$U_t U_s = U_{t+s}, \quad U_0 = 1. \tag{I.29$'$}$$

In addition, this group is unitary because of the conservative nature of equations (I.1′) and of the resultant invariance of the scalar product $(U_t f, U_t g)$. Thus, by applying one of Stone's theorems (Stone, 1930) we can define the spectral resolution of

U_t which can be written as

$$U_t = \int_{-\infty}^{+\infty} e^{i\omega t} \, dE(\omega). \tag{I.30}$$

By using the properties of the spectral operator $E(\omega)$ we can then prove von Neumann's theorem (von Neumann, 1932a):

For every given point $f(P)$ of the space $\mathscr{H}$, an average point $f^*(P)$ can be associated with the set of points $U_t f(P)$, such that

$$\lim_{T\to\infty} \left\| \frac{1}{T} \int_{t_0}^{t_0+T} U_t f(P) \, dt - f^*(P) \right\| \equiv \lim_{T\to\infty} \left\| \overline{f(P_t)}^T - f^*(P) \right\| = 0, \tag{I.31}$$

where $f^*(P)$ is invariant in the broad sense over Ω, or $f^*(P_t) \overset{\text{a.e.}}{=} f^*(P)$, for all given values of t. Thus, according to (I.28′), we have the relation

$$\|f^*\| \leqq \|f\|. \tag{1.32}$$

This theorem, which is mathematically a theorem of strong convergence in the Hilbert space $\mathscr{H}$ is analogous to Birkhoff's theorem but with a convergence in quadratic mean.

If, at the same time, we have $f(P) \in L_2(\Omega)$ and $f(P) \in L_1(\Omega)$, it can be shown that the two limit functions $f^*(P)$, defined respectively by Birkhoff's and von Neumann's theorems, are equal in as far as functions defined almost everywhere over Ω are concerned.

We note that if $f(P)$ is the characteristic function of a set $A \subset \Omega$, $f^*(P)$ represents the limit of the average time during which the point P_t is located in A. If we denote this mean time by $\overline{Z}^T$, (I.31) can be written as

$$\lim_{T\to\infty} \int_{\Omega} \left| \overline{Z}^T - Z(A, P) \right|^2 d\Gamma = 0, \tag{1.31′}$$

where $Z(A, P)$ corresponds to $f^*(P)$.

However, if (I.31) establishes property (a), property (b) and equation (I.26) will be satisfied only if $f^*(P)$ and $Z(A, P)$ are independent of the point P. For this, it is necessary again that the system possesses the property of metric transitivity; if this condition is fulfilled, we can write

$$f^*(P) = \frac{1}{\mu(\Omega)} \int_{\Omega} f(P)\, d\Gamma \tag{1.33}$$

and

$$Z(A, P) = \frac{\mu(A)}{\mu(\Omega)}. \tag{1.33'}$$

Thus, we have for a metrically transitive system

$$\lim_{T\to\infty} \left\| \overline{f(P_t)}^{T} - \overline{f(P)}^{\Omega} \right\| = 0. \tag{1.34}$$

Relation (I.26) is thus found to be satisfied, the proof of which constitutes the fundamental problem of statistical mechanics; moreover, it can be stated in a probabilistic form:

A value of $T = T(\varepsilon, \delta)$ can be found for every value of $\varepsilon > 0$ and $\delta > 0$, such that

$$\text{Prob}\left\{ \left| \frac{1}{\tau} \int_0^{\tau} f(P_t)\, dt - C \right| > \delta \right\} \leqq \varepsilon, \quad \text{provided} \quad \tau > T(\varepsilon, \delta). \tag{1.35}$$

It is obvious that the theorem is applicable without modification if the invariant set Ω is the hypersurface Σ; it is sufficient to replace Ω by Σ and $d\Gamma$ by $d\Sigma/\text{grad}\, E$ in the foregoing formulae.

4. *Hopf's theorem*

Hopf (1932b, c) has proposed a statement which is a little different from the previous theorem: it is important from the physical point of view since it enters into the framework of Gibbs' theory of ensembles and it is of interest in that it lends itself to a convenient comparison with quantum theory. The foregoing theorems can be derived by choosing as the invariant set Ω a shell of Γ-space, defined by the invariant volume contained between two adjacent hypersurfaces $H(P) = E$ and $H(P) = E + \delta E$; this corresponds more to the physical reality, where measurements are never absolutely precise.

On the other hand, Hopf's theorem is derived mathematically from that of von Neumann by transition from strong convergence to weak convergence. Physically, it introduces this transition by posing the ergodic problem in a different way: instead of justifying the substitution of time- by phase-averages, it defines an initial distribution $g(P)$ in Γ-space and it attempts to show that the evolution of the system has the consequence of making this density $g(P)$ tend towards a stationary and uniform limit density $g^*(P)$. In this form, Hopf's ergodic theorem is not without similarity to the H-theorems for ensembles of systems such as we shall be studying in Part II. If $g(P)$ is the initial distribution of the systems in Γ-space, the mean value of a function $f(P)$ is then given by the Hermitian scalar product

$$(f, g) = \int_V f(P)\, g(P)\, d\Gamma, \tag{1.36}$$

where V is the hypervolume enclosed between the two hypersurfaces.

At time t the initial distribution becomes $U_t g(P)$ and the problem is to study its evolution with time. If there is ergodicity, in the sense of a convergence in quadratic mean, $U_t g(P)$ will tend toward a limit distribution $g^*(P)$ with the result that we shall have

$$\lim_{T=\infty} \frac{1}{T} \int_{t_0}^{t_0+T} |(f, U_t g) - (f, g^*)|^2\, dt = 0. \tag{1.37}$$

This is the mathematical expression of Hopf's theorem; it denotes physically that the time-average of statistical fluctuations of the phase-average of $f(P)$ around its limit value (f, g^*) tends to zero, when T increases indefinitely.

In order that the tendency of the initial distribution $g(P)$ towards the limit distribution $g^*(P)$ be assured, it is necessary that on the shell δE a certain process of "mixing" of the initial conditions occurs; this is the process which Hopf has analysed and which we shall encounter again in studying the H-theorems. In order to state correctly the conditions which the system must fufill in order that the theorem (I.37) is true, we must introduce the Cartesian product of Γ-space with itself: $\Gamma \times \Gamma$; then, the hypersurface $\Sigma \times \Sigma'$ of the space $\Gamma \times \Gamma$ corresponds to a pair of hypersurfaces (Σ, Σ') of Γ-space. *Hopf's theorem is then satisfied if any hypersurface Σ in Γ-space is metrically transitive and if almost all the hypersurfaces $\Sigma \times \Sigma'$ in $\Gamma \times \Gamma$ have the same transitivity property.*

In order to facilitate comparison with quantum theory, we shall show briefly the mathematical elements of the proof of the foregoing theorem which rests, like that of von Neumann's theorem, on a theorem of harmonic analysis (Hopf, 1937):

If $g(t)$ is a function of the type $g(t) = \int_{-\infty}^{+\infty} e^{i\omega t}\, dv(\omega)$, where $v(\omega)$ is a distribution function with bounded variation, with both a continuous and a discontinuous part, we have

$$\lim_{T=\infty} \frac{1}{T} \int_{t_0}^{t_0+T} \left| g(t) - \sum_{\omega_\nu} [v(\omega_\nu + 0) - v(\omega_\nu)]\, e^{i\omega_\nu t} \right|^2 dt = 0, \tag{1.38}$$

where $\sum_{\omega_\nu}$ represents the discontinuous part of $v(\omega)$. For the application envisaged, this formula can be written also as

$$\lim_{T=\infty} \frac{1}{T} \int_{t_0}^{t_0+T} \left| (f, U_t\, g) - \left(f, \sum_{\omega_\nu} e^{i\omega_\nu t}[E(\omega_\nu + 0) - E(\omega_\nu)]\, g\right) \right|^2 dt = 0 \tag{1.38'}$$

which reduces to (I.37) when the only discontinuity of $v(\omega)$ is at the point $\omega = 0$, by putting

$$g^*(P) = (E(+0) - E(0))\, g(P). \tag{1.39}$$

The proof of (I.37) then becomes easy; in fact, we put $U'_t = U_t - (E(+0) - E(0))$ and

$$(f, U_t g) - (f, g^*) = (f, U'_t g) = \int_{-\infty}^{+\infty} e^{i\omega t} d(f, E'(\omega)\, g).$$

The spectral decomposition $E'(\omega)$ is now continuous everywhere and, by putting $(f, E'(\omega)\, g) = F(\omega)$, we show that

$$\lim_{T=\infty} \frac{1}{T} \int_{t_0}^{t_0+T} \left| \int_{-\infty}^{+\infty} e^{i\omega t} dF(\omega) \right|^2 dt = 0, \tag{1.40}$$

where $F(\omega)$ is continuous everywhere. By putting $x = \omega - \omega'$, we have

$$\left| \int_{-\infty}^{+\infty} e^{i\omega t} dF(\omega) \right|^2 = \int_{-\infty}^{+\infty} e^{ixt} \int_{-\infty}^{+\infty} dF(\omega)\, \overline{dF(\omega - x)}.$$

The function $G(x) = \int_{-\infty}^{+\infty} \overline{F(\omega - x)}\, dF(\omega)$ is continuous everywhere and we have

$$\left| \frac{1}{T} \int_{t_0}^{t_0+T} \int_{-\infty}^{+\infty} e^{ixt} dG(x)\, dt \right|$$

$$\leqq \int_{-\infty}^{+\infty} \left| \frac{e^{ix(t_0+T)} - e^{ixt_0}}{ixT} \right| d|G(x)| \leqq \int_{-\varepsilon}^{+\varepsilon} d|G(x)|$$

$$+ \frac{2}{\varepsilon T} \int_{-\infty}^{+\infty} d|G(x)\,| = |G(\varepsilon)| - |G(-\varepsilon)| + \frac{2}{\varepsilon T} |(f, g)|^2. \quad (1.41)$$

First of all, ε is chosen sufficiently small in order that the first term is less than $\delta/2$, then T sufficiently large so that the second term becomes, in its turn, smaller than $\delta/2$, which completes the proof of (I.40) and (I.37).

We can see that the result depends essentially on the continuity of $F(\omega)$ and therefore on the fact that $U_t g$ has a discontinuity only at the point $\omega = 0$, a condition that we shall meet again in quantum theory.

The last part of (I.41) enables us to give a mathematical estimate of the time T necessary that expression (I.40) is smaller than a given quantity δ. We see that the time T depends on the value of ε, which itself depends on the function $G(x)$ and which reflects the dynamic properties of the system; therefore, this estimate is in fact theoretical rather than practical and cannot lead to a precise evaluation of the relaxation times.

III. The Hypothesis of Metric Transitivity

1. *Properties of sum functions*

The ergodic theorems, which are essential for statistical mechanics, can thus be proved, provided the system possesses the property of metric transitivity. However, even though it is possible to construct models of ergodic systems, these models are far from real physical systems; there is, in fact, no criterion for recognising whether the usual models of macroscopic systems satisfy the ergodic hypothesis; systems are known, even, which are not ergodic, such as quasi-periodic systems which play an important role in physics. We note again that the hypothesis of metric transitivity just restates that over the invariant set Ω there is no integral of the equations of motion which is not constant almost everywhere over Ω.†

For all these reasons, we are led naturally to investigating whether or not it is possible to break away from this hypothesis in justifying the replacement of the time- by the phase-averages in more general cases. In this connection, we note immediately that the ergodic theorems stated previously are related to a convergence of a stochastic nature (almost certain convergence with the exception of a set of zero measure, or convergence in quadratic mean with the introduction of an initial distribution $g(P)$ in Hopf's hypothesis), with the result that the statistical element of the theory cannot be eliminated.

On the other hand, the theorems mentioned above are valid for any system, independently of the number of degrees of freedom of these systems: they correspond to results of general dynamics which are not specially adapted to the demands of statistical mechanics. However, macroscopic systems are systems with a very large number of degrees of freedom: this leads to special properties of the phase functions $f(P)$ and theorems can be established, which are very close to the ergodic theorems, without difficulty and without another hypothesis concerning the structure of the system. In a certain sense, they constitute a generalisation of the normal theorems and we shall call them "probability" ergodic theorems. We shall show some results obtained by Khinchin—results that we shall extend ultimately to quantum statistical mechanics.

The discussions which follow depend, on the one hand, on the very large number of degrees of freedom of the system (permitting the use of the central limit theorem of probability calculus) and, on the other hand, on the special form of the phase functions which

† Cf. Appendix I, § 3.

occur in statistical mechanics.† In general, these are "sum" functions, as Khinchin calls them (Khinchin, 1949; Truesdell and Morgenstern, 1958), that is to say, they are the sum of functions which each depend on the dynamic coordinates of a single particle. Such a phase function can thus be written as

$$f(P) = \sum_{i=1}^{n} f_i(P_i), \tag{1.42}$$

where each term is a function of the coordinates of a single particle.‡ By making use of the asymptotic formulae of the theory of probabilities†† we shall see that such functions possess certain remarkable properties:

(a) Their dispersion is of the order of magnitude $n^{\frac{1}{2}}$ (n denotes the number of particles here).
(b) The result is that these functions are constant over any hypersurface, except over a set of very small measure.
(c) It is legimitate to replace their time- by their phase-averages except over a set of very small measure.

The first property is easily shown by using the decomposition (I.42); we have, in fact,

$$\overset{-\Sigma}{f} = \sum_{i=1}^{n} \overset{-\Sigma}{f_i} \tag{1.43}$$

and for the dispersion of f:

$$\overline{\left(f - \overset{-\Sigma}{f}\right)^2}^{\Sigma} = \overline{\left\{\sum_{i=1}^{n} \left(f_i - \overset{-\Sigma}{f_i}\right)\right\}^2}^{\Sigma}$$

$$= \sum_{i=1}^{n} \overline{\left(f_i - \overset{-\Sigma}{f_i}\right)^2}^{\Sigma} + \sum_{i \neq k} \overline{\left(f_i - \overset{-\Sigma}{f_i}\right)\left(f_k - \overset{-\Sigma}{f_k}\right)}^{\Sigma}. \tag{1.44}$$

This dispersion consists thus of two sums: the first is the sum of n bounded terms and the second of n^2 terms of order $1/n$ [this

† We mention also that one of Hopf's theorems has been proved recently without a special hypothesis about the functions; see Chapter IV, § IV.4.

‡ We note that in the general case a phase function can be written in the form:

$$f(P) = \sum_{i=1}^{n} f_i^{(1)}(P_i) + \sum_{i,j=1}^{n} f_{ij}^{(2)}(P_i, P_j) + \cdots.$$

†† See Appendix I, § 4 and Khinchin (1949).

last result is obtained by applying the central limit theorem of probability calculus to the calculation of the averages and dispersions of phase functions of the type (I.42); see Khinchin, 1949]. We can then write

$$\overline{\left(f - \overline{f}^{\Sigma}\right)^{2}}^{\Sigma} = O(n), \tag{1.45}$$

or

$$\left\{\overline{\left(f - \overline{f}^{\Sigma}\right)^{2}}^{\Sigma}\right\}^{\frac{1}{2}} = O(n^{\frac{1}{2}}),$$

which proves property (a).

The dispersion can by definition be written as

$$\frac{1}{\mu(\Sigma)} \int_{\Sigma} \left[f(P) - \overline{f}^{\Sigma}\right]^{2} \frac{d\Sigma}{\text{grad } E} = S.$$

If, now, we consider the quantity

$$\frac{1}{\mu(\Sigma)} \int_{\Sigma} \left|f(P) - \overline{f}^{\Sigma}\right| \frac{d\Sigma}{\text{grad } E} = S', \tag{1.46}$$

we can see immediately that, according to Schwartz's inequality,

$$S' \leqq S^{\frac{1}{2}} = O(n^{\frac{1}{2}}), \tag{1.47}$$

which proves property (b).

2. *The probability ergodic theorem*

We are now in a position to justify the replacement of the time-averages by phase-averages. By giving the same meaning to $\overline{f(P_t)}^{T}$ and $\overline{f(P_t)}^{\infty}$ as before, a set of points A of the surface Σ can be defined for which $|\overline{f(P_t)}^{\infty} - \overline{f(P)}^{\Sigma}| > a$ and a set A^T for which $|\overline{f(P_t)}^{T} - \overline{f}^{\Sigma}| > a/2$, where a is a positive number. Because, according to Birkhoff's theorem, $\overline{f(P_t)}^{T} \to \overline{f(P_t)}^{\infty}$ as $T \to \infty$, we have for a sufficiently large value of T the obvious relation between the measures of the sets A and A^T:

$$\mu(A^T) > \frac{1}{2}\mu(A).$$

We derive from it:

$$\frac{a\mu(A)}{4} < \int_{A^T} \left| \overline{f(P_t)}^T - \overline{f}^{\Sigma} \right| \frac{d\Sigma}{\text{grad } E}$$

$$\leqq \frac{1}{T} \int_0^T dt \int_{A^T} \left| f(P_t) - \overline{f}^{\Sigma} \right| \frac{d\Sigma}{\text{grad } E}$$

$$= \frac{1}{T} \int_0^T dt \int_{A^T(t)} \left| f(P) - \overline{f}^{\Sigma} \right| \frac{d\Sigma}{\text{grad } E}$$

$$\leqq \frac{1}{T} \int_0^T dt \int_{\Sigma} \left| f(P) - \overline{f}^{\Sigma} \right| \frac{d\Sigma}{\text{grad } E} = S'\mu(\Sigma), \tag{I.48}$$

where we have used (I.46). Then, according to (I.48) we obtain

$$\frac{\mu(A)}{\mu(\Sigma)} < \frac{4S'}{a}. \tag{1.49}$$

If we choose, for example, $a = S'^{3/2}$, by using (I.49) and (I.47) we find that the relative measure of the set of points for which

$$\left| \overline{f(P_t)}^{\infty} - \overline{f(P)}^{\Sigma} \right| > Cn^{\frac{3}{4}}$$

is a quantity of the order of magnitude $n^{-\frac{1}{4}}$. This result can be expressed in the probabilistic form

$$\text{Prob} \left\{ \left| \overline{f(P_t)}^{\infty} - \overline{f(P)}^{\Sigma} \right| > Cn^{\frac{3}{4}} \right\} = O(n^{-\frac{1}{4}}). \tag{1.50}$$

Thus, we have an estimate of the order of magnitude of the error incurred by replacing the time- by the phase-averages. This error will be small since n will be large, which allows us to resolve practically the fundamental problem of statistical mechanics without resorting to the ergodic theorems of the foregoing section and, as a consequence, by freeing ourselves from the improvable hypothesis of metric transitivity. We shall say that (I.50) constitutes a probability ergodic theorem and that it is equivalent to (I.35) with an ε and a δ dependent on n; it follows that the equality

$\overline{f(P_t)}^{\infty} = \overline{f(P)}^{\Sigma}$ is valid over almost the entire surface Σ apart from a set of very small measure, corresponding to exceptional initial conditions.

In addition, a similar result can be obtained by using relation (I.32) which, in our case, can be written as

$$\int\limits_{\Sigma} \left| \lim \frac{1}{T} \int\limits_{0}^{T} f(P_t)\, dt \right|^2 d\mu(\Sigma) \leqq \int\limits_{\Sigma} |f|^2\, d\mu(\Sigma).$$

If, now, the constant $(\bar{f}^{\Sigma})^2$ is subtracted, we have

$$\frac{1}{\mu(\Sigma)} \int\limits_{\Sigma} \left\{ \left| \overline{f(P_t)}^{\infty} \right|^2 - \left| \bar{f}^{\Sigma} \right|^2 \right\} d\mu(\Sigma) \leqq \frac{1}{\mu(\Sigma)} \int\limits_{\Sigma} \left(|f|^2 - \left| \bar{f}^{\Sigma} \right|^2 \right) d\mu(\Sigma). \tag{I.51}$$

According to property (I.24′), we can write

$$\bar{f}^{\Sigma} = \frac{1}{\mu(\Sigma)} \int\limits_{\Sigma} f d\mu(\Sigma) = \frac{1}{\mu(\Sigma)} \int\limits_{\Sigma} f^* d\mu(\Sigma),$$

as a result of which (I.51) assumes the form

$$\frac{1}{\mu(\Sigma)} \int\limits_{\Sigma} \left(\overline{f(P_t)}^{\infty} - \bar{f}^{\Sigma} \right)^2 d\mu(\Sigma) \leqq \frac{1}{\mu(\Sigma)} \int\limits_{\Sigma} \left(f - \bar{f}^{\Sigma} \right)^2 d\mu(\Sigma) = O(n), \tag{I.52}$$

because of the assumption which we have accepted in this section. Thus, we find a result similar to (I.50) in a form which will be directly comparable with the results that we shall obtain in quantum theory.

The postulates of classical statistical mechanics are thus proved in ultimate analysis by considerations which involve the law of large numbers and especially the asymptotic geometrical properties of hypersurfaces of constant energy (or structure functions) in $6n$-dimensional space.† It must be emphasised also that even though in the foregoing proof a predominant role is played by statistical considerations based on the limit process $n \to \infty$, it

† See Appendix I, § 4 and Appendix II, § 1; see also Part II, Chapter V, § III.1.

does not impose any structure which is peculiar to the Hamiltonian of the system, apart from the canonical nature of the time-evolution; this point of view is contrary to that of Birkhoff's theorem, where the statistical element is reduced to eliminating a set of zero measure, but where the conditions imposed on the structure of the system play an essential role. The advantage of these results lies not only in avoiding the difficulties of the ergodic theory but also in extending the methods of statistical mechanics to systems which are not ergodic in the strict sense: this is the case, for example, with quasi-periodic systems composed of a large number of harmonic oscillators loosely coupled to one another (Terletskii, 1949). We shall not stress these developments, because we shall consider them more thoroughly in quantum theory.

In conclusion, it is important above all to point out that even if the ergodic theory enables the fundamental hypothesis of statistical mechanics to be proved, it does not constitute a derivation of the principles of statistical mechanics beginning with the laws of mechanics. Actually, the possibility of exceptional trajectories always exists and these can be eliminated only by neglecting either a set of zero measure in the exact theory or a set of very small measure in the approximate theory; the ergodic theory thus always maintains a statistical aspect. From this point of view, we could in addition break away from the hypothesis of microcanonical distributions over the hypersurface Σ, since it is sufficient to have a distribution function such that a set of very small measure has a very small probability.

It can be shown also that the ergodic theory is only a particular application of the general ergodic theorems of the theory of stochastic processes; for this, it is sufficient to note that a mechanical function $f(T_t P)$ appears as a strictly stationary random function of the time t, if we suppose that the point P is chosen at random in Γ-space with a probability law defined by $\text{Prob}\{P \in A\} = \mu(A)$, where A is a sub-set of Ω. Moreover, if $f(P)$ is quadratically summable, $f(T_t P)$ is a strictly stationary random function of the second order and we can apply to it all the known properties of groups of unitary transformations of Hilbert space; the ergodic theorems of Birkhoff and von Neumann can be derived from these properties, which appear as a particular case of the ergodic theorems dealing with stochastic processes represented by strictly stationary random functions of the second order.

3. *The role of primary integrals—Lewis's theorem*

In recent years Khinchin's work has given rise to statements and comments from various authors† referring at the same time to the validity and physical significance of his method; they have likewise given birth to a flow of similar ideas in quantum theory and we shall discuss these developments at length in Chapter IV of this book. In what follows, we shall confine ourselves to stressing two statements which appear to us to be essential: the first concerns the possibility of putting part of Khinchin's proof into the form of a theorem from the general theory of dynamics; the second concerns the difficulties encountered by Khinchin's theory because of the possible existence of primary integrals other than those of energy.

In order to be able to discuss usefully these various points we shall first of all in this section analyse the role of the primary integrals of a Hamiltonian system in ergodic theory. This method will lead us naturally to stating Lewis's theorem which, in a certain sense, generalises Birkhoff's theorem. Then, in the light of these latter developments we shall return in the next section to Khinchin's method and we shall then try to define more precisely its exact significance.

(a) *The role of primary integrals.* The connection between the ergodic hypothesis and the primary integrals of a system results from the following important property: if we consider a Hamiltonian system, which is metrically transitive on a constant energy hypersurface Σ, no other primary integrals of motion can exist which are not constant almost everywhere over Σ; in fact, if such an integral were to exist, we could then resolve Σ into two invariant sub-sets of non-zero measure.‡ Since the non-existence of primary integrals other than energy is thus a necessary condition for metric transitivity, it can be seen that the finding of the constants of motion of a Hamiltonian system is an essential problem of the ergodic theory. Morever, let us note that it is convenient to distinguish two types of primary integrals in this problem: the so-called "*global*" integrals which are constant at all times for an arbitrary trajectory of the system and the so-called

† Amongst these should be mentioned the contributions of Truesdell (1961) and Farquhar (1964).

‡ See § III.4 of this chapter and Appendix I, § 3.

"*local*" integrals which are constant only for a limited time and for certain regions of phase space. Only the global integrals are of interest for the ergodic theory, since $2n$-1 local integrals always exist, corresponding to a particular trajectory of the system.

Unfortunately, there is no general method available which enables us to determine all the global primary integrals of a Hamiltonian system; the only known integrals are in reality the following: the integral of the energy for an isolated system enclosed in a space of finite volume and the integrals of energy, linear and angular momentum for an isolated system which is free in space. We have already pointed out† that the necessary existence of walls leads to the disappearance of the conservation of linear and angular momentum, with the result that the only known remaining primary integral is the energy of the system. Nevertheless, the work of Prigogine and co-workers‡ enables us to establish that in the case of an infinite system other primary integrals with singular Fourier transforms exist which are associated with each of the invariants of the unperturbed Hamiltonian. However, since the ergodic theory only considers systems of finite volume, the wall effects also tend to destroy the invariants with singular Fourier transforms: these must be included, therefore, in the category of local integrals.

Apart from the cases which we have just mentioned, nothing is known for certain concerning the existence of other primary integrals. The only known result is due to a theorem by Poincaré (1892) which states that a dynamic Hamiltonian system of finite volume has no "uniform" primary integral other than that of energy and the components of linear and angular momentum. In order to understand the exact scope of this theorem, certain definitions are necessary; actually, Wintner (1941) has shown that in reality the theorem concerns not the "uniform" integrals in the mathematical sense of a function which assumes only a single value at each point, but the so-called "isolating" integrals which define a hypersurface in phase space and which enable us to reduce the number of dimensions of the set over which the trajectory moves. Since these are the only integrals in which we are interested from the point of view of the ergodic theory, it would

† Cf. Chapter I, § 2c.
‡ Cf. Chapter V, Section VI.

appear that Poincaré's theorem makes a decisive contribution to our problem. It turns out, however, that this is not the case, for the following reasons: Poincaré's theorem does not apply to *any* "isolating" integral, but only to the class of these integrals which are analytic in a parameter which measures the development. This parameter depends on the particle masses of the system which must therefore be considered as variables. Since they are actually fixed in any dynamic system, Wintner concludes that the domain of analyticity is reduced to a single point, which removes all dynamic significance from the theorem; other pertinent criticisms have been made also against the significance of Poincaré's result by Truesdell (1961) and by Cherry (1925).

Apart from these negative conclusions, it has not been possible to obtain any other result about the existence of "global" and "isolating" primary integrals other than energy. Similarly, tentative attempts, especially by Oxtoby and Ulam (1941), to establish the metric transitivity of general dynamic systems does not appear to lead to results which are directly applicable in physics (cf. Appendix I, § 1). Nevertheless, the number of primary integrals to be considered can be again reduced by noting, with Khinchin, that these integrals must necessarily belong to the class of so-called "normal" phase functions (Khinchin, 1949) [or "uniform" phase functions according to Rosenfeld's terminology (Rosenfeld, 1952)]; that is, functions which take only one value for all phases corresponding to the same physical state of the system (cf. Appendix I, § 3). Under these conditions, we must substitute for the concept of metric transitivity in the strict sense that of *metric transitivity in the extended sense* (or in the "physical" sense according to Rosenfeld); this is a property according to which an invariant set cannot be resolved into two sub-sets of non-zero measure, such that all points corresponding to the same physical state are contained in the same sub-set (cf. Appendix I, § 3). Our problem, therefore, reduces to that of finding all the global, "isolating" and normal primary integrals of a Hamiltonian system. But since it appears to be just as difficult to determine the normal integrals as those which are not normal, it can be seen that our problem remains as large as ever and that, despite a formulation which is as precise as possible, we are not actually able to establish a general criterion of metric transitivity in the extended sense.

(b) *Lewis's theorem.* As it is evident from the foregoing analysis that it is not possible to eliminate the *a priori* existence of primary integrals other than of energy, it is useful to study what happens with the ergodic theory in the general case where a Hamiltonian system possesses several global integrals which are independent of one another. This is the purpose of Lewis's theorem (Lewis, 1960) which enables us to calculate the time-average in the case where the hypersurface Σ is not metrically transitive.

Let us consider a global integral $y(P)$, or invariant of motion, which is by definition a constant function over almost all trajectories of Γ-space, that is to say we have: $y(T_t P) = y(P)$ except at most over a set of zero measure. Since, in physical cases, the number of dimensions of Γ-space is finite and equal to $2n$, at most $(2n - 1)$ independent integrals of motion can exist. Suppose now that we have k independent primary integrals $y_i(P)$ with $k \leqq 2n - 1$; we say that they constitute a *complete invariant* of motion which we shall denote by $Y(P)$ with:

$$Y(P) \equiv \{y_1(P), y_2(P), \ldots, y_k(P)\}. \tag{I.53}$$

From this definition, it follows necessarily that any measurable integral of motion depends functionally on $Y(P)$ almost everywhere in Γ-space. But we know, from Birkhoff's theorem, that the time-average $\overline{f(P_t)}^{\infty}$ of a phase function $f(P)$ exists in this case and that it is constant along almost all trajectories; it is thus an invariant of motion which depends functionally on $Y(P)$. We can write therefore

$$\overline{f(P_t)}^{\infty} \underset{\text{a.e.}}{=} F[Y(P)], \tag{I.54}$$

with the result that, since the complete invariant Y is assumed to be known, the problem of calculating $\overline{f(P_t)}^{\infty}$ is found to reduce to that of determining the function F.

Now, the values of the complete invariant $Y(P)$ defined in Γ-space constitute a set of points in the Euclidean space R_k with k dimensions. If we consider an arbitrary Borel set B in R_k, a measurable set of points of Γ-space corresponds to the average of the inverse function $Y^{-1}(B)$; because of the invariance of $Y(P)$, this set $Y^{-1}(B)$ differs from an invariant set in Γ-space only by a set of zero measure. Birkhoff's theorem can be applied therefore to the set $Y^{-1}(B)$ and, if we denote the invariant Lebesgue measure in Γ-space by $dm(P)$, we obtain from equation (I.24′)

$$\int_{Y^{-1}(B)} F[Y(P)]\, dm(P) = \int_{Y^{-1}(B)} \overline{f(P_t)}^{\infty}\, dm(P) = \int_{Y^{-1}(B)} f(P)\, dm(P). \quad (I.55)$$

However, the measure m in Γ-space introduces a measure M in R_k according to the formula:

$$M(B) = m[Y^{-1}(B)]. \quad (I.56)$$

Thus, for every measurable function F:

$$\int_{Y^{-1}(B)} F[Y(P)]\, dm = \int_{B} F(Y)\, dM, \quad (I.57)$$

whence, by comparison with equation (I.55),

$$\int_{B} F(Y)\, dM = \int_{Y^{-1}(B)} f dm. \quad (I.58)$$

Thus, we have found in terms of known quantities the integral of F over an arbitrary Borel set of R_k. In order to obtain F itself, it remains now for us to differentiate in the sense of the abstract theory of differentiation; this derivative must be calculated for a value of Y given by $Y(P) = u$, where $u \equiv \{u_1, u_2, \ldots, u_k)$ represents particular values, assumed to be known, of the integrals $y_1, y_2, \ldots, y_k$. According to a result from lattice theory, the concept of differentiation is generalised in such a way that the derivative of an integral of the function $F(Y)$ over an arbitrary set B of R_k-space is evaluated for the set $Y = u$ by:

$$\lim_{\delta \to 0} \frac{\int_{I_\delta(u)} F(Y)\, dM}{M[I_\delta(u)]} \quad (I.59)$$

where $I_\delta(u)$ is the set of points for which the invariants $y_i(P)$ satisfy the inequalities:

$$|y_i - u_i| \leqq \delta. \quad (I.60)$$

Applying this result to equation (I.58), we obtain finally:

$$F(u) \underset{\text{a.e.}}{=} \lim_{\delta \to 0} \frac{\int_{I_\delta(u)} F(Y)\, dM}{M[I_\delta(u)]} \quad (I.61)$$

and, returning to Γ-space,

$$F(u) \underset{\text{a.e.}}{=} \lim_{\delta\to 0} \frac{\int\limits_{Y^{-1}[I_\delta(u)]} f(P)\, dm}{m\{Y^{-1}[I_\delta(u)]\}}. \tag{I.62}$$

The set $S_\delta(u) \equiv Y^{-1}[I_\delta(u)]$ is a set of points of Γ-space which form, according to the definition of $I_\delta(u)$, a thin layer around the hypersurface $Y(P) = u$. Taking into account equations (I.24) and (I.54), we have finally:

$$\overline{f(P_t)}^{\infty} \underset{\text{a.e.}}{=} f^*[Y^{-1}(u)] \underset{\text{a.e.}}{=} \lim_{\delta\to 0} \frac{\int\limits_{S_\delta(u)} f dm}{m[S_\delta(u)]}. \tag{I.63}$$

This is Lewis's first theorem: it shows us that for almost all trajectories for which the complete invariant $Y(P)$ has the value u, the time-average of a phase function $f(P)$ is equal to the phase average as defined in equation (I.63), that is to say, with an equal *a priori* probability in the sense of the measure m for each phase of a thin layer around the hypersurface $Y(P) = u$. If this hypersurface is sufficiently "smooth" for analytical methods to be applicable, we can calculate the limit of equation (I.63) and thus obtain the time-averages.

In the case where the complete invariant $Y(P)$ is reduced to the energy of the system, we fall back on Birkhoff's theorem with the property of metric transitivity. If the motion were not metrically transitive, we could resolve the hypersurface Σ into two invariant subsets of positive measure, whose characteristic functions would then be invariants of motion, functionally independent of the energy. If, however, the complete invariant $Y(P)$ is not reduced to the energy of the system, Lewis's theorem goes much further than Birkhoff's result, in the sense that it shows us how the time-averages must be calculated in the case where the hypersurface Σ is no longer metrically transitive. We have still the equality of the time-averages and of the phase-averages, provided that we confine ourselves to that part of phase space which is contained in a thin layer around the hypersurface $Y(P) = u$, over which we can define an invariant measure. This procedure, therefore, has the effect of replacing the hypersurface Σ—which does not have the property of metric transitivity—by the hypersurface $Y(P) = u$, which has this property since it is defined by a *complete* invariant of motion; put in this form, Lewis's result could have been expected *a priori*. We note that Truesdell (1961) called these new phase-averages panta-microcanonical averages.

To conclude, we note also that Lewis has demonstrated a second theorem which is valid for "small" systems in weak interaction with a large thermostat. As before, it is assumed that we know one complete invariant of the total system, small system + thermostat; by taking into account the asymptotic properties of the thermostat [see, for example, § V.2 of Chapter IV and also Appendix I, § 4], it can then be shown that the time-averages of the phase functions of the small system are equal to the phase-averages taken over an ensemble which is a generalisation of the canonical distribution. In this distribution, which Truesdell called panta-canonical, the complete invariant Y plays the role of the energy E, with the result that the exponential $e^{-\beta E}$ is replaced by an expression of the form $e^{-(\beta_1 y_1 + \beta_2 y_2 + \cdots + \beta_k y_k)}$; we note that we can develop, starting from this new ensemble, a thermostatics which differs significantly from the normal thermostatics (Grad, 1952).

4. *Some remarks on Khinchin's asymptotic method*

Having thus defined the role of primary integrals in ergodic theory, we can now return to a discussion of the asymptotic results obtained by Khinchin. We point out first of all that Khinchin's proof consists of two essentially different stages. The first concerns the asymptotic properties of the sum functions which are expressed by formulae (I.45) and (I.47); this is a geometrical result which is related to the large number of degrees of freedom of the system, but which does not involve its dynamic behaviour. The second stage corresponds to the proofs of § 2 which show up a fundamental property of the phase functions contained in relation (I.52). This property can be stated as follows: *the phase dispersion of the time-average* $\overline{f(P_t)}^{\infty}$ *of a phase function* f *is always less than the phase dispersion of the function* f *itself.* Although this result has been proven in § 2 for a hypersurface Σ of constant energy, it naturally remains valid for any invariant set in Γ-space. Moreover, as its proof only involves Birkhoff's theorem and as the measure used does not come into the reasoning, this result can be extended to the case of an arbitrary measure on condition that it is conserved by the dynamic group T_t. This part of Khinchin's theory thus assumes a very general mathematical character which relates it to the theorems of general dynamics established by Birkhoff, von Neumann, and Hopf. If Ω denotes any invariant set, an identical reasoning to that which led to equation (I.52) shows that we have completely generally:

$$\overline{\left(\overline{f(P_t)}^{\infty} - \overline{f}^{-\Omega}\right)^2}^{\Omega} \leqq \overline{\left(f - \overline{f}^{-\Omega}\right)^2}^{\Omega}, \tag{I.64}$$

where the averages over the set Ω are taken with any measure which is invariant with respect to T_t. Equation (I.64) can be expressed in terms of probability by following Kurth (1958) and using Bienaymé and Chebichev's inequality which allows us to evaluate the probability for the deviation of a random function h from its average value $\bar{h}$ as a function of its dispersion; this inequality can be written as:

$$\text{Prob}\,\{|h - \bar{h}| \geqq \alpha\} \leqq \frac{1}{\alpha^2}\overline{(h - \bar{h})^2}. \tag{I.65}$$

By putting $h = \overline{f(P_t)}^{\infty}$ and by using equation (I.24′), we have according to equation (I.52)

$$\text{Prob}\left\{\left|\overline{f(P_t)}^{\infty} - \overline{f}^{-\Omega}\right| \geqq \alpha\right\} \leqq \frac{1}{\alpha^2}\overline{\left(f - \overline{f}^{-\Omega}\right)^2}^{\Omega}, \tag{I.66}$$

where α is an arbitrary positive number. If we put $\alpha = K\sqrt[4]{\overline{\left(f - \overline{f}^{-\Omega}\right)^2}^{\Omega}}$ we obtain the inequality:

$$\text{Prob}\left\{\left|\overline{f(P_t)}^{\infty} - \overline{f}^{-\Omega}\right| \geqq K\sqrt[4]{\overline{\left(f - \overline{f}^{-\Omega}\right)^2}^{\Omega}}\right\} \leqq \frac{1}{K^2}\sqrt{\overline{\left(f - \overline{f}^{-\Omega}\right)^2}^{\Omega}}, \tag{I.67}$$

which can be compared with equation (I.50).

Inequalities (I.66) and (I.67) express the complete dynamic content of Khinchin's method; they are valid for any definition of probability, provided it is invariant with respect to the group T_t. However, in order that these results are applicable to the ergodic theory, it is essential that this definition satisfies the following two conditions:

1. That the phase dispersion of the physically interesting functions f (i.e. of the sum functions in Khinchin's theory) is small.

2. That the probability for the occurrence of phases corresponding to a set of very small measure is itself very small.

The first condition involves the asymptotic geometrical properties of systems with a large number of degrees of freedom, established in § 1 in the case of a microcanonical distribution. In order to satisfy the second condition it suffices to limit oneself to probability distributions which are *absolutely continuous* relative to the invariant measure over the invariant set Ω considered.

We note now that the introduction of absolutely continuous probability distributions is already necessary in the strict ergodic theory, in order to eliminate the exceptional trajectories for which there is no equality of time and phase averages. In addition, this will raise a serious objection to the point of view which consists in founding statistical mechanics on a purely dynamic basis because, even by accepting the hypothesis of metric transitivity, it is necessary to give again an *a priori* significance to the probability concept of a set on the hypersurface Σ (see also below). Thus, we return to the statistical point of view of the theory of representative ensembles which the ergodic theory aims to prove; we note, however, that the condition of absolute continuity of the probability relative to the measure over Σ is much weaker than the hypothesis of equal probability of equal volumes of phase space which is the basis of the theory of representative ensembles.

This being so, Khinchin's method has essentially the aim of using the statistical element which is the basis of the ergodic theory in order to try to dispense with the hypothesis of metric transitivity relating to the dynamic structure of macroscopic systems; for this, it is sufficient to replace the sets of zero measure of the exact theory by sets whose measure approaches zero as $n \to \infty$. However, it is precisely on this point that the use of Khinchin's method encounters a serious difficulty which we shall now examine. Suppose, for example, that there exists a supplementary primary integral in addition to the energy. The geometrical result, according to which the dispersion of a sum phase function is very small relative to the microcanonical ensemble, always remains valid; we can again deduce from it that the time average of a sum function only differs from the microcanonical average over a set of very small measure. However, because of the existence of a supplementary integral, the trajectories are all located on a subset whose number of dimensions is smaller than that of Σ; this subset is therefore of zero measure with respect to the microcanonical distribution over Σ and, consequently, the set of all phases with a physical significance is also of zero measure. If, then, we define a distribution of probability over Σ such that a set of zero measure has a zero probability (absolutely continuous probability), it follows that these phases have a zero probability of being observed, although these are the only physically achievable ones. Thus, the use of Khinchin's method, which, in principle, would enable us to dispense with the hypothesis of metric transitivity, runs up against a contradiction in the case where the hypersurface Σ is not metrically transitive; such a paradox does not occur in the exact ergodic theory, since in this case we must *postulate the metric transitivity* of Σ, which is perfectly compatible with the elimination of a set of exceptional phases of zero measure.

Two ways are presented to us for trying to avoid this difficulty. The first consists in assuming that the supplementary invariants of motion have a very small microcanonical dispersion with the result that they are constant over almost the entire hypersurface Σ, except over a set of very small measure; this condition, which would be effectively fulfilled if the invariants of motion were sum functions, reverts in fact to making an assumption concerning the structure of the Hamiltonian which seems as difficult to prove as that of metric transitivity, even by restricting ourselves to "isolating" and "normal" integrals. Another method, proposed by Truesdell, starts by distinguishing the so-called "controllable" integrals, whose form is known explicitly and whose value can be fixed by experiment, and the non-controllable integrals—called "free" integrals by Khinchin, or "residual" integrals by Truesdell—whose form and value are not known to us (cf. Appendix I, § 3). We note that the controllable integrals are necessarily normal whilst the free integrals are not necessarily so and may therefore not have a physical significance. The value of the controllable integrals is always fixed by experiment: this is the case with the energy for all systems; this would also be the case for linear and angular momentum for systems in which these quantities are conserved. In fixing the value of these controllable integrals, we define a subset $\mathscr{M}$ of phase space over which we can define an invariant measure according to Lewis's method; in this way, we obtain a distribution which Truesdell called polymicrocanonical and which can be related to the microcanonical distribution in the case when the only controllable integral is the energy of the system. In order that Khinchin's method might then be applied, Truesdell suggests that the unknown values of the "free" integrals be considered as random variables defined on the subset $\mathscr{M}$, their values being distributed with equal *a priori* probability relative to the polymicrocanonical measure. In Truesdell's terminology this is the *hypothesis of the zero set*: the time averages are equal to the phase averages taken over the subset $\mathscr{M}$, except for a set of very small polymicrocanonical measure. In this framework, the introduction of a polymicrocanonical measure does not correspond to that of a representative ensemble of a collection of identical systems, but provides only a criterion for determining the trajectories which are of no importance in the application of Khinchin's ergodic theorem; this is the minimum concession that can be made to the theory of representative ensembles.

Let us quote yet another paradox of Khinchin's theory, called the paradox of weak interactions, which we also consider in Appendix I, § 4, and which leads to some interesting comments. The argument is as follows: Khinchin's theorem has been proved only for systems whose Hamiltonian is separable, that is to say of the form $H = \sum_{i=1}^{n} H_i$; however, in this case we have n integrals of motion $H_i = E_i$, and there can be no question of ergodicity with respect to the hypersurface Σ since there is no interaction between the various components of the system. The reply to this objection depends on the fact that the Hamiltonian is only separable in as far as the interaction term, of the order of magnitude of λ, is negligible compared with the Hamiltonian of a component. The results obtained with the separable Hamiltonian constitute, therefore, an approximation of more exact results which we would have if the interaction was not neglected. In reality, we are dealing with two limit processes, $n \to \infty$ and $\lambda \to 0$, whose order is not unimportant; the paradox arises from the fact that we have first taken the limit $\lambda \to 0$, which leads us to consider the double limit process $\lim_{n\to\infty} \lim_{\lambda\to 0}$. However, it seems that the limit $n \to \infty$ must be taken first, since an interaction which is itself very small compared with the energy of a component can involve phase transitions which are not negligible. In order to obtain physically acceptable results it is necessary to interchange the foregoing double limit and to consider the process $\lim_{\lambda\to 0} \lim_{n\to\infty}$.

This idea has been used recently by Mazur and van der Linden (1963) who have been able to show, by means of certain supplementary hypotheses, that the asymptotic properties of Khinchin's phase functions could be extended to systems formed of weakly interacting components. The fact that the two limit processes $\lambda \to 0$ and $n \to \infty$ are not commutative is an important result that we shall meet again in establishing a Master Equation, especially in the "weak coupling" limit (cf. Chapter V, § VI, and Chapter VI, § V).

In conclusion, we see that the ergodic theory is actually a long way from having attained its objectives. In fact, it encounters two major difficulties: the first, of a dynamic nature, is related to the hypothesis of metric transitivity or to hypotheses relating to the structure of the invariants of motion, for which there is no criterion which enables their justification to be proved, at least in certain cases; the second, perhaps more serious, is related to the inevitable introduction *a priori* of a statistical element into the theory, in the form of a hypothesis of absolute continuity of the probability distribution of the initial phases, which is necessary in order to eliminate exceptional trajectories. Although the statistical content of this hypothesis is much "weaker" than that of the theory of representative ensembles, nevertheless the possibility of justifying the methods of statistical mechanics on a purely dynamic basis appears to remain very improbable.

Nevertheless, there is a natural way to introduce statistical concepts into ergodic theory: this is the so-called "coarse-graining" of phase space which leads to replacing "fine-grained" by "coarse-grained" quantities. According to this procedure, the phase space is divided in phase cells, constructed in such a way that macroscopic observation—which is necessarily imprecise—does not enable us to distinguish between different phases in one cell; under these conditions, the observed physical quantities are no longer phase-functions themselves but only certain averages of these functions (see Chapter V, § II.1). Since the probability distribution used for calculating these averages obviously is not known, the hypothesis of absolute continuity can be assumed, in order that the exceptional phases of zero measure (or of very small measure) are automatically eliminated by the averaging process.

We are thus led to trying to replace the ergodic theorems, which have been established for fine-grained quantities, by the ergodic theorems dealing with coarse-grained quantities, the most convenient way of doing this being that offered by Hopf's theorem. This method, however, raises difficult problems: first of all, conditions for ergodicity must be found which correspond to the "mixing" processes of the initial conditions ("weakly mixing" processes in the case of Hopf's theorem, cf. Chapter I, § II.4), which imply very strong hypotheses concerning the dynamic structure of the Hamiltonian; moreover, it appears that these conditions are still insufficient for ensuring uniformity of the density and of the coarse-grained quantities (see the detailed discussion by Farquhar, 1964). On the other hand, the "coarse graining" method encounters in the same way the following important difficulty: actually, there is no precise indication which enables us to determine the size of the phase cells or the number of macroscopic quantities which must be measured in order to determine at every instant the macroscopic state of the system. Uhlhorn's attempt (1960, 1961), in which Γ-space is replaced by the "phase space of an experiment" associated with the measure of a "complete" ensemble of macroscopic observables, also encounters an objection of this kind. We point out, finally, that research which is actually being undertaken is tending to prove that the whole procedure of "coarse-graining", over time (see Kirkwood's method, described in Chapter V, § V. 3) or over phase space, is insufficient by itself for justifying the methods of statistical mechanics.

CHAPTER II

Quantum Mechanical Ensembles. Macroscopic Operators

WE shall develop quantum statistical mechanics by following a similar scheme to that of classical theory and it will be interesting to show the points which are common to both theories. In this chapter we shall define statistical ensembles of quantum systems by using the formalism of the density matrix and we shall study subsequently their evolution; having defined stationary ensembles, we shall then be able to calculate the statistical averages of observables for quantum systems, similar to the phase-averages of classical quantities. Then, having defined the macroscopic operators associated with macroscopic observations, we shall deal in Chapter III with the study of the ergodic quantum theory by following a course comparable with that of Hopf's theorem in classical mechanics. We shall then encounter the quantum analogy of the hypothesis of metric transitivity and we shall show, in Chapter IV, how we can avoid these difficulties by similar reasonings to those of Khinchin in classical mechanics.

I. Statistical Ensembles of Quantum Systems

1. *The statistical matrix in quantum mechanics*

A quantum system with n degrees of freedom is described by a wave function $\Psi(\dots q_l, \dots, t)$ over the configuration space of the system defined by the n coordinates q_l. The evolution of the system is determined by Schrödinger's equation

$$\hat{H}\Psi = i\hbar \frac{\partial \Psi}{\partial t}, \tag{II.1}$$

where $\hat{H}$ is the quantum Hamiltonian of the system; the Hamiltonian of an isolated system is defined by conditions similar to those used in classical theory. We shall accept, in what follows, that the system occupies a finite part of space and that the spectrum of $\hat{H}$ is a discrete spectrum which can be degenerate for certain energy values; these conditions correspond to those of classical theory relative to the finite measure of the hypersurface Σ.

Suppose, now, that we have expanded the function $\Psi(t)$ in a series of orthonormal functions φ_i†, which constitute a basis in Hilbert space; we have

$$\Psi(\ldots q_l, \ldots, t) = \sum_{i=1}^{\infty} c_i(t)\, \varphi_i(\ldots q_l \ldots). \tag{II.2}$$

The coefficients $c_i(t)$ define in Hilbert space the vector $\Psi(t)$ of unit length whose end moves over the hypersphere defined by

$$\sum_{i=1}^{\infty} |c_i(t)|^2 = 1. \tag{II.3}$$

The trajectory of $\Psi(t)$ over this hypersphere is determined by the solution of equation (II.1). As in classical mechanics, a group of U_t transformations can be associated with (II.1), which transform the hypersphere into itself in the course of time; we have the relations

$$\Psi(t) = \hat{U}_t \Psi(0) \tag{II.4}$$

and

$$\hat{U}_t \hat{U}_s = \hat{U}_{t+s}, \quad \hat{U}_0 = 1. \tag{II.5}$$

The U_t transformation represents the operator of evolution associated with the system (II.1) and it can be written, if the Hamiltonian is time-independent (we are interested in what follows only in conservative systems, as in classical mechanics), as

$$\hat{U}_t = \exp\left[-\frac{i}{\hbar}\hat{H}t\right]. \tag{II.6}$$

If we know the function $\Psi(t)$, we can calculate the quantum mechanical average of an observable defined by a Hermitian operator $\hat{A}$:

$$(\Psi, \hat{A}\Psi) = \sum_{i,j} c_j^*(t)\, c_i(t)\, (\varphi_j, \hat{A}\varphi_i). \tag{II.7}$$

† We shall assume always that the wave functions and the eigenfunctions satisfy the symmetry properties required by the nature of the particles considered.

By putting

$$c_j^*(t)\, c_i(t) = \varrho_{ij}(t), \tag{II.8}$$

equation (II.7) becomes

$$A(t) = (\Psi, \hat{A}\Psi) = \sum_{i,j} \varrho_{ij}(t)\, (\varphi_j, \hat{A}\varphi_i). \tag{II.9}$$

Equation (II.8) defines a matrix $\hat{\varrho}$ called the *density matrix* (or the *quantum statistical operator*) associated with the *pure case* $\Psi(t)$. It is easy to see that it has the following properties:

(a) This matrix is Hermitian since $\varrho_{ij} = \varrho_{ji}^*$.

(b) Its trace is equal to unity; we have, in fact,

$$\mathrm{Tr}\, \hat{\varrho} = \sum_{i=1}^{\infty} \varrho_{ii} = \sum_{i=1}^{\infty} |c_i(t)|^2 = 1. \tag{II.10}$$

(c) It satisfies the relation

$$\hat{\varrho}^2 = \hat{\varrho}, \tag{II.11}$$

since we have

$$(\varrho^2)_{ij} = \sum_{k=1}^{\infty} c_j^* c_k c_k^* c_i = c_j^* c_i = \varrho_{ij}.$$

According to (II.11), the matrix $\hat{\varrho}$ is a projection operator on the state Ψ and it can be written: $\hat{\varrho} = \hat{P}_\psi$.

With these definitions, the quantum mechanical average of an observable $\hat{A}$ is given by

$$A(t) = \mathrm{Tr}\, [\hat{\varrho}(t)\hat{A}]. \tag{II.12}$$

Thus, we see that the quantum properties of a system existing in a pure case are described completely by the statistical matrix $\hat{\varrho}$.

The equation of evolution of this matrix is obtained easily by starting with equation (II.1). In fact, it can be written as

$$\frac{\partial c_i}{\partial t} = -\frac{i}{\hbar} \sum_k H_{ik} c_k,$$

with $H_{ik} = (\varphi_i, \hat{H}\varphi_k)$; whence, by taking the complex conjugate of the preceding expression and by noting that

$$\frac{\partial \varrho_{ij}}{\partial t} = \frac{\partial c_j^*}{\partial t} c_i + c_j^* \frac{\partial c_i}{\partial t},$$

we arrive at

$$\frac{\partial \varrho_{ij}}{\partial t} = -\frac{i}{\hbar}\sum_k (H_{ik}\varrho_{kj} - \varrho_{ik}H_{kj}); \qquad \text{(II.13)}$$

equation (II.13) can be written in matrix language:

$$\frac{\partial \hat{\varrho}}{\partial t} = -\frac{i}{\hbar}(\hat{H}\hat{\varrho} - \hat{\varrho}\hat{H}) \qquad \text{(II.14)}$$

or, by introducing commutators which, apart from a factor, are the quantum equivalent of the Poisson brackets in classical mechanics,

$$\frac{\partial \hat{\varrho}}{\partial t} = \frac{i}{\hbar}[\hat{\varrho}, \hat{H}]_{-}. \qquad \text{(II.15)}$$

The solution of (II.14) can be written in operator form as

$$\hat{\varrho}(t) = \hat{U}_t \hat{\varrho}(0)\, \hat{U}_t^* \qquad \text{(II.15')}$$

with

$$\varrho_{ij}(0) = c_j^*(0)\, c_i(0).$$

The matrix $\hat{\varrho}(t)$ represents the maximum knowledge we can have about a quantum system in the pure case $\Psi(t)$: it comprises the probabilities of the various states (diagonal elements) and the correlations between these states. The probabilities which it involves are the specifically quantum probabilities which follow from the nature of the description of a physical system in wave mechanics. These probabilities must be carefully distinguished from the probabilities in the classical sense which occur in the definition of ensembles of quantum systems. [In addition, we know that the probabilities of quantum mechanics do not obey the normal rules of probability calculations (de Broglie, 1948, 1957).]

2. *Definition of ensembles of quantum systems*

For similar reasons to those explained in classical theory we are led, in quantum statistical mechanics, to study ensembles of systems. As before, it is the incomplete nature of the macroscopic observation, only giving us limited information about a system dependent on a very large number of parameters, which leads us to replace the real system by an ensemble of systems conveniently constructured and corresponding to various microscopic states compatible with our observation [*accessible states*†].

Suppose, now, that to a physical system studied through macroscopic observation is associated an ensemble of $\mathcal{N}$ identical systems which are independent of one another and each described in a pure case by a wave function $\Psi^{(\alpha)}(t)$ ($\alpha = 1, 2, \ldots, \mathcal{N}$):

$$\Psi^{(\alpha)}(t) = \sum_{i=1}^{\infty} c_i^{(\alpha)}(t)\,\varphi_i. \tag{II.16}$$

An elementary statistical matrix

$$\varrho_{ij}^{(\alpha)}(t) = c_j^{(\alpha)*}(t)\,c_i^{(\alpha)}(t) \tag{II.17}$$

corresponds to each of these functions $\Psi^{(\alpha)}(t)$.

The quantum mechanical average of an observable $\hat{A}$ for a system in the state $\Psi^{(\alpha)}(t)$ can, according to (II.12), be written as

$$A^{(\alpha)}(t) = \mathrm{Tr}\,[\hat{\varrho}^{(\alpha)}(t)\,\hat{A}] \tag{II.18}$$

and the mean value of $\hat{A}$, over the ensemble of the $\mathcal{N}$ systems considered, assumes the form

$$\bar{A}(t) = \frac{1}{\mathcal{N}} \sum_{\alpha=1}^{\mathcal{N}} \mathrm{Tr}\,[\hat{\varrho}^{(\alpha)}(t)\,\hat{A}] = \frac{1}{\mathcal{N}} \sum_{\alpha=1}^{\mathcal{N}} \sum_{i,j} \varrho_{ij}^{(\alpha)} A_{ji}. \tag{II.19}$$

$$= \sum_{i,j} \left(\frac{1}{\mathcal{N}} \sum_{\alpha=1}^{\mathcal{N}} \varrho_{ij}^{(\alpha)} \right) A_{ji}.$$

Equation (II.19) shows us that the mean value of a quantity $\hat{A}$ taken over such an ensemble of systems can be written in a form which is identical to equation (II.12), provided that an average statistical matrix is defined

$$\bar{\varrho}_{ij} = \frac{1}{\mathcal{N}} \sum_{\alpha=1}^{\mathcal{N}} \varrho_{ij}^{(\alpha)} = \overline{c_j^{(\alpha)*}(t)\,c_i^{(\alpha)}(t)}. \tag{II.20}$$

This matrix characterises the statistical ensemble used: we have defined it by supposing that the ensemble contains a finite number of systems. However, one is more often led to using infinite sets

† We note that in quantum theory, the concept of accessibility must take account of the properties of symmetry or antisymmetry of the wave functions.

of systems; in this case, the sums of (II.19) and (II.20) must be replaced by integration in the functional space with a suitable weight (in the sense of classical probabilities, naturally): the statistical ensemble considered is thus represented by a cluster of points on the hypersphere (II.3) of the complete Hilbert space.

Since we cannot define a measure in Hilbert space with a countable infinity of dimensions, we are led to assume that our data concerning the system are such that the various possible states compatible with our observation form a discrete and finite series (the case where it is known that a quantity exists in a given interval comprising n possible states). The hypersphere (II.3) has, in this case, a finite number of dimensions and a measure can be defined on this hypersphere, and therefore a weight for our statistical ensembles. We shall see in the next section that it is always possible to return to this case and we shall apply this method, ultimately, to the microcanonical ensemble with a view to establishing the quantum ergodic theorems in quadratic mean. For this purpose, we shall use the statistical ensemble defined by a uniform distribution over the unit-hypersphere of $2n$-dimensional space;† in addition, we shall encounter other statistical ensembles in Chapter VI during the proof of a quantum kinetic equation.‡

The matrix (II.20) is thus the quantum parallel of the classical distributions (I.9) in Γ-space; we must not, however, lose sight of the fact that statistical ensembles represented by $\hat{\bar{\varrho}}$ are defined in functional space and that they are, in particular, much richer in possibilities than the corresponding classical ensembles (we shall have occasion to return to this point in Chapter IV, § IV.4). Taking account of this difference, the matrix $\hat{\bar{\varrho}}$ possesses, in addition, similar properties to those previously established for the classical $\varrho(P, t)$. They are derived from the relations (II.10) and (II.11), proved for the statistical matrices associated with a single system. We have the following properties for $\hat{\bar{\varrho}}$:

(a) It is Hermitian, since it is the sum of Hermitian matrices:

$$\bar{\varrho}_{ij} = \bar{\varrho}^*_{ji}. \qquad \text{(II.21)}$$

(b) Its trace is equal to unity, or

$$\operatorname{Tr} \hat{\bar{\varrho}} = 1; \qquad \text{(II.22)}$$

† See Appendix II.
‡ Cf. Chapter VI, § II.

this relation is, of course, the quantum parallel of equation (I.10). We note further that the eigenvalues of $\hat{\varrho}$ are necessarily positive (or zero) and that their sum is equal to unity by (II.22), since $\bar{\varrho}_{ii}$ represents the probability that a system chosen at random in the set is found in the state i.

On the other hand, it no longer has, in general, the property (II.11): $\hat{\varrho}^2 = \hat{\varrho}$. This relation is valid only for statistical matrices describing a pure case and it is easy to show that (II.11) is the necessary and sufficient condition for a given statistical matrix to represent a pure case. The relationship $\hat{\varrho}^2 = \hat{\varrho}$ then characterises the case of maximum knowledge of the state of a quantum system.

The statistical matrices $\hat{\varrho}$ describe the most general *mixtures* (von Neumann, 1932f; Fano, 1957) of pure cases. The spectrum of $\hat{\varrho}$ is generally a point spectrum; it can be written $\hat{\varrho} = \sum_i w_i \hat{P}_{\varphi_i}$, where w_i is the relative weight of the pure case $\hat{P}_{\varphi_i}$. In such a mixture, statistics is involved twice: first of all in the form of specifically *quantum* probabilities (associated with the pure cases $\hat{P}_{\varphi_i}$) and then by the introduction of *classical* probabilities translating our incomplete knowledge of the dynamic state of the system. The operators $\hat{\varrho}$ involve, as special cases, mixtures which can be associated with pure cases by cancelling the phase relationships existing between the various eigenstates of a quantity. For example, if the system is described by a wave function $\Psi(t)$, expanded in a series of energy eigenfunctions ψ_i, we can write

$$\Psi(t) = \sum_{i=1}^{\infty} r_i e^{i(\alpha_i - iE_i t/\hbar)} \psi_i. \tag{II.23}$$

The mixture corresponding to this pure case is that in which we have a set of systems distributed over various states i with probabilities given by r_i^2. It is easy to see that we obtain such a mixture by considering the ensemble of systems described by functions similar to $\Psi(t)$ but with arbitrary initial phases α_i, distributed uniformly between 0 and 2π. The statistical matrix associated with this ensemble is then

$$\bar{\varrho}_{ij}^{\alpha}(t) = r_i r_j e^{-i(E_i - E_j)t/\hbar} \frac{1}{4\pi^2} \iint e^{i(\alpha_i - \alpha_j)} \, d\alpha_i \, d\alpha_j = r_i^2 \delta_{ij} = \varrho_{ii}(0). \tag{II.24}$$

These results are ultimately used in the quantum ergodic theory (especially in the first quantum ergodic theorem) and in the quantum theory of measurement.

The evolution of the statistical matrix $\hat{\bar{\varrho}}(t)$ as a function of time is obtained easily by starting from equation (II.15) for an elementary statistical matrix. Actually, the matrix $\hat{\bar{\varrho}}(t)$ is derived from $\hat{\varrho}(t)$ by a linear operation (summation or integration); equation (II.15), which is itself linear, can then be written immediately as

$$\frac{\partial \hat{\bar{\varrho}}}{\partial t} = \frac{i}{\hbar} [\hat{\bar{\varrho}}, \hat{H}]_{-}. \tag{II.25}$$

This relationship is the *quantum parallel of Liouville's theorem*; by using the definition and the properties (Aeschlimann, 1952) of the time-derivative of an operator in quantum mechanics, we note that (II.25) can be written, in fact, in the form

$$\frac{\partial \hat{\bar{\varrho}}}{\partial t} - \frac{i}{\hbar} [\hat{\bar{\varrho}}, \hat{H}]_{-} = \frac{d\hat{\bar{\varrho}}}{dt} = 0. \tag{II.26}$$

Thus, the comparison between the statistical matrices $\hat{\bar{\varrho}}(t)$ and the distributions $\varrho(P, t)$ in the phase space of classical mechanics is accomplished. We have thus been able to define statistical ensembles without having recourse to a phase space, which is forbidden to us in quantum mechanics because of the rules of non-commutation between canonically conjugate quantities.

The solution of (II.25) is analogous to (II.15′); we have:

$$\hat{\bar{\varrho}}(t) = \hat{U}_t \hat{\bar{\varrho}}(0)\, \hat{U}_t^* \tag{II.25′}$$

and the mean value of a quantum observable taken over the ensemble described by the matrix $\hat{\bar{\varrho}}(t)$ can then be written as

$$\bar{A}(t) = \mathrm{Tr}\,[\hat{\bar{\varrho}}(t)\, \hat{A}] = \mathrm{Tr}\,[\hat{U}_t \hat{\bar{\varrho}}(0)\, \hat{U}_t^* \hat{A}]. \tag{II.27}$$

This is the quantum parallel of the phase averages of classical statistical mechanics given by (I.21). Because of the properties of invariance of the trace under unitary transformations, this expression is invariant under these transformations as its physical nature requires.

3. *Stationary ensembles*

As in classical mechanics, we shall now define stationary ensembles; these are ensembles for which we have $\partial\hat{\bar{\varrho}}/\partial t = 0$. As in classical mechanics, one can satisfy this relation either by a constant matrix

$$\hat{\bar{\varrho}} = \hat{\bar{\varrho}}_0, \quad \bar{\varrho}_{ij} = \bar{\varrho}_0 \, \delta_{ij}, \tag{II.28}$$

or by matrices for which $[\hat{\bar{\varrho}}, \hat{H}]_- = 0$, i.e. with integrals of motion In the general case, the statistical matrix of a stationary ensemble is thus a function of the Hamiltonian $\hat{H}$; we must have the operator relationship

$$\hat{\bar{\varrho}} = f(\hat{H}), \quad \text{or} \quad \bar{\varrho}_{ij} = [f(\hat{H})]_{ij} \tag{II.29}$$

for the matrix elements. If the matrix is written in the energy representation, we have

$$\bar{\varrho}_{ij} = f(E_i) \, \delta_{ij}. \tag{II.30}$$

According to the choice of the function $f(E_i)$, we shall have particular statistical ensembles; as in classical mechanics, three ensembles are of physical interest: these are the microcanonical, canonical, and grand-canonical ensembles.

(a) *Uniform ensemble.* This is defined by equation (II.28); $\hat{\bar{\varrho}}_0$ here denotes the relative probability since it is not possible to normalise the probability, as this ensemble contains an infinite number of states. This ensemble has an important property: that of invariance of the density matrix under unitary transformations; we have, in fact,

$$\bar{\varrho}'_{ij} = \sum_{n,m} \bar{\varrho}_{mn} S^*_{mj} S_{ni} = \sum_{n,m} \bar{\varrho}_0 \, \delta_{mn} S^*_{mj} S_{ni} = \bar{\varrho}_0 \sum_m S^*_{mj} S_{mi} = \bar{\varrho}_0 \, \delta_{ij} \tag{II.28'}$$

according to the properties of unitary transformations. This property is the quantum parallel of the invariance of the uniform ensemble in classical mechanics under canonical transformations. This uniform ensemble will be useful to us later for stating the fundamental postulate of quantum statistical mechanics.

(b) *Microcanonical ensembles.* This corresponds to a uniform distribution for an ensemble of systems whose energy levels are contained in a narrow band δE and, as a result, it permits us to describe an isolated physical system (whose definition is similar to the classical definition given in Chapter I); the energy of such a system must, in fact, be considered as constant but, since it cannot be known with precision,† we should know simply that the system

† In quantum theory, such a precision is anyway excluded by the fourth Heisenberg relation.

is found in an energy interval $(E, E + \delta E)$. By accepting that there should be n eigenvalues of the energy within this interval, this isolated system will be described by an ensemble of systems uniformly distributed over the various levels of the interval δE; the corresponding statistical matrix will thus be obtained with a function $f(E_i)$ which is constant within the interval considered, or

$$\bar{\varrho}_{ij} = \begin{cases} \bar{\varrho}_0 \, \delta_{ij} & \text{if} \quad E \leqq E_i < E + \delta E, \\ 0 & \text{for other values of } E_i, \end{cases} \tag{II.31}$$

with $\bar{\varrho}_0 = 1/n$ if the levels E_i are not degenerate [otherwise, a degenerate level must be reckoned as often as its degree of degeneracy†]. It can be written in operator language as

$$\hat{\bar{\varrho}} = \frac{1}{n} \sum_{i=1}^{n} \hat{P}\psi_i . \tag{II.31'}$$

Since, according to the fourth Heisenberg relation there is an uncertainty in the energy ΔE for a duration of observation Δt such that $\Delta E \geqq h/\Delta t$, the magnitude of the interval δE defining the microcanonical ensemble will be, in general, much larger than ΔE, since it corresponds to a macroscopic observation of the system. We shall thus have

$$\delta E \gg \Delta E . \tag{II.32}$$

(c) *Canonical ensemble.* This corresponds to a system in thermal equilibrium with a thermostat and its statistical matrix is defined by the operator relation

$$\hat{\bar{\varrho}} = e^{(\Psi - \hat{H})/\theta}, \tag{II.33}$$

where the parameters Ψ and θ have meanings similar to those in classical theory. In the energy representation, the function $f(E_i)$ of (II.30) assumes the form

$$f(E_i) = e^{(\Psi - E_i)/\theta}$$

† In the case of a precise observation of the system, extending over a very long duration, we should have a single eigenstate and the microcanonical ensemble would reduce to the uniform distribution over the various degenerate states corresponding to the observed eigenvalue.

and the canonical ensemble is described in this case by the statistical matrix

$$\bar{\varrho}_{ij} = e^{(\Psi - E_i)/\theta}\,\delta_{ij}. \qquad \text{(II.34)}$$

(d) *Grand canonical ensemble.* Grand ensembles are introduced into quantum statistical mechanics for the same reasons as in classical theory; moreover, their definition is simpler in quantum theory since the property of indistinguishability of particles of the same species in wave mechanics makes the distinction between specific and generic phases useless.

If we assume, as in Chapter I, that the system consists of h kinds of particles (with a variable number n_i of particles of the ith kind), the energy levels of the whole system are functions of the numbers n_i: we write them as $E_n(n_i)$; we must introduce, therefore, the particle number operators $\hat{N}_i$ having n_i for eigenvalues (Jordan and Klein, 1927; Jordan and Wigner, 1928). The statistical matrix of the quantum mechanical grand canonical ensemble must then commute with $\hat{H}$ and also with all the $\hat{N}_i$; it is given by

$$\hat{\varrho} = \exp\left[\left(\Omega + \sum_{i=1}^{h} \mu_i \hat{N}_i - \hat{H}\right)\Big/\theta\right], \qquad \text{(II.35)}$$

where the parameters Ω, θ and μ_i have the same interpretation as in classical theory. In particular, we have the relation

$$e^{-\Omega/\theta} = \mathrm{Tr}\,\exp\left[\left(\sum_{i=1}^{h} \mu_i \hat{N}_i - \hat{H}\right)\Big/\theta\right]. \qquad \text{(II.36)}$$

If we use an orthonormal base which simultaneously diagonalises $\hat{H}$ and all $\hat{N}_i$, equation (II.36) is written in this base

$$e^{-\Omega/\theta} = \sum_{n_1 \dots n_h, n} \exp\left[\left(\sum_{i=1}^{h} \mu_i n_i - E_n(n_i)\right)\Big/\theta\right], \qquad \text{(II.36')}$$

which is similar to the classical formula (I.19').

4. *Fundamental hypothesis of quantum statistical mechanics*

Exactly as in classical mechanics we have just obtained an ensemble of systems, which have the same structure as the actual system and which are independent of one another, to correspond to macroscopic observations on a physical system; this ensemble is weighted by a quantum statistical matrix $\hat{\varrho}$. Nevertheless, in order to choose this distribution $\hat{\varrho}$, we must have, as in classical theory, a starting assumption concerning *a priori* probabilities to be assigned to various states of the system compatible with our information.

The required statistical hypothesis in quantum statistical mechanics is that of equal *a priori* probability and of randomly distributed phases for the non-degenerate quantum states of a system. This hypothesis implies that in the absence of precise information concerning the probability amplitudes $c_i = r_i e^{i\alpha_i}$, we choose a representa tiveensemble such that the probabilities $\bar{\varrho}_{ii} = \overline{r_j^2}$ are equal and such that the phases α_i for all systems in the same state i are uniformly distributed. We shall then have

$$\bar{\varrho}_{ij} = r_i r_j \overline{e^{i(\alpha_i - \alpha_j)}} = \begin{cases} \overline{r_i^2} & \text{if} \quad i = j, \\ 0 & \text{if} \quad i \neq j, \end{cases} \tag{II.37}$$

with $\overline{r_i^2} = \overline{r_j^2} = \varrho_0$, for all i and j.

This corresponds to the uniform statistical ensemble defined earlier for the various possible states of a system; as in classical mechanics, equation (II.26) shows that this fundamental statistical hypothesis cannot contradict the laws of quantum mechanics. Moreover, as the uniform ensemble $\varrho_0 \delta_{ij}$ is invariant under unitary transformations according to (II.28′), it follows that the condition of equal *a priori* probabilities and of randomly distributed phases for the various quantum states is *invariant both under the evolution of the system and under all changes of representation*; this hypothesis thus plays a similar role to that of equal probability of equal volumes of extension in phase in classical statistical mechanics.

On the other hand, we must note here a difference with classical theory: in classical mechanics, it is sufficient to assume equal probability of equal volumes in Γ-space in order to obtain the invariance with respect to the equations of motion, whilst in quantum mechanics the assumption of equal probability for a set of energy states is not sufficient for establishing invariance with respect to the unitary operator of evolution; this can easily be seen by putting, for example, $\bar{\varrho}_{ij} = \varrho_0 \delta_{ij} + A_{ij}(1 - \delta_{ij})$ which is neither invariant nor stationary. This is due to the fact that (II.37) can be satisfied in numerous different ways because of the phase relationships and, consequently, because of the new possibilities for superposition of the quantum states; therefore it is necessary to refine the two parts of the fundamental statistical assumption referring to the amplitudes and phases of the c_i coefficients.†

As in classical statistical mechanics, the fundamental problem of quantum statistical mechanics is to justify the use of these representative ensembles and of the statistical averages to which they are related: the quantum ergodic theorem is intended to accomplish this programme by attempting to define in a suitable manner, starting from the wave function $\Psi(t)$ of the observed system, the macroscopically observed quantities. However, before passing on to this study, we must yet develop a particular point of quantum theory—that of macroscopic operators.

II. Macroscopic Operators

1. *Macroscopic energy*

We know that in quantum mechanics any two physical observables are not generally measurable simultaneously. If $\hat{A}$ and $\hat{B}$ denote respectively the Hermitian operators corresponding to these quantities, we have the Heisenberg relations

$$\Delta A \, \Delta B \geqq \tfrac{1}{2} \, |\overline{[\hat{A}, \hat{B}]_-}|, \tag{II.38}$$

where ΔA and ΔB are defined by

$$\Delta A = \sqrt{\overline{(\hat{A} - \overline{\hat{A}})^2}} \quad \text{and} \quad \Delta B = \sqrt{\overline{(\hat{B} - \overline{\hat{B}})^2}} \tag{II.39}$$

(the mean values here, obviously, denote the quantum mechanical averages). We note the well-known fact, that two events are measurable simultaneously if the two corresponding operators $\hat{A}$ and $\hat{B}$ commute; this property of quantum observables constitutes one of the essential differences between wave mechanics and classical mechanics; it involves a difficulty which is inherent in quantum statistical mechanics, since quantities observed macroscopically are measurable simultaneously.

This is why numerous authors (von Neumann, 1929; Watanabe, 1935; Pauli and Fierz, 1937; van Kampen, 1954), in order to take into account this essential difference between microscopic observables and quantities observed macroscopically, have defined—

† The importance of the uniform ensemble has been stressed in particular by Dirac (1929) and by von Neumann (1927a).

starting from microscopic operators—the so-called macroscopic operators which have the required properties of commutability. The introduction of these operators is not without analogy to those of phase cells, which enable a coarse-grained density to be defined, in the sense of the Ehrenfests, in classical statistical mechanics. We shall study the role of these macroscopic operators, together with the definition of coarse-grained quantum densities in the second part of this book, but we must introduce them at this stage because they play an important role in quantum ergodic theory; we shall draw inspiration here from the intuitive method of van Kampen, while referring the reader to the works of von Neumann for the mathematical aspects of this problem.

The definition of these operators begins with that of the macroscopic energy. We note that a macroscopic measurement of energy of a system always occurs with an inaccuracy of δE, identical to that which we have used to define the microcanonical ensemble. Since the system has a large number of degrees of freedom, there will be always a very large number of eigenvalues of the energy spectrum contained in the interval δE. As we have mentioned already, the conditions for macroscopic observation are such that always $\delta E \gg \Delta E$, where ΔE is the uncertainty in the energy according to the fourth Heisenberg relation. We are thus led to divide the energy spectrum into cells $e^{(1)}, e^{(2)}, \ldots, e^{(\alpha)}, \ldots$, in such a way that a macroscopic measurement of energy could only indicate to us that the system has an energy belonging to one of the cells $e^{(\alpha)}$. Therefore, we must consider all the eigenvalues of the same cell as equal to the same value $\mathscr{E}^{(\alpha)}$ lying between $E^{(\alpha)}$ and $E^{(\alpha)} + \delta E^{(\alpha)}$. Corresponding to the microscopic energy operator

$$\hat{H} = \sum_i E_i \hat{P}_{\psi_i}, \tag{II.40}$$

where $\hat{P}_{\psi_i}$ are the projection operators associated with the energy eigenvectors, we then have a macroscopic energy operator

$$\hat{\mathscr{H}} = \sum_\alpha \mathscr{E}^{(\alpha)} \hat{P}^{(\alpha)}, \tag{II.41}$$

where we have put

$$\hat{P}^{(\alpha)} = \sum_{i=1}^{S(\alpha)} \hat{P}_{\psi_i}. \tag{II.42}$$

The macroscopic operator $\hat{\mathscr{H}}$ is obviously much more degenerate than the corresponding microscopic operator $\hat{H}$. Each energy cell $e^{(\alpha)}$ contains $S^{(\alpha)}$ eigenvalues of the microscopic energy, with the result that each eigenvalue $\mathscr{E}^{(\alpha)}$ of the macroscopic operator is $S^{(\alpha)}$ times degenerate and we have

$$S^{(\alpha)} = \text{Tr}\, \hat{P}^{(\alpha)}. \tag{II.43}$$

2. *Macroscopic observables in general*

Having thus defined a macroscopic energy operator, we shall construct other macroscopic quantities commuting with $\hat{\mathscr{H}}$. Suppose that $\hat{A}$ is a microscopic operator corresponding to a quantum observable which is not measurable simultaneously with $\hat{H}$. We should have for $\hat{A}$ and $\hat{H}$ a relationship similar to (II.38): we shall use it here in its approximate form, that is to say, by keeping to orders of magnitude and by replacing the standard deviations of (II.38) by the uncertainties themselves. We shall then be able to write

$$\Delta E \,\Delta A \sim \left| \overline{[\hat{A}, \hat{H}]_-} \right|. \tag{II.44}$$

However, macroscopic observation of this quantity is made necessarily with an inaccuracy of δA, which will also satisfy the relation $\delta A \gg \Delta A$. The product of the inaccuracies of the quantities will then be much larger than that of the uncertainties of a quantum origin. We shall then have†

$$\delta E \,\delta A \gg \Delta E \,\Delta A. \tag{II.45}$$

Let us consider now the matrix elements A_{ij} in the microscopic energy representation ψ_i and let us write out the matrix element of the commutator $[A, H]_-$; this becomes

$$[\hat{A}, \hat{H}]_{-ij} = (E_i - E_j)\, A_{ij}. \tag{II.46}$$

Then, according to (II.44), by considering only orders of magnitude, we can write

$$(E_i - E_j)\, A_{ij} \sim \Delta E \,\Delta A \ll \delta E \,\delta A. \tag{II.47}$$

If, now, we chose i and j in such a way that $E_i - E_j \sim \delta E$, we have

$$A_{ij} \sim \frac{\Delta E}{\delta E} \Delta A \ll \delta A. \tag{II.48}$$

Thus, it can be seen that the matrix elements A_{ij} corresponding to differences $E_i - E_j$ of the order of δE can be neglected; this in-

† This condition can be obtained also by supposing that A is a "slowly varying" quantity, which satisfies the relation $\delta A \gg \dot{A}\,\Delta t$.

volves the grouping of all non-negligible elements of the matrix A_{ij} into a band along the principal diagonal, a band which is narrow compared with the size of the $e^{(\alpha)}$ cells according to (II.48). By neglecting moreover the matrix elements involving two adjacent energy zones, the matrix associated with the microscopic operator $\hat{A}$ will be reduced to a set of sub-matrices, each corresponding to a single energy cell $e^{(\alpha)}$.

In denoting the operator constructed in this way by $\hat{\mathscr{A}}$, it can be seen that it commutes with the operator $\hat{\mathscr{H}}$ and answers completely to the definition of macroscopic operators stated previously. Actually, according to (II.46) the commutator $[\hat{A}, \hat{H}]_-$ is equal to $A_{ij}(E_i - E_j)$: if E_i and E_j belong to the same energy cell, the bracket $(E_i - E_j)$ is zero, since macroscopically E_i and E_j are considered to be equal to $\mathscr{E}^{(\alpha)}$ and, if E_i and E_j belong to different energy cells, then the element A_{ij} is zero, according to what was said earlier.

In particular, we can find a unitary change of variables in each cell $e^{(\alpha)}$ which enables us to make $\hat{\mathscr{H}}$ and $\hat{\mathscr{A}}$ simultaneously diagonal; this change of variables can be written for the cell α as:

$$\chi_j^{(\alpha)} = \sum_{i=1}^{S(\alpha)} d_{ji}^{(\alpha)} \psi_i . \tag{II.49}$$

The set of these changes of variables for the different cells $e^{(\alpha)}$ defines an orthonormal base in Hilbert space relative to which the operators $\hat{\mathscr{H}}$ and $\hat{\mathscr{A}}$ are simultaneously diagonal. In each of the energy cells we have

$$(\chi_n^{(\alpha)}, \hat{A}\chi_m^{(\alpha)}) = A_m^{(\alpha)}\, \delta_{mn} . \tag{II.50}$$

As there are errors $(\delta A)_\beta$ in the macroscopic observation [we put $(\delta A)_\beta$ since, although of the same order of magnitude, they are not necessarily identical for all cells of the spectrum of $\hat{A}$], the set of the $A_m^{(\alpha)}$ can be divided into $N^{(\alpha)}$ groups for which the eigenvalues of $\hat{\mathscr{A}}$ must be considered as equal to the same value $a_\beta^{(\alpha)}$. We are thus led to a finer sub-division of Hilbert space into cells, to which correspond the projection operators

$$\hat{P}_\beta^{(\alpha)} = \sum_{m=1}^{s_\beta^{(\alpha)}} \hat{P}\chi_m^{(\alpha)} , \tag{II.51}$$

the degeneracy of the value $a_\beta^{(\alpha)}$ being given by

$$\text{Tr}\,\hat{P}_\beta^{(\alpha)} = s_\beta^{(\alpha)}. \tag{II.52}$$

These numbers will always be very large for a system with a large number of degrees of freedom. According to the foregoing definitions we have the following relations:

$$S^{(\alpha)} = \sum_{\beta=1}^{N(\alpha)} s_\beta^{(\alpha)} \quad \text{and} \quad \hat{P}^{(\alpha)} = \sum_{\beta=1}^{N(\alpha)} \hat{P}_\beta^{(\alpha)}, \tag{II.53}$$

and the macroscopic operator $\hat{\mathscr{A}}$ assumes the form

$$\hat{\mathscr{A}} = \sum_\alpha \sum_{\beta=1}^{N(\alpha)} a_\beta^{(\alpha)} \hat{P}_\beta^{(\alpha)}. \tag{II.54}$$

Having thus defined the macroscopic operators for the energy and for a quantity $\hat{A}$, it is easy to extend this definition to other quantities $\hat{B}$, $\hat{C}$, ... The arguments are identical and lead us to the definition each time of a new set of cells, after making a change of variables, appropriate for each energy cell $e^{(\alpha)}$. (If we consider, for example, a second quantity $\hat{B}$, we have the relation $\delta A\,\delta B \gg \Delta A \Delta B$; the result, as before, is that the elements of the matrix $B_{\beta\beta'}$ involving two cells β are negligible, and that the operator $\hat{B}$ can be divided up in its turn into a system of sub-matrices, each operating in one cell β.) Thus, it can be assumed that the macroscopic state of the observed system is defined completely by the macroscopic energy and a set of macroscopic operators ($\hat{\mathscr{A}}$, $\hat{\mathscr{B}}$, $\hat{\mathscr{C}}$, ...) which set up a system of cells over the Hilbert space subtended by a suitably chosen orthonormal base $\Omega_k^{(\alpha)}$. This base is obtained from the energy eigenfunctions by a unitary transformation of the form†

$$\Omega_k^{(\alpha)} = \sum_{i=1}^{S(\alpha)} C_{ki}^{(\alpha)} \psi_i. \tag{II.55}$$

A system of cells, defined by their projection operators $\hat{P}_\nu^{(\alpha)}$, corresponds to this, with

$$\hat{P}_\nu^{(\alpha)} = \sum_{k=1}^{s_\nu^{(\alpha)}} \hat{P}_{\Omega_k^{(\alpha)}}, \tag{II.56}$$

† (II.55) obviously is only an approximate relation, with the same approximations as before; it would only be exact if the summation were taken over all states ψ_i of Hilbert space.

where $s_\nu^{(\alpha)}$ is the size of the cell. In supposing that there are $N^{(\alpha)}$ cells in each cell $e^{(\alpha)}$ obviously we have the relationship

$$S^{(\alpha)} = \sum_{\nu=1}^{N(\alpha)} s_\nu^{(\alpha)}, \quad \text{and} \quad \hat{P}^{(\alpha)} = \sum_{\nu=1}^{N(\alpha)} \hat{P}_\nu^{(\alpha)}. \tag{II.57}$$

Since the transformation matrix $C_{ki}^{(\alpha)}$ is unitary for each cell $e^{(\alpha)}$, we have in this case

$$\sum_{i=1}^{S(\alpha)} C_{kj}^{(\alpha)*} C_{ki}^{(\alpha)} = \delta_{ij}, \tag{II.58}$$

which can also be written as

$$\sum_{i=1}^{S(\alpha)} C_{ki}^{(\alpha)*} C_{li}^{(\alpha)} = \delta_{kl}. \tag{II.58$'$}$$

Other relations can be deduced from (II.58′) which will be useful later on. In fact, we can derive two relations from (II.58′) by taking $k = l$ and $k \neq l$:

$$\sum_i{}^{(\alpha)} |C_{ki}^{(\alpha)}|^2 = 1 = \delta_{kk}, \tag{II.59}$$

$$\sum_i{}^{(\alpha)} C_{ki}^{(\alpha)*} C_{li}^{(\alpha)} = 0 \quad (k \neq l), \tag{II.60}$$

where the symbol $\Sigma^{(\alpha)}$ denotes summation over the $S^{(\alpha)}$ states of the cell $e^{(\alpha)}$.

By squaring (II.59) we obtain

$$\Big(\sum_i{}^{(\alpha)} |C_{ki}^{(\alpha)}|^2\Big)^2 = \sum_i{}^{(\alpha)} |C_{ki}^{(\alpha)}|^4 + \sum_{i\neq j}{}^{(\alpha)} |C_{ki}^{(\alpha)}|^2\, |C_{kj}^{(\alpha)}|^2 = 1 = \delta_{kk}. \tag{II.61}$$

Likewise, by multiplying (II.59) by the identity $\sum_i{}^{(\alpha)} |C_{li}^{(\alpha)}|^2 = \delta_{ll}$ it also becomes

$$\sum_{i,j}{}^{(\alpha)} |C_{ki}^{(\alpha)}|^2\, |C_{lj}^{(\alpha)}|^2 = \sum_i{}^{(\alpha)} |C_{ki}^{(\alpha)}|^2\, |C_{li}^{(\alpha)}|^2 + \sum_{i\neq j}{}^{(\alpha)} |C_{ki}^{(\alpha)}|^2\, |C_{lj}^{(\alpha)}|^2 = \delta_{kk}\delta_{ll}. \tag{II.62}$$

We obtain another type of relation by starting from (II.60); by multiplying by $\sum_j^{(\alpha)} C_{kj}^{(\alpha)} C_{lj}^{(\alpha)*} = 0$, it becomes

$$\sum_{i,j}^{(\alpha)} C_{ki}^{(\alpha)*} C_{li}^{(\alpha)} C_{kj}^{(\alpha)} C_{lj}^{(\alpha)*} = \sum_j^{(\alpha)} |C_{ki}^{(\alpha)}|^2 \, |C_{li}^{(\alpha)}|^2 + \sum_{i \neq j}^{(\alpha)} C_{ki}^{(\alpha)*} C_{li}^{(\alpha)} C_{kj}^{(\alpha)} C_{lj}^{(\alpha)*} = 0 \quad (k \neq l), \tag{II.63}$$

or even

$$\sum_{i \neq j}^{(\alpha)} C_{ki}^{(\alpha)*} C_{li}^{(\alpha)} C_{kj}^{(\alpha)} C_{lj}^{(\alpha)*} = - \sum_i^{(\alpha)} |C_{ki}^{(\alpha)}|^2 \, |C_{li}^{(\alpha)}|^2 \quad (k \neq l). \tag{II.63'}$$

Finally, we point out that we shall see expressions of the type

$$D_{ij}^{(\nu)} = \sum_{k=1}^{s_\nu^{(\alpha)}} C_{ki}^{(\alpha)*} C_{kj}^{(\alpha)}, \tag{II.64}$$

appear in the calculation of the macroscopic averages, defined in a cell ν with $s_\nu^{(\alpha)}$ dimensions. These quantities satisfy the following relations:

$$D_{ji}^{(\nu)} = \sum_k^{(\nu)} C_{kj}^{(\alpha)*} C_{ki}^{(\alpha)} = D_{ij}^{(\nu)*}, \tag{II.65}$$

$$\sum_i^{(\alpha)} D_{ii}^{(\nu)} = \sum_{k=1}^{s_\nu^{(\alpha)}} \sum_i^{(\alpha)} |C_{ki}^{(\alpha)}|^2 = \sum_{k=1}^{s_\nu^{(\alpha)}} \delta_{kk} = s_\nu^{(\alpha)}. \tag{II.66}$$

In the foregoing, we now have the necessary equipment to deal with the various aspects of the ergodic theory in wave mechanics and we shall see, in the second part of this book, that these macroscopic operators enable us to define the coarse-grained statistical densities necessary for proving a quantum H-theorem.

CHAPTER III

The Ergodic Theorem in Quantum Statistical Mechanics

I. The Ergodic Problem in Quantum Mechanics

As in classical mechanics, where the ergodic theory presents two aspects, two ergodic theorems can be distinguished in quantum mechanics, which correspond respectively to those of Birkhoff and Hopf in classical theory. The evolution of an isolated physical system is determined by the motion of the end-point of the vector $\Psi(t)$ over the unit hypersphere of Hilbert space; this motion is defined by equations (II.1)–(II.6); to each observable represented by a Hermitian operator $\hat{A}$ there corresponds a quantum average value depending on $\Psi(t)$, or $A(t) = (\Psi, \hat{A}\Psi)$, and $A(t)$ is completely defined if the trajectory of $\Psi(t)$ is known. As in classical mechanics, we accept that macroscopic observation does not permit the instantaneous value of $A(t)$ to be found, but only its time-average over a time interval T; thus, we study the quantity

$$\overline{A(t)}^{T} = \frac{1}{T}\int_{0}^{T} A(t)\, dt \qquad \text{(III.1)}$$

which are the quantum parallel of (I.23). Under these conditions, the fundamental hypothesis of statistical mechanics will be justified if we can prove the equality of the time-averages (III.1) and of the statistical averages (II.27), calculated with the microcanonical distribution (II.31). We shall see that this programme, in quantum theory, encounters additional difficulties due to the special properties of stationary quantum states.

The first ergodic theorem consists in proving the existence of a limit of $\overline{A(t)}^T$, when $T \to \infty$ and in establishing under what conditions this limit is independent of the initial state. It is presented, therefore, as the quantum parallel of Birkhoff's theorem and the condition for the absence of degeneracy in the spectrum of the Hamiltonian corresponds to the classical hypothesis of metric transitivity.

The second quantum ergodic theorem, more useful for physical applications, is analogous to Hopf's theorem in classical mechanics; in this case, it will be necessary for us to have recourse to a certain definition of macroscopic observables (in the sense of G. Ludwig) in order to try to establish a convergence towards averages which are time independent. Moreover, we shall have to assume not only the non-degeneracy of the Hamiltonian spectrum but also the absence of resonance frequencies in this spectrum: these conditions are analogous to those stated by Hopf's theorem, but we shall see that they appear to be even more restrictive in quantum mechanics than in classical mechanics.

Finally, we shall show that it has not been possible to prove rigorously the equality of the time-averages and the microcanonical averages for one system but only statistically, by resorting to an average over a set of possible macroscopic observers, following von Neumann; this latter restriction modifies considerably the real significance of the quantum ergodic theorem of von Neumann's, the statistical nature of which thus becomes obvious.

We shall try to break away from the two foregoing ergodic theorems, which have a range of application more restricted than in classical theory because of the nature of the assumptions concerning the Hamiltonian spectrum and because of the difficulties which are inherent to the quantum ergodic theory; we shall see that we can succeed by taking account of the statistical aspect of the theory. This is why, after developing the two usual theorems of the quantum ergodic theory in this chapter, we shall study in Chapter IV an approximate form of the ergodic theorems called, as in classical theory, the "probability" ergodic theorems and, at the same time, we shall rely on the quantum parallel of von Neumann's condition (I.26) and on the asymptotic behaviour of certain quadratic averages (when the number of degrees of freedom of the system becomes very large). We shall use from the start ensemble theory, employing the quantum ensembles of Chapter II and we shall show that ergodic theorems in quadratic mean can then be proved without needing the assumptions made above. This result, which depends on the macroscopic nature of the system considered, will be compared with Khinchin's asymptotic formulae in classical mechanics.

II. The First Quantum Ergodic Theorem

In this theorem, we consider an isolated system as being in a pure case described by the wave function $\Psi(t)$; we seek, therefore, to prove the existence of a limit of $\overline{A(t)}^T$ when $T \to \infty$, or

$$\lim_{T=\infty} \overline{A(t)}^T = \lim_{T=\infty} \frac{1}{T} \int_0^T A(t)\, dt. \tag{III.2}$$

and to deduce the conditions under which this limit is independent of the initial phases: thus, we obtain the quantum parallel of Birkhoff's theorem.

If ψ_i and E_i are the eigenfunctions and eigenvalues of the Hamiltonian $\hat{H}$ and if $\Psi(0)$ is the wave function of the system at $t = 0$, we have

$$\Psi(0) = \sum_{i=1}^{\infty} c_i(0)\, \psi_i, \tag{III.3}$$

with

$$c_i(0) = r_i e^{i\alpha_i}, \tag{III.4}$$

where r_i is the modulus and α_i is the phase of $c_i(0)$. At time t, $\Psi(0)$ is transformed into $\Psi(t)$ according to equation (II.4), and it can be written as

$$\Psi(t) = \sum_{i=1}^{\infty} r_i\, e^{i\alpha_i} e^{-iE_i t/\hbar} \psi_i. \tag{III.5}$$

It can be seen that the transformation $\Psi(0) \to \Psi(t)$ leaves the probabilities for various states ψ_i which are equal to r_i^2 unchanged. On the contrary, the phases of the various states are functions of time and, in the ψ_i representation, the statistical matrix of the pure case $\Psi(t)$ can be written as

$$\varrho_{ij}(t) = r_i r_j\, e^{i(\alpha_i - \alpha_j)}\, e^{-i(E_i - E_j)t/\hbar} = \varrho_{ij}(0)\, e^{-i(E_i - E_j)t/\hbar}. \tag{III.6}$$

With the foregoing expressions, we have for the quantum average value $A(t)$ of an observable A

$$A(t) = \sum_{i,j} r_i r_j \, e^{i(\alpha_i - \alpha_j)} \, e^{-i(E_i - E_j)t/\hbar} \, A_{ji}, \tag{III.7}$$

where A_{ji} is the matrix element $(\psi_j, A\psi_i)$.

If the series $\sum_{i=1}^{\infty} r_i$ is convergent, expression (III.1) can then be written as

$$\overline{A(t)}^T = \frac{1}{T}\int_0^T A(t)\,dt = \sum_{i,j} r_i r_j A_{ji} \, e^{i(\alpha_i - \alpha_j)} \left\{ \frac{1}{T}\int_0^T e^{-i(E_i - E_j)t/\hbar} \, dt \right\} \tag{III.8}$$

and, if now we take the limit of this expression as $T \to \infty$, it can be seen that we revert (always in the case where the series of the values r_i is convergent) to evaluating the limits

$$\lim_{T=\infty} \left\{ \frac{1}{T}\int_0^T e^{-i(E_i - E_j)t/\hbar} dt \right\}. \tag{III.9}$$

However, these expressions are equal to 0 or 1 according to whether $E_i \neq E_j$ or $E_i = E_j$. It follows that the limit $\overline{A(t)}^{\infty}$ of $\overline{A(t)}^T$ as $T \to \infty$ always exists: this result is *the quantum equivalent of Birkhoff's theorem.*

This limit, however, depends generally on the initial phases α_i: it is independent of them only if there is no degeneracy in the spectrum of the Hamiltonian. In this case, we have

$$\lim_{T=\infty} \overline{A(t)}^T = \overline{A(t)}^{\infty} = \sum_{i=1}^{\infty} r_i^2 A_{ii}. \tag{III.10}$$

If, for example, an eigenvalue E_j had a degeneracy of order 2, the eigenfunctions ψ_{j1} and ψ_{j2}, the amplitudes r_{j1} and r_{j2} and the phases α_{j1} and α_{j2} would correspond to it. Then, in the expansion of $\overline{A(t)}^{\infty}$ we should have, in addition to the diagonal terms $r_{j1}^2 A_{j1j1} + r_{j2}^2 A_{j2j2}$, expressions of the form

$$\begin{aligned} & r_{j1} r_{j2} A_{j2j1} \, e^{i(\alpha_{j1} - \alpha_{j2})} + r_{j2} r_{j1} A_{j1j2} \, e^{i(\alpha_{j2} - \alpha_{j1})} \\ &= 2 r_{j1} r_{j2} \operatorname{Re}(A_{j2j1} \, e^{i(\alpha_{j1} - \alpha_{j2})}), \end{aligned} \tag{III.11}$$

which depend, obviously, on the initial phases α_i.

If we introduce the mean statistical matrix defined by (II.24), equation (III.10) can then be written as

$$\overline{A(t)}^{\infty} = \lim_{T=\infty} \frac{1}{T} \int_0^T \mathrm{Tr}(\hat{\varrho}(t)\,\hat{A})\,dt = \mathrm{Tr}(\hat{\varrho}^{\alpha}\hat{A}). \qquad \text{(III.12)}$$

Thus, we see that the time-average taken over the pure case represented by the matrix (III.6) has the effect of replacing this pure case by the mixture defined by the matrix (II.24): the second term of (III.12) is thus independent of the initial phases α_i. We can say, by using the terminology of quantum statistical ensembles, that *the time-average of a quantity $A(t)$ is equal to the statistical average taken over the pure case ensemble $\Psi(t)$ corresponding to (III.6) with the phases distributed uniformly over the interval* $(0, 2\pi)$.

This result has been obtained at the price of an assumption concerning the nature of the system, namely that the Hamiltonian spectrum is non-degenerate. This corresponds to the hypothesis of metric transitivity in classical theory: actually, this latter hypothesis, in classical mechanics, can be defined by saying that the trajectory covers completely the hypersurface of constant energy (by passing as close as we wish to every point of the hypersurface Σ); in the same way, we can say in wave mechanics, that the hypothesis of non-degeneracy of the spectrum of $\hat{H}$ has the effect of allowing the trajectory of $\Psi(t)$, over the unit hypersphere, to cover the statistical ensemble corresponding to the matrix which has the r_i^2 as weights: stated in this form, the first ergodic theorem is then completely the quantum equivalent of Birkhoff's theorem.

Their physical significance is different, however: actually, although knowledge of the initial phase is sufficient to determine the energy surface in classical mechanics over which the movement takes place, the knowledge of the initial vector $\Psi(0)$, according to (III.3), defines only a statistical distribution of the quantum system over all possible energy states. Moreover, and this is the essential fact, the previous theorem does not establish the equality of $\overline{A(t)}^{\infty}$ with the microcanonical mean $((1/n) \sum_i A_{ii})$, since it does not deal with the probabilities r_i^2, which remain unchanged during the evolution of the system: this is a characteristic property of wave mechanics; it is associated with the fact that, since the energy is an integral of motion in a conservative system, the distribution in probability of any quantity is independent of time when the system is in a stationary state (we have $\psi = a_i\psi_i$ and the time-dependent exponential, of modulus 1, is contained in the factor a_i). This property constitutes an additional difficulty of the quantum ergodic theorem which will make it necessary to introduce averages over macroscopic observers in the second ergodic theorem; moreover, we shall see in Chapter IV that the elimination of this difficulty rests, in fact, on the asymptotic properties of the quantities $\sum_i r_i^2 A_{ii}$ on the unit sphere in $2n$-dimensional space.†

We note also that we have had to assume convergence of the series $\sum_i r_i$; this convergence cannot be associated with any physical significance. Apart from this difficulty of a mathematical nature [which we shall avoid by showing that in quantum statistical mechanics we can always restrict ourselves to a finite number of terms in (III.8)], the hypothesis of non-degeneracy of the spectrum of the Hamiltonian does not seem to be acceptable, since in the majority of simple quantum physical systems (atoms, molecules), the existence of degenerate levels is the general rule and not the exception; moreover, there is little likelihood that such is not the case also for the more complex systems envisaged in statistical mechanics. Thus, it appears that the quantum ergodic theory encounters a difficulty at this point which is greater than in classical theory, where the assumption of metric transitivity does not clash at first sight with objections of the same kind: thus, we can say that the existence of ergodic systems in the foregoing sense is much more conjectural in wave mechanics than in classical mechanics. Since we shall encounter this difficulty again in the quantum parallel of Hopf's theorem, we can conclude that the justification of the principles of quantum statistical mechanics by the ergodic theory is even more slender than in classical mechanics.

III. The Second Quantum Ergodic Theorem

1. *Statement of the theorem*

This is the quantum parallel of Hopf's theorem. If $\hat{\varrho}(t)$ is the matrix representing a statistical ensemble of systems, the average

† Another method of overcoming this difficulty consists in the introduction of non-stationary ensembles (cf. Chapter VI, Section III, § 2).

value over this ensemble of the observable A is given, according to (II.27), by $\mathrm{Tr}(\hat{A}\hat{\bar{\varrho}}(t))$; if we define the operator $\hat{W}_t$ by the relationship

$$\hat{\bar{\varrho}}(t) = \hat{W}_t\hat{\bar{\varrho}}(0) = \hat{U}_t\hat{\bar{\varrho}}(0)\,\hat{U}_t^*, \tag{III.13}$$

this average value can be written as

$$\mathrm{Tr}\,(\hat{A}\hat{\bar{\varrho}}(t)) = \mathrm{Tr}\,(\hat{A}\hat{W}_t\hat{\bar{\varrho}}(0)). \tag{III.14}$$

We shall have an ergodic theorem similar to that of Hopf if we can prove that the distribution $\hat{W}_t\hat{\bar{\varrho}}(0)$ tends in the quadratic mean towards a limit distribution $\hat{\bar{\varrho}}^+(0)$, in such a way that

$$\lim_{T=\infty}\frac{1}{T}\int\limits_0^T |\mathrm{Tr}\,(\hat{A}\hat{W}_t\hat{\bar{\varrho}}(0)) - \mathrm{Tr}\,(\hat{A}\hat{\bar{\varrho}}^+(0))|^2\,dt \to 0. \tag{III.15}$$

However, the statistical matrix of a mixture $\hat{\bar{\varrho}}(0)$ can always be written as the weighted sum of elementary matrices $\hat{\varrho}_{\Phi_\nu}(0)$ corresponding to pure cases described by the functions Φ_ν, or

$$\hat{\bar{\varrho}}(0) = \sum_\nu w_\nu\hat{\varrho}_{\Phi_\nu}(0). \tag{III.16}$$

Expression (III.15) can then be written as

$$Z = \lim_{T=\infty}\frac{1}{T}\int\limits_0^T \left|\sum_\nu w_\nu\{\mathrm{Tr}(\hat{A}\hat{W}_t\hat{\varrho}_{\Phi_\nu}(0)) - \mathrm{Tr}(\hat{A}\hat{\varrho}^+_{\Phi_\nu}(0))\}\right|^2 dt, \tag{III.17}$$

since the statistical weights w_ν are unaffected by the operator $\hat{W}_t$. According to Schwartz's inequality we have

$$Z \leq \lim_{T=\infty}\frac{1}{T}\int\limits_0^T \sum_\nu w_\nu|\mathrm{Tr}\,(\hat{A}\hat{W}_t\hat{\varrho}_{\Phi_\nu}(0)) - \mathrm{Tr}\,(\hat{A}\hat{\varrho}^+_{\Phi_\nu}(0))|^2\,dt$$

$$= \lim_{T=\infty}\sum_\nu w_\nu\left\{\frac{1}{T}\int\limits_0^T |\mathrm{Tr}(\hat{A}\hat{W}_t\hat{\varrho}_{\Phi_\nu}(0)) - \mathrm{Tr}(\hat{A}\hat{\varrho}^+_{\Phi_\nu}(0))|^2\,dt\right\}, \tag{III.18}$$

with the result that it is sufficient to prove theorem (III.15) for a pure case represented by the statistical matrix $\hat{\varrho}_\Phi(0) = \hat{P}_\Phi$.

As before, we shall assume that we are dealing with a system whose Hamiltonian is time-independent, with a discrete spectrum and that it can be written $\hat{H} = \sum_i E_i \hat{P}_{\Psi_i}$, where $\hat{P}_{\Psi_i}$ are the projection operators corresponding to the eigenvectors of the energy; we shall assume further that the spectrum of the Hamiltonian is non-degenerate. The initial state of the system is represented by the matrix $\hat{\varrho}_\Phi(0)$ [which we denote by $\hat{\varrho}(0)$ for simplicity], and the quantum average value $A(t)$ of an observable $\hat{A}$ in the pure case described by $\hat{P}_\Phi$ can be written [in the Heisenberg representation, which will be more convenient here for us†]:

$$A(t) = \mathrm{Tr}(\hat{W}'_t\hat{A}\hat{\varrho}(0)) = \sum_{i,j} e^{i(E_i-E_j)t/\hbar}(E_i|\hat{A}|E_j)\,(E_j|\hat{\varrho}(0)|E_i), \quad \text{(III.19)}$$

with $\hat{W}'_t\hat{A} = \hat{U}^*_t\hat{A}\hat{U}_t$; by putting $E_i - E_j = h\omega/2\pi$, we obtain for (III.19), from the fact that the spectrum of H is non-degenerate,

$$A(t) = \sum_\omega e^{i\omega t} \sum_{E_i} (E_i|\hat{A}|E_i - \hbar\omega)\,(E_i - \hbar\omega|\hat{\varrho}(0)\,|E_i). \quad \text{(III.20)}$$

[Expression (III.19) is obviously identical with (III.7) because of (III.6), but the notation used here is more convenient for stressing the role of degeneracies and of resonance frequencies; cf. (III.20) and (III.25).]

The problem is to show that $A(t)$ tends in quadratic mean towards a limit which is time-independent. The time-independent terms in $A(t)$ are of the form

$$\sum_{E_i} (E_i|\hat{A}|E_i)\,(E_i|\hat{\varrho}(0)|E_i),$$

and so we shall put

$$\mathrm{Tr}\,(\hat{A}\hat{\varrho}^+(0)) = \sum_{E_i} (E_i|\hat{A}|E_i)\,(E_i|\hat{\varrho}(0)|E_i). \quad \text{(III.21)}$$

We note that for the same reasons as before, $\mathrm{Tr}(\hat{\varrho}^+(0)\hat{A}) = \sum_i r_i^2 A_{ii}$ (in the notation of Section II) is not equal to the microcanonical average; this difficulty will be raised only with von Neumann's theorem and the "probability" theorems of Chapter IV.

† Here, we follow Ludwig (1954).

We shall now try to show that

$$Z_\Phi = \lim_{T=\infty} \frac{1}{T} \int_0^T A|(t) - \mathrm{Tr}(\hat{A}\hat{\varrho}^+(0))|^2 \, dt$$

$$= \lim_{T=\infty} \frac{1}{T} \int_0^T \left| \sum_\omega e^{i\omega t} g(\omega) \right|^2 dt \to 0, \qquad \text{(III.22)}$$

where, according to (III.20) and (III.21), we have put

$$g(\omega) = \begin{cases} 0 \quad \text{when} \quad \omega = 0, \\ \sum_{E_i} (E_i|\hat{A}|E_i - \hbar\omega)\,(E_i - \hbar\omega|\hat{\varrho}(0)|E_i) \quad \text{when} \quad \omega \neq 0. \end{cases} \qquad \text{(III.23)}$$

According to the foregoing definition we have $g^*(\omega) = g(-\omega)$, with the result that (III.22) can be written as

$$Z_\Phi = \lim_{T=\infty} \frac{1}{T} \int_0^T \left| \sum_\omega e^{i\omega t} g(\omega) \right|^2 dt = \sum_\omega |g(\omega)|^2 \to 0; \qquad \text{(III.24)}$$

Z_Φ can become very small only if all the moduli of $g(\omega)$ are small: it is sufficient that a single value of $|g(\omega)|$ is large in order that $A(t)$ does not tend towards a limit. The necessary assumption for proving that Z_Φ tends towards zero concerns the Hamiltonian of the system: it is assumed that there are no resonance frequencies ω, i.e. that unless $i = i'$ and $j = j'$, $E_i - E_j \neq E_i' - E_j'$, for every pair ij, $i'j'$; applying this condition to (III.23) we obtain

$$g(\omega) = \begin{cases} 0 \quad \text{if} \quad \omega = 0, \\ (E_i|\hat{A}|E_j)\,(E_j|\hat{\varrho}(0)|E_i) \\ \text{for one single term for which} \quad E_i - E_j = \hbar\omega. \end{cases} \qquad \text{(III.25)}$$

We can now evaluate Z_Φ for a pure case; actually, we have—with the previous notation—$(E_j|\hat{\varrho}(0)|E_i) = (E_j|\Phi)\,(\Phi|E_i)$,† whence

$$Z_\Phi = \sum_\omega |g(\omega)|^2 \leqq \sum_{E_i, E_j} |(E_i|\hat{A}|E_j)\,(E_j|\Phi)\,(\Phi|E_i)|^2. \qquad \text{(III.26)}$$

† With the notation of Chapter II, we should have for example $(E_j|\Phi) = c_j(0)$.

The form of (III.26) (where we have the inequality sign because we should drop terms where $E_i = E_j$) suggests that we compare Z_Φ with the average value of $\hat{A}^2$ taken with respect to $\hat{\varrho}^+(0)$, or $\overline{\hat{A}^2} = \text{Tr}(\hat{A}^2\hat{\varrho}^+(0))$. Putting

$$m = \underset{E_i, E_j}{\text{Max}} \, |(E_i|\hat{A}|E_j)\,(E_j|\Phi)|, \tag{III.27}$$

equation (III.26) can be written as

$$Z_\Phi \leqq m \sum_{E_i, E_j} |(E_i|\hat{A}|E_j)\,|.|\,(E_j|\Phi)\,|.|\,(\Phi|E_i)|^2$$

and, by applying Schwartz's inequality, it becomes

$$Z_\Phi \leqq m \Big[\sum_{E_i, E_j} |(E_i|\hat{A}|E_j)|^2\,|(\Phi|E_i)|^2\Big]^{\frac{1}{2}} \Big[\sum_{E_i, E_j} |(E_j|\Phi)|^2\,|(\Phi|E_i)|^2\Big]^{\frac{1}{2}}.$$

Since we have $\sum_{E_i} |(\Phi|E_i)|^2 = 1$, the previous relation can be written as

$$Z_\Phi \leqq m[\text{Tr}\,(\hat{A}^2\hat{\varrho}^+(0))]^{\frac{1}{2}} \tag{III.28}$$

or even

$$Z_\Phi \leqq \delta\,\text{Tr}(\hat{A}^2\hat{\varrho}^+(0)), \tag{III.28$'$}$$

with

$$\delta^2 = \frac{m^2}{\sum_{E_i, E_j} |(E_i|\hat{A}|E_j)|^2\,|(\Phi|E_i)|^2}. \tag{III.29}$$

In order that δ be very small compared with 1, the matrix elements $(E_i|\hat{A}|E_j)$ must be very numerous; this condition can be defined more precisely by noting that if E_0 and E_0' are the values of the energy corresponding to m, it follows by (III.29) that

$$\delta^2 \leqq \frac{|(E_0|\hat{A}|E_0')|^2}{\sum_{E_i} |(E_i|\hat{A}|E_0')|^2}. \tag{III.30}$$

If, on the other hand, S^ν is the number of elements $(E_i|\hat{A}|E_0')$ for which

$$\frac{1}{\nu - 1}(E_0|\hat{A}|E_0')| > |(E_i|\hat{A}|E_0')| \geqq \frac{1}{\nu}|(E_0|\hat{A}|E_0')|, \tag{III.31}$$

we shall have according to (III.30):

$$\delta \leqq \left(\sum_\nu \frac{S^\nu}{\nu^2}\right)^{-\frac{1}{2}}. \tag{III.30$'$}$$

Thus, we can see that according to (III. 30′), the number S^{ν} must be very large in order that δ be very small, with the result that among the matrix elements $(E_i|\hat{A}|E_j)$ there will be a large number of the same order of magnitude; this condition may be compared with that used in Chapter II to define macroscopic observables and it has been generalised in a recent paper (Ludwig, 1958a, b): it leads to the concept of *strongly ergodic* observables and of macroscopic observables in the sense of Ludwig.

Having established the conditions for which theorem (III.22) is valid for the pure case $\hat{P}_{\Phi_\nu}$, by putting this result in equation (III.18) we have

$$Z \leqq \sum_{\nu} w_\nu \delta \operatorname{Tr}(\hat{A}^2 \varrho^{+}_{\Phi_\nu}(0)) = \delta \operatorname{Tr}(\hat{A}^2 \hat{\bar{\varrho}}^{+}(0)), \qquad \text{(III.32)}$$

with the result that (III.15) is satisfied if $\delta \to 0$; thus, we have more precisely defined the method which is to be used for a rigorous proof of the second quantum ergodic theorem, which requires that the average of the variations of $A(t)$ around $\operatorname{Tr}(\hat{A}\hat{\bar{\varrho}}^{+}(0))$ are very small compared with $\operatorname{Tr}(\hat{A}^2\hat{\bar{\varrho}}^{+}(0))$.

2. *Existence conditions and comparison with Hopf's theorem*

We have seen in the preceding section that the proof of theorem (III.15) requires the following three conditions:

(a) Non-degeneracy of the spectrum of $\hat{H}$: this assumption has already been analysed in the first quantum ergodic theorem.

(b) Absence of resonance frequencies in the spectrum of H, which is the quantum parallel of the assumption necessary for proving Hopf's theorem, dealing with the metric transitivity of the coupled hypersurface $\Sigma \times \Sigma'$. Actually, if these conditions were not fulfilled, the time-averages Z would depend on the initial phases α_l, and we should have terms similar to (III.11). Ludwig has, in fact, shown, as an example, that in the case where $A(t)$ takes the form of a Fourier series and where the motion is periodic, every possibility of convergence would be excluded unless $\hat{A}$ commutes with $\hat{H}$.

(c) Condition (III.31) imposed on the observable $\hat{A}$, which amounts to considering its spectrum as strongly degenerate from the macroscopic point of view. This condition, which is necessary for having $\delta \to 0$, has no classical equivalent and thus constitutes an additional difficulty which is inherent in the quantum ergodic theory. Although Ludwig (1958a, b) has attempted, in his recent papers† to derive such a condition for the intrinsic properties of observed macroscopic systems, we may say that the precise proof of the second quantum ergodic theorem is not always realised in practice. In order to avoid this difficulty, we can try to accept that the correct statement of Hopf's theorem in quantum mechanics must make use of the macroscopic observables defined in Chapter II; we are thus led to the statement of the theorems of von Neumann and of Pauli-Fierz which we are now going to study.

Before doing this, we shall show by relying on the unitarity of the operator of evolution that we can extend more widely the mathematical analogy between the classical and quantum forms of Hopf's theorem. Indeed, because of the unitary nature of the operator W_t', defined by (III.19), we can define its spectral resolution as follows

$$\hat{W}_t' = \int_{-\infty}^{+\infty} e^{i\omega t} d\hat{F}(\omega), \tag{III.33}$$

where the $\hat{F}(\omega)$ are the projection operators defined by

$$\hat{F}(\omega)\,\hat{A} = \sum_{E_i, E_j}^{\omega} \hat{P}_{\psi_i}\hat{A}\hat{P}_{\psi_j}; \tag{III.34}$$

the symbol $\sum^{\omega}$ denotes here a sum over all the E_i and E_j such that $E_i - E_j \leqq \hbar\omega$. With (III.34) we have a spectral resolution corresponding to that given by Stone's theorem (I.30) in classical mechanics.

At the point $\omega = 0$, $\hat{F}(\omega)\hat{A}$ has a discontinuity defined by

$$(\hat{F}(+0) - \hat{F}(0))\,\hat{A} = \sum_{E_i} \hat{P}_{\psi_i}\hat{A}\hat{P}_{\psi_i}. \tag{III.35}$$

If the assumptions of the ergodic theorem are fulfilled (non-degeneracy and absence of resonance frequencies), to some value of ω which is non-vanishing there corresponds a single matrix element which defines the discontinuity in $\hat{F}(\omega)\,\hat{A}$ at the point ω, or $(\hat{F}(+\omega) - \hat{F}(\omega))\,\hat{A} = \hat{P}_{\psi_i}\hat{A}\hat{P}_{\psi_i}$, where naturally we have $E_i - E_j = \hbar\omega$. In addition, if the observable $\hat{A}$ satisfies the requirements of a macroscopic observable in Ludwig's sense,

† We shall return to the proof of the quantum equivalent of Hopf's theorem: see Chapter VI, § III, 2.

i.e. (III.31), the sum on the right-hand side of (III.35) includes a large number of terms, with the result that the discontinuity $\hat{P}_{\psi_i}\hat{A}\hat{P}_{\psi_i}$ is negligible compared with the discontinuity in $\hat{F}(\omega)\,\hat{A}$ at the origin. If follows that $\mathrm{Tr}\,(\hat{W}_t'\hat{A}\hat{\varrho}(0))$ is a function whose spectral decomposition is almost continuous and involves only a single discontinuity at the origin; moreover, this is given by (III.35). Thus, we can apply the mathematical theorem (I.38) to $\mathrm{Tr}\,(\hat{W}_t'\hat{A}\hat{\varrho}(0))$ which has helped us to prove Hopf's classical theorem. We then have

$$\lim_{T=\infty}\frac{1}{T}\int_0^T |\mathrm{Tr}\,(\hat{W}_t'\hat{A}\hat{\varrho}(0)) - \mathrm{Tr}\,(\hat{\varrho}(0)\,[\hat{F}(+0) - \hat{F}(0)]\hat{A})|^2\,dt \to 0 \tag{III.36}$$

which assumes the form (III.22) if we note that, according to (III.35) and (III.21), we have

$$\mathrm{Tr}\,(\hat{\varrho}(0)\,[\hat{F}(+0) - \hat{F}(0)]\,\hat{A}) = \mathrm{Tr}\,(\hat{A}\hat{\varrho}^{+}(0)).$$

In conclusion, we see that there is a close parallelism between the classical or quantum expression of the ergodic theorems in quadratic mean and that they are both based on the same mathematical foundations. It is in order to be able to use the mathematical theorem (I.38) that we have had to make the assumptions concerning the spectrum of the Hamiltonian on the one hand and to introduce, on the other hand, macroscopic quantities in the sense of Ludwig.

We have mentioned already that the assumptions concerning the spectrum of $\hat{H}$ are very limited from the point of view of quantum mechanics; in the present case, the assumption of non-degeneracy of the eigenstates of $\hat{H}$ is still inadequate: it is necessary to add that of the absence of resonance frequencies in the spectrum of $\hat{H}$. This additional assumption will even reinforce the hypothetical nature of the existence of ergodic systems in wave mechanics.

With regard to the last condition (c) referring to macroscopic observables it is, as we have seen, an additional obstacle to the accurate proof of the second quantum ergodic theory. We can endeavour to satisfy it by recourse to the macroscopic observables of Chapter II; by way of justification, we can say that their introduction is made necessary at the same time by the special nature of quantum observables, which are not measurable simultaneously in general and by the time-evolution of the quantum average values—the probability amplitudes being time-independent in the energy representation—that we have mentioned already on the occasion of the first ergodic theorem. By using the macroscopic operators already defined in (II.54)–(II.58) we shall prove, in the next section, an ergodic theorem in quadratic mean, either in the form which von Neumann gave to it, or in the form of Pauli–Fierz; we shall then be able to define more precisely the exact role played in the proof by these macroscopic operators and we shall see that we can obtain, in reality, only a convergence in probability by resorting to the concept of an average over an ensemble of possible macroscopic observables.

IV. The Proofs of von Neumann and Pauli–Fierz

1. *Von Neumann's method* (Von Neumann, 1929)

This consists in the first place in replacing the microscopic operator $\hat{A}$ in expressions (III.19) and (III.22) by a suitably constructed macroscopic operator: we shall use, for this purpose, the operator defined by equations (II.54)–(II.58). Thus, we obtain for the average value of the macroscopic operator $\hat{\mathscr{A}}$ in the pure case $\Psi(t)$:

$$\mathscr{A}(t) = \mathrm{Tr}\,(\hat{P}_{\Psi(t)}\hat{\mathscr{A}}) = \sum_{\alpha}\sum_{\nu=1}^{N(\alpha)} a_\nu^{(\alpha)} w_\nu^{(\alpha)}(t), \tag{III.37}$$

with

$$w_\nu^{(\alpha)}(t) = \mathrm{Tr}\,(\hat{P}_{\Psi(t)}\,\hat{P}_\nu^{(\alpha)}). \tag{III.38}$$

In order to calculate the expression for the statistical average, we define in the second place the operator $\hat{\bar{\varrho}}^{+}(0)$ so as to obtain the microcanonical distribution; we then determine it by assuming that each state on the same energy shell $e^{(\alpha)}$ has the same probability, which corresponds to a generalisation of the microcanonical ensemble. If $\hat{P}_{\Psi(0)}$ represents the pure case at $t = 0$, we shall have:

$$\hat{\bar{\varrho}}^{+}(0) = \sum_{\alpha} \frac{\mathrm{Tr}\,(\hat{P}_{\Psi(0)}\hat{P}^{(\alpha)})}{\mathrm{Tr}\,\hat{P}^{(\alpha)}}\,\hat{P}^{(\alpha)}. \tag{III.39}$$

the quantity $W^{(\alpha)} = \mathrm{Tr}\,(\hat{P}_{\Psi(0)}\hat{P}^{(\alpha)})$ is the probability of finding the system at $t = 0$ in the energy shell $e^{(\alpha)}$. With these definitions, the statistical average $\overline{\mathscr{A}}$ can be written as

$$\overline{\mathscr{A}} = \mathrm{Tr}\,(\hat{\varrho}^{+}(0)\,\hat{\mathscr{A}}) = \sum_{\alpha}\sum_{\nu=1}^{N(\alpha)} a_\nu^{(\alpha)} s_\nu^{(\alpha)} \frac{W^{(\alpha)}}{S^{(\alpha)}}. \qquad \text{(III.40)}$$

We can then form the expression

$$\frac{1}{T}\int_0^T |\mathscr{A}(t) - \overline{\mathscr{A}}|^2\, dt, \qquad \text{(III.41)}$$

in place of (III.22), where we have for the term to be integrated

$$|\mathscr{A}(t) - \overline{\mathscr{A}}|^2 = \left|\sum_{\alpha}\sum_{\nu=1}^{N(\alpha)} a_\nu^{(\alpha)} \left(w_\nu^{(\alpha)} - \frac{s_\nu^{(\alpha)} W^{(\alpha)}}{S^{(\alpha)}}\right)\right|^2. \qquad \text{(III.42)}$$

As before, the form of this expression suggests a comparison with the average of $\hat{\mathscr{A}}^2$, which can be written as

$$\overline{\mathscr{A}^2} = \sum_{\alpha}\sum_{\nu=1}^{N(\alpha)} (a_\nu^{(\alpha)})^2\, s_\nu^{(\alpha)} \frac{W^{(\alpha)}}{S^{(\alpha)}}. \qquad \text{(III.43)}$$

By using Schwartz's inequality in equation (III.42), we obtain, using (III.43),

$$|\mathscr{A}(t) - \overline{\mathscr{A}}|^2 \leqq \overline{\mathscr{A}^2} \sum_{\alpha}\sum_{\nu=1}^{N(\alpha)} \frac{S^{(\alpha)}}{s_\nu^{(\alpha)} W^{(\alpha)}} \left(w_\nu^{(\alpha)} - \frac{s_\nu^{(\alpha)} W^{(\alpha)}}{S^{(\alpha)}}\right)^2 \equiv \lambda \overline{\mathscr{A}^2}. \qquad \text{(III.44)}$$

Thus, we shall have proved theorem (III.15) if we are able to prove that the time-average $\bar{\lambda}^{\infty}$ of the factor λ tends to zero as T tends to infinity: we shall see that, in fact, we can obtain only a convergence in probability.

In order to proceed to this proof, it is necessary to study the term $w_\nu^{(\alpha)}(t)$ which is time-dependent; this can be written as

$$w_\nu^{(\alpha)}(t) = \sum_{k=1}^{s_\nu^{(\alpha)}} |(\Omega_k^{(\alpha)}, \Psi(t))|^2, \qquad \text{(III.45)}$$

because of the definition (II.55) of the functions $\Omega_k^{(\alpha)}$, which subtend macroscopic cells with index ν. If $\Psi(t)$ is the pure case defined by (III.5), (III.45) assumes the form

$$w_\nu^{(\alpha)}(t) = \sum_{k=1}^{s_\nu^{(\alpha)}} \left|\sum_i r_i e^{i\alpha_i} e^{-iE_i t/h} (\Omega_k^{(\alpha)}, \psi_i)\right|^2, \qquad \text{(III.46)}$$

or again, by expanding the square,

$$w_\nu^{(\alpha)}(t) = \sum_{i,j} r_i r_j \, e^{i(\alpha_i-\alpha_j)} \, e^{-i(E_i-E_j)t/\hbar} \sum_{k=1}^{s_\nu^{(\alpha)}} (\psi_j, \Omega_k^{(\alpha)}) \, (\Omega_k^{(\alpha)}, \psi_i). \tag{III.46'}$$

The last sum in this expression can be written as

$$\sum_{k=1}^{s_\nu^{(\alpha)}} (\psi_j, \Omega_k^{(\alpha)}) \, (\Omega_k^{(\alpha)}, \psi_i) = \sum_{k=1}^{s_\nu^{(\alpha)}} C_{ki}^{(\alpha)*} C_{kj}^{(\alpha)} \equiv D_{ij}^{(\nu)}, \tag{III.47}$$

by using the transformation matrix $C_{ki}^{(\alpha)}$ of (II.55) and the definition (II.64).

Because of the definition of the functions $\Omega_k^{(\alpha)}$ and equation (II.58), the sum (III.47) vanishes if ψ_i and ψ_j do not both belong to the energy range $e^{(\alpha)}$; the result is that the sum (III.46′) over i and j involves only the indices corresponding to the range $e^{(\alpha)}$: the quantity $w_\nu^{(\alpha)}(t)$ is thus expressed as a finite sum of terms belonging to a single energy shell. In order to prove our theorem, we must now evaluate the time-average as follows:

$$\overline{z_\nu^{(\alpha)}(t)}^T = \frac{1}{T} \int_0^T \left(w_\nu^{(\alpha)}(t) - \frac{s_\nu^{(\alpha)} W^{(\alpha)}}{S^{(\alpha)}} \right)^2 dt$$

$$= \frac{1}{T} \int_0^T \left\{ \sum_{i,j}^{\alpha} r_i r_j \, e^{i(\alpha_i-\alpha_j)} \, e^{-i(E_i-E_j)t/\hbar} \left(D_{ij}^{(\nu)} - \frac{s_\nu^{(\alpha)}}{S^{(\alpha)}} \delta_{ij} \right) \right\}^2 dt, \tag{III.48}$$

where $\sum_{i,j}^{\alpha}$ represents a sum over states ψ_i, ψ_j, belonging to the energy shell $e^{(\alpha)}$.

We can see immediately that the previous expression can only be reduced to terms independent of the phases α_i if the two conditions of absence of degeneracy and resonance frequencies in the spectrum of the Hamiltonian are fulfilled simultaneously: these are the assumptions which we have discussed already.

If these two conditions are fulfilled, we can see now how it can be proved that the time-average of the factor λ of (III.44) tends to zero. We calculate the time-average (III.48), which can be written with the foregoing assumptions as

$$\overline{z_\nu^{(\alpha)}(t)}^{\infty} = \left\{\sum_i{}^{\alpha} r_i^2\left(D_{ii}^{(\nu)} - \frac{s_\nu^{(\alpha)}}{S^{(\alpha)}}\right)\right\}^2 + \sum_{i\neq j}{}^{\alpha} r_i^2 r_j^2 \, |D_{ij}^\nu|^2 . \quad \text{(III.49)}$$

By applying Schwartz's inequality to the first term and by noting that $\sum_i{}^{\alpha} r_i^2 = W^{(\alpha)}$, we have

$$\overline{z_\nu^{(\alpha)}(t)}^{\infty} \leqq W^{(\alpha)} \sum_i{}^{\alpha} r_i^2\left(D_{ii}^{(\nu)} - \frac{s_\nu^{(\alpha)}}{S^{(\alpha)}}\right)^2 + \sum_{i\neq j}{}^{\alpha} r_i^2 r_j^2 \, |D_{ij}^{(\nu)}|^2 .$$

(III.50)

Thus, according to (III.44), we obtain an upper limit of $\overline{\lambda}^{\infty}$ by writing

$$\overline{\lambda}^{\infty} = \sum_\alpha \sum_{\nu=1}^{N(\alpha)} \frac{S^{(\alpha)}}{s_\nu^{(\alpha)} W^{(\alpha)}} \overline{z_\nu^{(\alpha)}(t)}^{\infty}$$

$$= \sum_\alpha \sum_{\nu=1}^{N(\alpha)} \frac{S^{(\alpha)}}{s_\nu^{(\alpha)} W^{(\alpha)}} \left\{\left[\sum_i{}^{(\alpha)} r_i^2\left(D_{ii}^{\nu} - \frac{s_\nu^{(\alpha)}}{S^{(\alpha)}}\right)\right]^2 + \sum_{i\neq j}{}^{(\alpha)} r_i^2 r_j^2 \, |D_{ij}^\nu|^2\right\}$$

$$\leqq \sum_\alpha \left\{\sum_i{}^{\alpha} r_i^2 \mu_{ii}^{(\alpha)} + \sum_{i\neq j}{}^{\alpha} \frac{r_i^2 r_j^2}{W^{(\alpha)}} \mu_{ij}^{(\alpha)}\right\}, \quad \text{(III.51)}$$

where we have put

$$\mu_{ij}^{(\alpha)} = \sum_{\nu=1}^{N(\alpha)} \frac{S^{(\alpha)}}{s_\nu^{(\alpha)}} \left| D_{ij}^{(\nu)} - \frac{s_\nu^{(\alpha)}}{S^{(\alpha)}} \delta_{ij} \right|^2 . \quad \text{(III.52)}$$

As in the previous section, we verify again that the assumptions of ergodicity, depending on the spectrum of $\hat{H}$, are insufficient for making $\overline{\lambda}^{\infty}$ approach zero and that it is necessary to add a supplementary condition. However, we can see easily that the terms of expression (III.51), which depend on changing the variables $C_{ki}^{(\alpha)}$ (i.e. on a change in the definition of the macroscopic cells) by means of $\mu_{ij}^{(\alpha)}$, are separated from the terms which define the pure case considered, that is, r_i^2 and $W^{(\alpha)}$. If we can show that all the $\mu_{ij}^{(\alpha)}$ are less than ε, we can deduce easily that $\overline{\lambda}^{\infty}$ is less than 2ε.

The method used by von Neumann and Pauli–Fierz to prove this latter point consists in considering the set of possible subdivisions of Hilbert space into macroscopic cells (or, in other words the set of possible macroscopic observers). By assuming *the equi-*

probability of all macroscopic observers we can show that the probability of a subdivision of the set providing a value of $\mu_{ij}^{(\alpha)}$ which is larger than a certain given value is very small.

The subdivisions of Hilbert space are defined by the matrix $C_{ki}^{(\alpha)}$ which occurs in the preceding expressions through $D_{ij}^{(\nu)}$; we are thus led to studying the distribution in probability of the expressions $D_{ij}^{(\nu)}$ for an energy shell $e^{(\alpha)}$. If the coefficients $C_{ki}^{(\alpha)}$ are considered as the components of unitary vectors in a Hilbert space with $S^{(\alpha)}$ dimensions, each value of $D_{ij}^{(\nu)}$ represents the scalar product of the projections of two such vectors on a cell with $s_\nu^{(\alpha)}$ dimensions. Because of the assumption of the equiprobability of macroscopic observers, the probability of finding subdivisions corresponding to elements of $C_{ki}^{(\alpha)}$ located inside a given solid angle of Hilbert space is proportional to the magnitude of this solid angle (or, what amounts to the same, to the area of surface cut out of the unit sphere by this solid angle); the calculation of the probability $P(D)\,dD$ that the quantities $D_{ii}^{(\nu)}$ lie between D and $D + dD$ is thus reduced to a problem of geometrical probabilities dealt with by von Neumann and Pauli–Fierz in particular. The following expression is found for $P(D)$:

$$P(D) = KD^{s_\nu^{(\alpha)}}(1 - D)^{S^{(\alpha)} - s_\nu^{(\alpha)}}, \qquad \text{(III.53)}$$

where K is a normalising constant. It can be seen from (III.53) that the most probable value of $D_{ii}^{(\nu)}$ is $s_\nu^{(\alpha)}/S^{(\alpha)}$ and, on the other hand, that the maximum is extremely sharp: this is a consequence of the asymptotic geometrical properties of the unit hypersphere in $2n$-dimensional space. We can deduce from this result the probability that the upper limit of the $\mu_{ij}^{(\alpha)}$ is larger than a number ε_0 fixed in advance (always over an interval $e^{(\alpha)}$);† we find for this probability an expression of the form

$$K' \exp\left[-K''\sqrt{\frac{S^{(\alpha)}}{N^{(\alpha)}}}\,\varepsilon_0 + 2\ln S^{(\alpha)}\right], \qquad \text{(III.54)}$$

where K'' is a number of order unity; this result is valid only if $\varepsilon_0 > \dfrac{2N^{(\alpha)}}{(S^{(\alpha)} - 2)}$. We can see that in this case, for values of ε_0

† We refer to Appendix II for all intermediate calculations.

which are not too small, the probability (III.54) becomes very small provided that

$$\frac{S^{(\alpha)}}{N^{(\alpha)}} \gg (2 \ln S^{(\alpha)})^2. \tag{III.55}$$

This condition expresses simply that the average number of eigenstates per macroscopic cell with suffix ν must be very large, which is completely acceptable if we are dealing with a macroscopic physical system.

Thus, we can say that it is very probable that a macroscopic observation attributes values to $D_{ij}^{(\nu)}$ such that

$$D_{ij}^{(\nu)} = \frac{S_\nu^{(\alpha)}}{S^{(\alpha)}} \delta_{ij}, \quad \mu_{ij}^{(\alpha)} = 0, \tag{III.56}$$

with the result that we can write

$$\text{Prob}\left\{ \lim_{T=\infty} \frac{1}{T} \int_0^T (\mathscr{A}(t) - \overline{\mathscr{A}})^2 \, dt > \eta \right\} < \delta, \tag{III.57}$$

(the probability having the classical meaning which we have just assigned to it); this inequality is satisfied when the conditions concerning the spectrum of the Hamiltonian and those relating to the definition of macroscopic operators are fulfilled. We can see that it is only possible, in fact, to prove the second quantum "probability" ergodic theorem by invoking the macroscopic nature of the system and the assumption of equiprobability of all macroscopic observers; this constitutes an additional difficulty which is inherent in the quantum ergodic theory, because in classical mechanics the hypothesis of metrical transitivity is sufficient to ensure an almost certain convergence or a convergence in quadratic mean. We shall show, moreover, in the next chapter, that the average over the macroscopic observers actually plays a fundamental role in the foregoing proof and that it renders useless any assumption concerning the spectrum of the Hamiltonian.

Finally, we note that the distribution (III.53) also enables us to calculate the average of the factor $\overline{\lambda}^{\infty}$ for all macroscopic observers; for this purpose we use formulae (36) to (43) of Appendix II, which give the averages $\langle |C_{kl}^{(\alpha)}|^2 \rangle$, $\langle |C_{kl}^{(\alpha)}|^4 \rangle$, $\langle |C_{kl}^{(\alpha)}|^2 \, |C_{kj}^{(\alpha)}|^2 \rangle$ and $\langle |C_{kl}^{(\alpha)}|^2 \, |C_{k'j}^{(\alpha)}|^2 \rangle$, where the symbol $\langle \, \rangle$ denotes the average over different macroscopic observers. If we assume that the system occupies a single energy shell $e^{(\alpha)}$, according to (III.51) we have:

$$\overset{-\infty}{\lambda} = \sum_{\nu=1}^{N(\alpha)} \frac{S^{(\alpha)}}{s_i^{(\alpha)}} \left\{ \sum_i^{(\alpha)} r_i^4 \left(D_{ii}^{(\nu)} - \frac{s_\nu^{(\alpha)}}{S^{(\alpha)}} \right)^2 \right.$$

$$\left. + \sum_{i \neq j}^{(\alpha)} r_i^2 r_j^2 \left[\left(D_{ii}^{(\nu)} - \frac{s_\nu^{(\alpha)}}{S^{(\alpha)}} \right) \left(D_{jj}^{(\nu)} - \frac{s_\nu^{(\alpha)}}{S^{(\alpha)}} \right) + |D_{ij}^{(\nu)}|^2 \right] \right\}, \tag{III.51'}$$

whence

$$\left\langle \overset{-\infty}{\lambda} \right\rangle = \sum_{\nu=1}^{N(\alpha)} \frac{S^{(\alpha)}}{s_\nu^{(\alpha)}} \left\{ \sum_i^{(\alpha)} r_i^4 \left\langle \left(D_{ii}^{(\nu)} - \frac{s_\nu^{(\alpha)}}{S^{(\alpha)}} \right)^2 \right\rangle \right.$$

$$\left. + \sum_{i \neq j}^{(\alpha)} r_i^2 r_j^2 \left\langle \left[\left(D_{ii}^{(\nu)} - \frac{s_\nu^{(\alpha)}}{S^{(\alpha)}} \right) \left(D_{jj}^{(\nu)} - \frac{s_\nu^{(\alpha)}}{S^{(\alpha)}} \right) + |D_{ij}^{(\nu)}|^2 \right] \right\rangle \right\}.$$

By replacing the averages by their value, we obtain finally

$$\left\langle \overset{-\infty}{\lambda} \right\rangle = \frac{N^{(\alpha)} - 1}{S^{(\alpha)} + 1} = O\left(\frac{N^{(\alpha)}}{S^{(\alpha)}} \right) \tag{III.57'}$$

which, put into (III.44), enables us to obtain a result which is equivalent to (III.57); we shall encounter also a similar formula in the next chapter [cf. (IV.77)].

2. *The Pauli–Fierz method*

This method can be reduced easily to that of von Neumann, because the definition of macroscopic cells which it uses comes within the framework of the general scheme that we have indicated. Since the probability $w_\nu^{(\alpha)}(t)$ of finding the system in the νth cell is defined by (III.38),t he entropy of the system can be written as

$$S(t) = -\sum_\alpha \sum_{\nu=1}^{N(\alpha)} w_\nu^{(\alpha)}(t) \ln \frac{w_\nu^{(\alpha)}(t)}{s_\nu^{(\alpha)}} \tag{III.58}$$

[putting $k = 1$ in formula (26) of Appendix III.B] and for the corresponding microcanonical entropy we have

$$\overset{-m}{S} = -\sum_{\alpha=1}^{\omega} W^{(\alpha)} \ln \frac{W^{(\alpha)}}{S^{(\alpha)}}, \tag{III.59}$$

where $W^{(\alpha)} = \mathrm{Tr}\,(\hat{P}_{\Psi 0)}\hat{P}^{(\alpha)})$ is the probability of finding the system in the shell $e^{(\alpha)}$. In order to prove an ergodic theorem we consider

$$\overline{S(t)}^{\infty} = \lim_{T=\infty} \frac{1}{T} \int_0^T S(t)\, dt \tag{III.60}$$

and we evaluate the difference $\overline{S(t)}^{\infty} - \overline{S}^{m}$ by using a well-known relation satisfied by the function $L(x, y) = x(\ln x - \ln y) - x + y$, namely that we have for x and $y > 0$,

$$0 \leqq L(x, y) \leqq \frac{1}{y}(x - y)^2.$$

If we put $x = w_{\nu}^{(\alpha)}/s_{\nu}^{\alpha}$ and $y = W^{(\alpha)}/S^{(\alpha)}$, we obtain

$$\begin{aligned} 0 &\leqq -W^{(\alpha)} \ln \frac{W^{(\alpha)}}{S^{(\alpha)}} + \sum_{\nu=1}^{N(\alpha)} w_{\nu}^{(\alpha)} \ln \frac{w_{\nu}^{(\alpha)}}{s_{\nu}^{(\alpha)}} \\ &\leqq \sum_{\nu=1}^{N(\alpha)} \frac{S^{(\alpha)}}{s_{\nu}^{(\alpha)} W^{(\alpha)}} \left(w_{\nu}^{(\alpha)} - \frac{s_{\nu}^{(\alpha)} W^{(\alpha)}}{S^{(\alpha)}} \right)^2, \end{aligned} \tag{III.61}$$

by restricting ourselves to a single shell $e^{(\alpha)}$.

Putting this result in (III.60), we obtain for the difference $\overline{S(t)}^{\infty} - \overline{S}^{m}$:

$$\overline{S(t)}^{\infty} - \overline{S}^{m} \leqq \sum_{\alpha} \sum_{\nu=1}^{N(\alpha)} \frac{S^{(\alpha)}}{s_{\nu}^{(\alpha)} W^{(\alpha)}} \overline{\left(w_{\nu}^{(\alpha)}(t) - \frac{s_{\nu}^{(\alpha)} W^{(\alpha)}}{S^{(\alpha)}} \right)^2}^{\infty}. \tag{III.62}$$

Thus, we have encountered an expression which is similar to the factor $\overline{\lambda}^{m}$ of (III.51): starting from here, the reasoning of the previous section applies in the same way. We deduce that for almost all macroscopic observers the limit of the time-average of the entropy of the system (III.58) is equal to the microcanonical entropy, which can be written as

$$\mathrm{Prob}\left\{\left(\overline{S(t)}^{\infty} - \overline{S}^{m}\right) > \eta\right\} < \delta. \tag{III.63}$$

3. *Fierz's criticism*

This criticism is concerned first of all with the *a priori* assumption of equiprobability of macroscopic observers; Fierz (1955) draws attention to the fact that this assumption is unfounded physically and that it tends to associate the property of ergodicity not with the system observed but with the macroscopic observer. In addition, we shall encounter in Chapter IV another inconvenience of this type of averages so that it will be preferable to change to ensemble averages such as we defined in Chapter II.

Moreover, Fierz simplifies the proof of the previous section, on the one hand by using directly in the definition of $S(t)$ the fact that the system has a large number of degrees of freedom and, on the other hand, by using separately the first quantum ergodic theorem. We note in this case that the number of cells $N^{(\alpha)}$ contained within an energy shell $e^{(\alpha)}$ must be at least of order of magnitude 10^{20}, since $N^{(\alpha)}$ is the number of states which can be distinguished by a macroscopic observer; furthermore, the entropy of the system in units of k (Boltzmann's constant) is also of order 10^{20}, which means that $s_\nu^{(\alpha)}$ and $S^{(\alpha)}$ are approximately of order $\exp(10^{20})$. The result is that $w_\nu^{(\alpha)}$ is of order 10^{-20} and $\ln w_\nu^{(\alpha)}$ is negligible compared with $\ln s_\nu^{(\alpha)}$, so that the entropy (III.58) can then be written more simply as

$$S(t) \cong \sum_{\alpha} \sum_{\nu=1}^{N(\alpha)} w_\nu^{(\alpha)} \ln s_\nu^{(\alpha)}. \tag{III.64}$$

On the other hand, if we are restricted to a single energy shell $e^{(\alpha)}$ and if we consider the quantity

$$\mathfrak{S} = \sum_{\nu=1}^{N(\alpha)} \frac{s_\nu^{(\alpha)}}{S^{(\alpha)}} \ln s_\nu^{(\alpha)}, \tag{III.65}$$

it can be shown easily that $\mathfrak{S}$ is almost equal to the microcanonical entropy; in fact, the sum $\sum_\nu s_\nu^{(\alpha)} \ln s_\nu^{(\alpha)}$, which satisfies the condition $\sum_\nu s_\nu^{(\alpha)} = S^{(\alpha)}$, is a minimum for $s_\nu^{(\alpha)} = \text{constant} \simeq S^{(\alpha)}/N^{(\alpha)}$; we have, therefore,

$$\mathfrak{S} \geqq \sum_\nu \frac{1}{N^{(\alpha)}} \ln\left(\frac{S^{(\alpha)}}{N^{(\alpha)}}\right) = \ln S^{(\alpha)} - \ln N^{(\alpha)}.$$

In addition, by using the property $0 < s_\nu^{(\alpha)}/S^{(\alpha)} < 1$, we can also show that

$$\mathfrak{S} = \sum_\nu \frac{s_\nu^{(\alpha)}}{S^{(\alpha)}} \ln S^{(\alpha)} + \sum_\nu \frac{s_\nu^{(\alpha)}}{S^{(\alpha)}} \ln \frac{s_\nu^{(\alpha)}}{S^{(\alpha)}} \leqq \ln S^{(\alpha)}.$$

By comparing these two inequalities, we have finally

$$\ln S^{(\alpha)} - \ln N^{(\alpha)} \leqq \mathfrak{S} \leqq \ln S^{(\alpha)}.$$

Since $\ln S^{(\alpha)}$ is of order 10^{20} whilst $\ln N^{(\alpha)}$ is of order 50, we can neglect $\ln N^{(\alpha)}$ in comparison with $\ln S^{(\alpha)}$, whence

$$\mathfrak{S} \cong \ln S^{(\alpha)} = \overline{S}^{m}. \tag{III.66}$$

Because of the approximate formulae (III.66) and (III.64) we shall have established the existence of an ergodic theorem provided we prove the equality

$$\mathfrak{S} = \lim_{T=\infty} \frac{1}{T} \int_0^T \sum_{\nu=1}^{N(\alpha)} w_\nu^{(\alpha)}(t) \ln s_\nu^{(\alpha)} dt, \tag{III.67}$$

or

$$\sum_{\nu=1}^{N(\alpha)} \frac{s_\nu^{(\alpha)}}{S^{(\alpha)}} \ln s_\nu^{(\alpha)} = \sum_{\nu=1}^{N(\alpha)} \overline{w_\nu^{(\alpha)}(t)}^{\infty} \ln s_\nu^{(\alpha)};$$

the ergodic theorem will thus be proved if

$$\overline{w_\nu^{(\alpha)}(t)}^{\infty} = \frac{s_\nu^{(\alpha)}}{S^{(\alpha)}}. \tag{III.68}$$

According to equations (III.46′) and (III.47), we have at once

$$w_\nu^{(\alpha)}(t) = \sum_{i,j}^{\alpha} r_i r_j e^{i(\alpha_i - \alpha_j)} e^{-i(E_i - E_j)t/\hbar} D_{ij}^{(\nu)}, \tag{III.69}$$

and, by applying the first ergodic theorem to this expression, it becomes

$$\overline{w_\nu^{(\alpha)}(t)}^{\infty} = \sum_i^{\alpha} r_i^2 D_{ii}^{(\nu)}. \tag{III.70}$$

In order to obtain (III.70) we must assume that the spectrum of $\hat{H}$ is non-degenerate (on the other hand, the condition $\sum_i r_i < \infty$ has become superfluous, since we are dealing in general with a finite number of terms in the shell $e^{(\alpha)}$).

In order to prove (III.68) we now evaluate

$$\left| \overline{w_\nu^{(\alpha)}(t)}^{\infty} - \frac{s_\nu^{(\alpha)}}{S^{(\alpha)}} \right| = \left| \sum_i^{\alpha} r_i^2 \left(D_{ii}^{(\nu)} - \frac{s_\nu^{(\alpha)}}{S^{(\alpha)}} \right) \right| < \max \left| D_{ii}^{(\nu)} - \frac{s_\nu^{(\alpha)}}{S^{(\alpha)}} \right|, \tag{III.71}$$

in which we have used the relationship $\sum_i^\alpha r_i^2 = 1$, valid for the shell $e^{(\alpha)}$. Thus, it is sufficient to prove that the quantity on the right-hand side of (III.71) is practically zero for all changes of the variables $C_{ki}^{(\alpha)}$: it is thus related to the problem of geometrical probabilities which we have already resolved earlier. It can be deduced that (III.71) can be considered as always zero except for extremely improbable macroscopic observers.

The foregoing proof is simpler than the proofs of von Neumann and Pauli–Fierz for the following reasons: on the one hand it derives the maximum advantage in the definition of $S(t)$ from the large number of degrees of freedom of the system considered; on the other hand, it depends solely on the absolute value of the difference between the time-average and the statistical average, instead of depending on the time-average of the square of the difference between the actual quantity and its statistical average. This latter simplification enables us to use the first ergodic theorem without having recourse to the supplementary assumption of the absence of resonance frequencies in the spectrum of the Hamiltonian.

We note, finally, that expression (III.71) highlights the important role played by strongly degenerate macroscopic operators in the second part of the proof: in fact, the use of the first ergodic theorem is insufficient [depending only on the phases of $w_\nu^{(\alpha)}(t)$] and it is still necessary to prove the convergence of the square of the probability amplitudes (which are unaffected by the evolution operator) towards $s_\nu^{(\alpha)}/S^{(\alpha)}$. It is the introduction of macroscopic observers which enables us to use statistical considerations which are necessary for proving the theorem in probability.

In the following chapter, we shall deal with certain of these considerations which, combined with the systematic use of quantum statistical ensembles, will enable us to obtain the probability ergodic theorems without a restrictive assumption concerning the nature of the system. We must, therefore, develop in quantum theory a method which is similar to that proposed by Khinchin in classical statistical mechanics, and introduce averages over ensembles of systems (or, which amounts to the same thing, averages over ensembles of initial conditions). We shall then see that the ergodic theorems established by von Neumann and by Pauli–Fierz do, in fact, come within the framework of these probability ergodic theorems, the averages over ensembles of systems being replaced here by the averages over the ensemble of macroscopic observers. Moreover, we shall show that these two types of averages are mathematically equivalent, but that there are strong physical reasons for preferring the methods which use averages over statistical ensembles of systems.

These reasons depend on two types of argument: first of all, there are the criticisms already mentioned by Fierz relating to the hypothesis of equal *a priori* probability of all macroscopic observers. These criticisms are based on the fact that the thermodynamic properties of a macroscopic system are completely defined by a single complete system of macroscopic observables, which commute with one another. It follows that not all the macroscopic observables can be measured and that the introduction of the ensemble of all the macroscopic observers has no physical significance. In this connection, we shall note moreover with Farquhar (1964) that in practice, it is not the hypothesis of equal *a priori* probability of all macroscopic observers that is used, but a hypothesis of equal probability for the different unitary transformations of the sub-space of $S^{(\alpha)}$ dimensions corresponding to a given energy shell. However, as there is no one-to-one correspondence† between macroscopic observers and the bases $\{\Omega_k^{(\alpha)}\}$, this latter hypothesis is not equivalent to that of equal *a priori* probabilities for all macroscopic observers; in fact, the *a priori* probabilities for different macroscopic observers are not equal, but they are such that we have an equal probability for the bases $\{\Omega_k^{(\alpha)}\}$ which subtend the unitary sub-space considered; this is the assumption made implicitly in all calculations of averages over the "ensemble of all macroscopic observers".

Another type of objection to von Neumann's procedure is based on the fact that the ergodicity of any macroscopic observable can be established independently of the time evolution of the system. This result, which we shall be discussing in detail in the next chapter, shows us that von Neumann's theorem only expresses the geometrical properties of the set of unitary transformations of a space with a large number of dimensions; under these conditions, the ergodicity no longer appears to be connected in any way with the dynamic structure of the system. We shall see that these two kinds

† In fact, many equivalent unitary transformations $C_{ki}{}^{(\alpha)}$ can be associated with each macroscopic observer.

of difficulty disappear when we express it within the framework of the probability ergodic theorems, by adopting a point of view similar to that of Khinchin in classical theory.

4. *Return to "individual" ergodic theorems*

Before dealing with the study of the quantum probability ergodic theorems, it would appear to be useful, nevertheless, to return to the "individual" quantum ergodic theorems, for which the equality of the time averages and of the phase averages can only result from the properties of the dynamic evolution of a *single* system. The need for this return to the "individual" theorems is due to two reasons. On the one hand, this point of view is the only one that permits us to relate the fundamental principles of statistical mechanics to the dynamic properties of a system and to the structure of its Hamiltonian. On the other hand, the probability ergodic theorems which we shall obtain in Chapter IV themselves encounter difficulties which are completely comparable with those that we have already discussed in Chapter I, § III.3, when discussing the corresponding classical theorems. We can, in fact, argue along similar lines: the theorems in question should in principle be applicable whatever the dynamic structure of the system, provided a uniform probability distribution is taken over the whole of an energy shell; on the other hand, however, the existence of constants of motion other than energy, associated with degeneracies of the Hamiltonian, has the effect of making certain states of the given energy shell "inaccessible". There is therefore a conflict between the possible existence of primary integrals other than the energy of the system, and the hypothesis of equal probability of all states of an energy shell; it follows that the probability ergodic theorems, just like the von Neumann and Pauli–Fierz theorems, are too general to allow ergodic systems to be distinguished from those that are not.

Thus, there remains the problem of finding the conditions for ergodicity. We recall in this connection that the conditions for the absence of degeneracy and of resonance frequencies in the spectrum of the Hamiltonian are not sufficient, as we have already emphasized previously. In fact, this can be seen directly by constructing counter-examples, where manifestly non-ergodic systems satisfy these conditions. This is the case with the model proposed by Fierz in which a fluid rotates in a cylindrical container: the angular momentum around the axis of the cylinder is then a constant of motion, as a result of which not all the states of the energy shell are accessible. Such a system is therefore not ergodic, although the conditions for the absence of degeneracy and of resonance frequencies in the energy spectrum could be satisfied if intermolecular forces are taken into account. We shall also see in Chapter VI that other conditions of ergodicity have been obtained by Prosperi and Scotti by modifying the definition of the statistical ensembles considered; instead of calculating the averages by using a uniform probability distribution over the whole of an energy shell, we limit ourselves to a uniform distribution over that single cell, which is occupied initially by the system according to the macroscopic observation. Unfortunately, it appears to be difficult to define more precisely the exact significance of the conditions of ergodicity which one obtains in this way (cf. for example, the inequalities (E_2') in Chapter VI, § I.2) and especially to determine what class of physical systems satisfy these conditions.

Since these works have so far not given satisfactory solutions to these problems, it would appear to be of interest for us to give a brief account of two attempts which are capable of opening up new ways for research: these are recent papers by Ludwig (1960, 1961) and the ergodic theorem proposed by Golden and Longuet-Higgins (1960).

(a) *Ludwig's macroscopic observables.* Ludwig's point of view consists essentially in a criticism of the concept of macroscopic observable, as introduced by von Neumann and van Kampen. According to Ludwig, the definition of such observables should result objectively from the intrinsic properties of a system with a large number of degrees of freedom, and should not be connected in a subjective way with the imprecise nature of macroscopic observation. How to define macroscopic observables as intrinsic properties of the system is the problem set by Ludwig and he attempts to resolve it by introducing the concepts of "discernibility" of two states and the "measure" of this discernibility.

The basic idea is as follows: Consider two states of a system represented by the statistical operators $\hat{\varrho}_1$ and $\hat{\varrho}_2$; the average values of a microscopic Hermitian observable $\hat{A}$ are different for the two statistical ensembles, except obviously if $\hat{\varrho}_1 = \hat{\varrho}_2$. In this latter case, we say that the two ensembles $\hat{\varrho}_1$ and $\hat{\varrho}_2$ are microscopically "indiscernible". This is no longer true on the macroscopic scale in which, because of its strong degeneracy, a macroscopic observable can have its average values equal for two ensembles $\hat{\varrho}_1$ and $\hat{\varrho}_2$ which are microscopically discernible. This leads us to attempt to relate the problem of the objective definition of macroscopic observables to finding a criterion which enables us to distinguish two "macroscopically discernible" states.

In order to treat this idea mathematically, we consider first of all the microscopic case and we shall consider the Hilbert space formed by the class of completely continuous Hermitian operators. We define in this space the scalar product of two operators $\hat{A}$ and $\hat{B}$ by:

$$(\hat{A}, \hat{B}) = \mathrm{Tr}\,(\hat{A}^{\dagger}\hat{B}), \tag{III.72}$$

and the norm of some operator $\hat{A}$ by:

$$\|\hat{A}\| = (\hat{A}, \hat{A})^{\frac{1}{2}}; \tag{III.73}$$

this must be finite in accordance with the condition of continuity. In particular, the average value of any Hermitian operator $\hat{A}$ for a statistical ensemble $\hat{\varrho}$ can be written as follows:

$$\bar{A} = \mathrm{Tr}\,(\hat{\varrho}\hat{A}) = (\hat{\varrho}, \hat{A}). \tag{III.74}$$

Let us consider now two statistical ensembles represented by the operators $\hat{\varrho}_1$ and $\hat{\varrho}_2$. Ludwig then defines a "distance" between these two ensembles by:

$$d(\hat{\varrho}_1, \hat{\varrho}_2) \equiv \frac{1}{\sqrt{2}} \left\| \sqrt{\hat{\varrho}_1} - \sqrt{\hat{\varrho}_2} \right\|; \tag{III.75}$$

this has two important properties: on the one hand, it satisfies the triangular inequality; on the other hand, it is such that the average values of an observable $\hat{A}$ in these two ensembles satisfies the inequality

$$\left|\bar{A}_1 - \bar{A}_2\right| = \left|\mathrm{Tr}\,(\hat{\varrho}_1\hat{A}) - \mathrm{Tr}\,(\hat{\varrho}_2\hat{A})\right| \leqq 2\sqrt{2}\left|\hat{A}\right| d(\hat{\varrho}_1, \hat{\varrho}_2). \tag{III.76}$$

This latter property shows us that the "distance" $d(\hat{\varrho}_1, \hat{\varrho}_2)$ can be interpreted as a "microscopic measure of discernibility" of the two ensembles $\hat{\varrho}_1$ and $\hat{\varrho}_2$. In fact, if

$$d(\hat{\varrho}_1, \hat{\varrho}_2) \ll 1, \tag{III.77}$$

the average values A_1 and A_2 will be very close and it could be said that the states $\hat{\varrho}_1$ and $\hat{\varrho}_2$ are almost indiscernible; moreover, if $\hat{\varrho}_1 = \hat{\varrho}_2$, we have $\bar{A}_1 = \bar{A}_2$ and the two states are completely "indiscernible".

Having obtained these conditions, Ludwig's procedure then consists in attempting to generalise the concept of "measure of discernibility" of two ensembles to arrive at a definition of macroscopic observables. We begin by introducing more general "distances", $d_\alpha(\hat{\varrho}_1, \hat{\varrho}_2)$, which possess properties similar to the distance d (they satisfy especially the triangular inequality). We can then associate with these generalised distances, the so-called "α-continuous" classes of Hermitian operators, which are those which satisfy a relation of the form (III.76), with d replaced by d_α; these α-continuous operators correspond to observables whose measure only enables us to distinguish two statistical ensembles to an extent represented by the distance $d_\alpha(\hat{\varrho}_1, \hat{\varrho}_2)$.

We can apply this method to the definition of macroscopic observables; in this framework, it is necessary first of all to introduce a "macroscopic measure of discernibility" of two ensembles $d_M(\hat{\varrho}_1, \hat{\varrho}_2)$. We next define the macroscopic observables as being represented by the class of commuting operators which are M-continuous with respect to d_M. If $\{\Omega_k\}$ denotes a base of eigenvectors of these operators and if $\hat{P}_{\Omega_k}$ are the corresponding projection operators, the "distance" between two eigenvectors is given by:

$$d_M(\Omega_k, \Omega_l) = d_M(\hat{P}_{\Omega_k}, \hat{P}_{\Omega_l}). \qquad \text{(III.78)}$$

It follows that if we have $d_M(\hat{P}_{\Omega_l}, \hat{P}_{\Omega_k}) \ll 1$, we define a group of vectors Ω_l which are associated with the values of almost identical macroscopic measures; we note, moreover, that this property is analogous to that which results from the introduction of "cells" in the usual theory. The foregoing definitions therefore lead to macroscopic observables which are almost equivalent to those of von Neumann and van Kampen but which, by their construction, are independent of the observer considered and therefore no longer have a subjective character. Ludwig's procedure would therefore allow us to obtain, at least formally, the macroscopic observables as intrinsic properties of the system, on the one condition that we must know how to determine the "measure of macroscopic discernibility" $d_M(\hat{\varrho}_1, \hat{\varrho}_2)$. Unfortunately, it has not been possible to establish this essential point, even in the case of systems with a large number of degrees of freedom. Ludwig's programme has not, therefore, been completely accomplished, with the result that the theory which we have just outlined remains for the present just an attempt.

Leaving aside this problem, we now come to the ergodic theorem established by Ludwig for his macroscopic observables. It can be obtained by imposing on the M-continuous operators, which we have just introduced, a condition identical to the one we already used in Section III of this chapter; in our notation, this can be written as:

$$\overline{\left(A(t) - \overline{A(t)}^{\infty}\right)^2}^{\infty} \gg \overline{A^2(t)}^{\infty} ; \qquad \text{(III.79)}$$

it indicates that the relative time-variation of the average quantum value $A(t)$ of a macroscopic observable $\hat{A}$ must be very small. In order to satisfy equation (III.79), we show that any eigenvector Φ_i of the Hamiltonian $\hat{H}$ must satisfy the ergodicity condition

$$\|\Phi_i - \hat{P}_k^{(m)}\Phi_i\| \ll 1, \qquad \text{(III.80)}$$

where $\hat{P}_k^{(m)}$ denotes the projection operator onto a phase cell, characterised by a sharp entropy maximum, the macroscopic integrals of motion having fixed values. If we return to the notation of von Neumann's macroscopic observables, it can be shown that relation (III.80) leads to ergodicity conditions comparable with those proposed by Prosperi and Scotti (cf. Chapter VI, § III.2); in particular, we note that the quantities

$$\left|\left(\Omega_k^{(\alpha)}, \Phi_i\right)\right|^2 = \left|C_{ki}^{(\alpha)}\right|^2$$

must all be of the same order of magnitude. Nevertheless, this analogy cannot be pushed too far, because of the difference in structure between Ludwig's observables and those of von Neumann. In quite the same way, Ludwig's ergodicity condition comes up against the same objections as those of Prosperi and Scotti, namely, that its physical significance is not clear and that it does not provide any criterion which will allow us to recognise whether or not a Hamiltonian system is ergodic.

(b) *The ergodic theorem of Golden and Longuet-Higgins.* We shall now study a very different attempt, proposed by Golden and Longuet-Higgins, which results in the formulation of an ergodic theorem which is valid for a macroscopic system. The essential feature of this theory lies in the fact that it introduces neither macroscopic operators nor "coarse-graining" processes: the macroscopic nature of the system appears only in the spectrum of the Hamiltonian, which is assumed to be continuous. The theory thus obtained, is applicable therefore to a situation which is comparable to that envisaged by Van Hove in his theory of irreversible processes.

The fundamental argument of the work of Golden and Longuet-Higgins is that the time evolution of a quantum system is very different when the Hamiltonian spectrum is discrete from when it is continuous. Let us deal first with the case in which the Hamiltonian has a discrete spectrum with eigenvalues which can be degenerate. It is known that the average quantum value of any observable A will fluctuate indefinitely with time and that it will not tend to any limit unless it is constant at every instant. Moreover, a strict demonstration of this result can be given by considering, with Golden and Longuet-Higgins, the time average of the variation of $A(t)$,

$$\overline{\left(A(t) - \overline{A(t)}^{\infty}\right)^2}^{\infty}, \tag{III.81}$$

and by noting that this expression would become zero if $A(t)$ tended asymptotically towards a limit which, if it existed, would be necessarily equal to $\overline{A(t)}^{\infty}$. In order to calculate the time averages, these authors introduce the expression:

$$z\int_0^\infty dt\, e^{-zt} A(t) = z\int_0^\infty dt\, e^{-zt}\, \mathrm{Tr}\, [\hat{\bar{\varrho}}(t)\, \hat{A}]; \tag{III.82}$$

this is a time average weighted by e^{-zt} which, for $z = 0$, reduces to the ordinary time average of $A(t)$. Thus, we have the result:

$$\overline{A(t)}^{\infty} = \lim_{z\to 0} z\int_0^\infty dt\, e^{-zt}\, \mathrm{Tr}\, [\hat{\bar{\varrho}}(t)\, \hat{A}] = \sum_{\substack{mn \\ (E_m = E_n)}} \bar{\varrho}_{mn}(0) A_{nm}. \tag{III.83}$$

Similarly, we have:

$$\overline{\left(A(t) - \overline{A(t)}^{\infty}\right)^2}^{\infty} = \lim_{z\to 0} z\int_0^\infty dt\, e^{-zt} \left[A(t) - \overline{A(t)}^{\infty}\right]^2$$

$$= \sum_{\omega \neq 0} |g(\omega)|^2, \tag{III.84}$$

where we have put:

$$g(\omega) = \sum_{(E_n = E_m - \hbar\omega)} \bar{\varrho}_{nm}(0) A_{mn}; \tag{III.85}$$

we note in passing that this function $g(\omega)$ is reduced to that used by Ludwig (Chapter III, § III.1) if the spectrum of the Hamiltonian were non-degenerate. It can be seen from equation (III.84) that expression (III.81) will only vanish if all the $g(\omega)$ are zero for $\omega \neq 0$. However, according to equation (III.85), we have the two relations:

$$\overline{A(t)}^{\infty} = g(0), \quad A(0) = g(0) + \sum_{\omega \neq 0} g(\omega). \tag{III.86}$$

It follows that if $A(t)$ tends asymptotically towards a limit $\overline{A(t)}^{\infty}$, we must have $A(0) = \overline{A(t)}^{\infty}$, which involves $A(t)$ already having the value $\overline{A(t)}^{\infty}$ at $t = 0$ and that it remains constant at every subsequent instant. Actually, this result only expresses the properties of almost-periodicity of the state vector $\Psi(t)$ (cf. Chapter VI, § I.1) and of the statistical operator (cf. Ono, 1949 and Percival, 1961).

Moreover, it is easy to show that this result remains valid if we introduce "time-smoothing" of the statistical operator. The argument is the same as that used in Chapter V for Kirkwood's method (Chapter V, § V.3). We define a new statistical operator by:

$$\hat{\hat{\varrho}}(t) = \int_{-\infty}^{+\infty} d\tau f(\tau)\, \hat{\varrho}(t + \tau), \tag{III.87}$$

where $f(\tau)$ is a real non-negative function whose integral with respect to τ is equal to unity. By forming the time derivative $\partial\hat{\hat{\varrho}}(t)/\partial t$, it can be seen easily from equation (III.87) that $\hat{\hat{\varrho}}(t)$ obeys the equation of evolution:

$$\frac{\partial \hat{\hat{\varrho}}}{\partial t} = \frac{2\pi i}{h} [\hat{\hat{\varrho}}, \hat{H}]_- \tag{III.88}$$

As this equation is identical with equation (II.25) satisfied by $\hat{\varrho}$, the theorem that we have just proved for $\hat{\varrho}$ will be applicable in the same way to $\hat{\hat{\varrho}}$.

On the other hand, since the observable A is considered to be arbitrary, we can equally apply the foregoing result to a "coarse-grained" property of the system. It can be concluded, therefore, from this study that the two processes of "time-smoothing" and "coarse-graining" are incapable of freeing us from the almost periodic nature of evolution and enabling us to obtain an asymptotic convergence of $A(t)$ towards a limit value. This is the fundamental obstacle encountered by the ergodic theory of finite systems, whose Hamiltonian has a discrete spectrum; in order to attempt to avoid this difficulty, it appears necessary therefore to take into account the macroscopic nature of the system being considered.

For this reason, we now come to examine—with Golden and Longuet-Higgins—the case of an infinite system whose Hamiltonian has a purely continuous spectrum. In this case, the eigenstates are represented by the set of numbers (E', α'), where E' denotes the energy of the system and α' represents the values of another set of variables of the system; we note that these quantities α' can always be omitted if the system is not degenerate. With this notation, $\langle E'\alpha'|\hat{\varrho}|E''\alpha''\rangle$ and $\langle E'\alpha'|\hat{A}|E''\alpha''\rangle$ are the matrix elements corresponding to the operators $\hat{\varrho}$ and $\hat{A}$ in the energy representation and the representation associated with the quantities α'. We shall assume in what follows that the energy E' is measured by starting from the lowest eigenvalue of the Hamiltonian.

As before, the operators $\hat{A}$ and $\hat{H}$ do not explicitly depend on the time (isolated system case), as a result of which we have:

$$A(t) = \int_0^\infty dE' \int_0^\infty dE'' \sum_{\alpha',\alpha''} \langle E'\alpha'|\hat{\varrho}|\, E''\alpha''\rangle \langle E''\alpha''|\hat{A}|E'\alpha'\rangle \times \\ \times \exp[i(E'' - E')\, t/\hbar]; \qquad \text{(III.89)}$$

we note that this average value must be an even function of t because of the dynamic reversibility of the equations of motion. By putting

$$x = \tfrac{1}{2}(E'' + E'), \quad y = \tfrac{1}{2}(E'' - E'), \qquad \text{(III.90)}$$

equation (III.89) can be written as

$$A(t) = 2\int_{-\infty}^{+\infty} dy \int_{|y|}^{\infty} dx \Big[\sum_{\alpha',\alpha''} \langle x - y, \alpha'|\hat{\varrho}|\, x + y, \alpha''\rangle \times \\ \times \langle x + y, \alpha''|\hat{A}|\, x - y, \alpha'\rangle \Big] \exp(2iyt/\hbar)$$

$$= 2\int_{-\infty}^{+\infty} dy\, g(y) \exp(2iyt/\hbar), \qquad \text{(III.91)}$$

in which, following a procedure already used, we have put:

$$g(y) = \int_{|y|}^{\infty} dx \left[\sum_{\alpha'\alpha''} \langle x-y,\alpha'|\hat{\varrho}|x+y,\alpha''\rangle \langle x+y,\alpha''|\hat{A}|x-y,\alpha'\rangle \right]. \tag{III.92}$$

This expression shows us that $g(y)$ must be real in order that $A(t)$ is an even function of t. But we have also $g(-y) = g^*(y)$ because of the hermiticity of the operators $\hat{\varrho}$ and $\hat{A}$. It follows that $g(y)$ is also an even function of y.

With these definitions, the time average of $A(t)$ can, according equation (III.83), be written as

$$\begin{aligned}\overline{A(t)}^{\infty} &= 2 \lim_{z\to 0} z \int_{-\infty}^{+\infty} dt\, e^{-zt} \int_{-\infty}^{+\infty} dy g(y) \exp(2iyt/\hbar) \\ &= 2 \lim_{z\to 0} z \int_{-\infty}^{+\infty} dy g(y) \left[\frac{1}{z - 2iy/\hbar}\right] \\ &= 2 \lim_{z\to 0} \int_{-\infty}^{+\infty} dy g(y) \left[\frac{z^2}{z^2 + 4y^2/\hbar^2}\right] \end{aligned} \tag{III.93}$$

in which we have used the fact that $g(y)$ is real.

We consider now the function $\left[\dfrac{z^2}{z^2 + 4y^2/\hbar^2}\right]$: we note that it is very small for $|y| \gg z$ and that it is close to unity if $|y| \ll z$. As we are interested in the limit case as $z \to 0$, it can be seen that the integral (III.93), and therefore $\overline{A(t)}^{\infty}$, vanishes in the limit as $z \to 0$ if $g(y)$ is not singular at the point $y = 0$. If, now, we assume the following general form for $g(y)$:

$$g(y) = G\,\delta(y) + h(y), \tag{III.94}$$

where $h(y)$ is not singular at the point $y = 0$, we obtain for the time average $\overline{A(t)}^{\infty}$:

$$\overline{A(t)}^{\infty} = \lim_{z\to 0} \int_{-\infty}^{+\infty} dy G\delta(y) \left[\frac{2z^2}{z^2 + 4y^2/\hbar^2}\right] = 2G, \tag{III.95}$$

The same reasoning can be applied for calculating the time average of $[A(t)]^2$. We have, first of all:

$$[A(t)]^2 = \left| 2 \int_{-\infty}^{+\infty} dy g(y) \exp(2iyt/\hbar) \right|^2$$

$$= 4 \int_{-\infty}^{+\infty} dy \int_{-\infty}^{+\infty} dy' g(y)\, g(y') \exp[2i(y - y')\, t/\hbar], \qquad \text{(III.96)}$$

whence we have for the time average $\overline{[A(t)]^2}^{\infty}$:

$$\overline{[A(t)]^2}^{\infty} = 4 \lim_{z\to 0} \int_{-\infty}^{+\infty} dy \int_{-\infty}^{+\infty} dy' g(y)\, g(y') \left[\frac{z^2}{z^2 + 4(y - y')^2/\hbar^2} \right]. \qquad \text{(III.97)}$$

If we repeat the previous argument, it can be seen that integration over y' leads to zero, unless $g(y')$ has a singularity at $y' = y$. However, according to (III.94) the only singularity occurs at $y' = 0$; thus, we obtain:

$$\overline{[A(t)]^2}^{\infty} = 4 \lim_{z\to 0} \int_{-\infty}^{+\infty} dy g(y) \left[\frac{Gz^2}{z^2 + 4(y - y')^2/\hbar^2} \right],$$

which reduces in turn to $4G^2$ after integration over y. Finally, therefore, we have:

$$\overline{[A(t)]^2}^{\infty} = 4G^2 = \left(\overline{A(t)}^{\infty}\right)^2, \qquad \text{(III.98)}$$

which shows that the variations of $A(t)$ vanish completely in the course of time.

Naturally, it may be that the function $g(y)$ has other singularities than the one at $y = 0$. It can easily be seen in this case that the result (III.95) always remains valid whereas the value of $\overline{[A(t)]^2}^{\infty}$ is no longer given by equation (III.98). Thus, we now have $\overline{[A(t)]^2}^{\infty} \neq (\overline{A(t)}^{\infty})^2$, with the result that $A(t)$ continues to fluctuate indefinitely. It follows therefore that the necessary condition for satisfying that the system should have an ergodic behaviour is that the function $g(y)$ should have no other singularity than that located at $y = 0$, that is to say for $E' = E''$.

It will be noted that the theorem of Golden and Longuet-Higgins is thus stated in a form which is completely analogous to that of the theorems of harmonic analysis, the theory of which we have outlined in Chapter I, § II.4 in connection with the classical ergodic theory. The condition of ergodicity at which we have arrived is comparable with the hypothesis of metric transitivity but unfortunately it has the same disadvantage as the latter: it is, at present, impossible to obtain a criterion which allows us to recognise whether a real physical system satisfies this condition or not.

As we have mentioned, it follows from this study that the time behaviour of the average values $A(t)$ is qualitatively different for infinite systems. It is not without interest to note that this result apparently leads to a paradox, since an infinite system can only be expressed as the limit of a finite system whose dimensions are increased indefinitely: we could expect, therefore, to obtain final results in agreement with one another. In order to resolve this paradox, we are led to point out that the method used in fact involves two limiting processes, namely the processes $t \to \infty$ and V (or N) $\to \infty$. In the case of a finite system, we take the limit $t \to \infty$, then we make the dimensions of the system subsequently tend to infinity; in the case of the infinite system, on the contrary, transition to the limit V(or N) $\to \infty$ is carried out first of all, *before* letting $t \to \infty$. The fact that the results obtained by these two processes do not agree expresses simply that the two limiting processes considered do not commute with one another: this is the one idea which appears to us to be fruitful and to which we shall be returning (cf. for example, Chapter IV, § V.4). According to what we have said previously, it seems as though the correct way of dealing with the theory of irreversible processes consists in making the dimensions of the system tend to infinity before calculating the time averages of the observed quantities; in this way, the duration of the Poincaré cycles become infinite and we can rid ourselves of the almost-periodicity of the time evolution of finite systems.

In conclusion, we point out that the theorem of Golden and Longuet-Higgins can be compared with the result obtained by Van Hove in his work on the Master Equation. Van Hove studies the asymptotic behaviour of an infinite system whose Hamiltonian is the sum of an unperturbed Hamiltonian having a continuous spectrum and a perturbation term which is assumed to be small (hypothesis of weak coupling). The condition which ensures that such a system exhibits the desired irreversible behaviour, is that δ-type singularities exist for the diagonal matrix elements describing the transitions between the eigenstates of the unperturbed Hamiltonian. Although this result is completely similar to that of Golden and Longuet-Higgins, it is difficult to make a more detailed comparison because the two conditions which have been considered depend on functions which are defined differently. Moreover, the situations considered are appreciably different, since the theory of Golden and Longuet-Higgins is based on the total Hamiltonian of the system and does not depend on any perturbation calculations.

CHAPTER IV

Probability Quantum Ergodic Theorems

I. Comments on the Quantum Ergodic Theory

We have seen in the previous chapter that the quantum ergodic theory raises additional difficulties as compared with the classical ergodic theorem. In fact, even if we had been able to prove an ergodic theorem similar to that of Birkhoff, this would not apply to probability amplitudes (which are time-invariant in the energy representation) with the result that for applications to quantum statistical mechanics we must use the parallel of Hopf's theorem.

In the proof of this second theorem we have encountered the assumptions stated earlier concerning the spectrum of $\hat{H}$, which are the quantum parallel of the hypothesis of metrical transitivity, which is needed in classical mechanics; however, even though it is difficult to verify that a classical physical system actually possesses this property, the existence of ergodic systems is not altogether improbable in classical mechanics (it has been possible to construct mathematically some actual examples) (Hopf, 1937). On the other hand, we have mentioned already that by the very nature of quantum systems the assumption of non-degeneracy and the absence of resonance frequencies in the spectrum of the Hamiltonian restrict considerably the real applicability of the quantum ergodic theory; it can be said that the existence of quantum ergodic systems seems to be not very probable. Moreover, these assumptions do not suffice to ensure a convergence towards microcanonical averages: in fact, we have been able to prove the second probability ergodic theorem only by introducing the concept of macroscopic observables and of averages over a set of possible macroscopic observers. These concepts enable us to resolve the difficulties raised by the special nature of quantum observables (non-commutable with one another) and, above all, to devote ourselves to statistical considerations justified by the large number of degrees of freedom of systems observed macroscopically. Thus, even if we can avoid the inconvenience associated with the time-invariance of the probability amplitudes in the energy representation, we introduce, by means of averages over macroscopic observers, a statistical element which is foreign to quantum theory.

That is why we shall try, by relying on the definition of ensembles of quantum systems similar to Gibbs ensembles, to prove the probability ergodic theorems which allow us to break away from the restrictive assumptions concerning the spectrum of the Hamiltonian. Thus, we shall be able to justify the fundamental hypothesis of statistical mechanics, namely the replacement of time-averages by statistical averages (similar to phase-averages), with the help of arguments comparable with those of Khinchin in classical mechanics.

For this purpose, we shall let correspond to an isolated physical system whose energy is known macroscopically a microcanonical ensemble of systems represented by all points of the unit hypersphere of $2S^{(\alpha)}$ dimensions, which is invariant under the equations of motion and over which we define a uniform distribution in accordance with the fundamental assumption of quantum statistical mechanics. The time-average of an observable $\hat{A}$ is thus represented as a random variable defined over the unit hypersphere and, if the spectrum of the Hamiltonian is non-degenerate, it is independent of the initial phases (first ergodic theorem). We can then show that the dispersion of this random variable is of order $1/S^{(\alpha)}$, if we accept that the matrix elements A_{ll} are all bounded. Subsequently, we shall extend this first result to the case of macroscopic observables $\hat{\mathscr{A}}$, the division of Hilbert space into "cells" enabling us to dispense with any assumption concerning the nature of the spectrum of $\hat{H}$: thus, a probability ergodic theorem will be proved, which suffices to serve as a basis of quantum statistical mechanics.

Finally, in Section IV we shall discuss the significance of these results by establishing first of all an asymptotic property of the standard deviation of the macroscopic observables themselves from which the ergodic theorems are derived. In this way, we shall rederive the results of recent work by Bocchieri and Loinger and we shall show that because of the commutability of the operations of the time- and microcanonical-averages, von Neumann's quantum ergodic theorem becomes again a probability theorem, the assumptions concerning the spectrum of the Hamiltonian having become in fact useless. To conclude, we shall study the analogy between the classical and quantum aspects of the probability ergodic theorem and we shall discuss the properties of a model of quantum systems similar to that of Khinchin in classical mechanics.

The foregoing thus justifies the use of microcanonical ensembles for quantum statistical mechanics: it still remains to establish the relations between microcanonical, canonical, and grand-canonical ensembles, which play an essential role in the thermodynamic applications of statistical mechanics. As in classical mechanics (cf. Appendix I), we shall see that it is possible to show that a system in thermostatic equilibrium must be represented by an ensemble of canonically distributed systems, by starting from the fact that the total system (observed system + thermostat) is an isolated system represented by a microcanonical ensemble; moreover, the same method can be extended to grand-canonical ensembles. Thus, even if the microcanonical ensemble is little used in practice, its theoretical importance remains very great since, according to the foregoing, it is at the basis of the whole of statistical mechanics.

II. First Probability Ergodic Theorem (Jancel, 1955a)

1. *Status of the problem*

We shall suppose in what follows that the energy of the isolated system is known macroscopically, i.e. within an interval of δE; in other words, that the system is located in a single energy shell $e^{(\alpha)}$. We note without difficulty that this condition in no way restricts the generality because we have seen in the calculations of the previous chapter that all the arguments refer to a definite shell $e^{(\alpha)}$ and that their extension to many shells changes in no way the nature of the problem. Let, as before,

$$\Psi(t) = \sum_{i=1}^{S(\alpha)} r_i \, e^{i\alpha_i} e^{-iE_i t/\hbar} \, \psi_i \tag{IV.1}$$

be the wave function of a system corresponding to a pure case compatible with our macroscopic data. Let us now consider a microscopic observable $\hat{A}$ whose quantum average in the pure case $\Psi(t)$ is given by

$$A(t) = \sum_{i,j} r_i r_j e^{i(\alpha_i - \alpha_j)} e^{-i(E_i - E_j)t/\hbar} (\psi_j, \hat{A}\psi_i) = \sum_{i,j} \varrho_{ij}(0)\, e^{-i(E_i - E_j)t/\hbar} A_{ji}. \tag{IV.2}$$

To begin with, let us apply the first ergodic theorem, which leads us to assume that the Hamiltonian of the system is non-degenerate; with this assumption, the limit of the time-average of $A(t)$ exists and, according to (III.10), it is given by

$$\overline{A(t)}^{\infty} = \sum_{i=1}^{S(\alpha)} r_i^2 A_{ii}. \tag{IV.3}$$

We note that here we can manage without an assumption concerning the convergence of $\sum_i r_i$ since, usually, the number of eigenstates in a shell $e^{(\alpha)}$ is finite. On the other hand, since the system is non-degenerate, the microcanonical distribution corresponds to an equal probability for different eigenstates and, if $\overline{A}^m$ denotes the microcanonical average of $\hat{A}$, according to (II.31′) we have

$$\overline{A}^m = \frac{1}{S^{(\alpha)}} \sum_{i=1}^{S(\alpha)} A_{ii}. \tag{IV.4}$$

As we have noted already, we have generally $\overline{A}^{\infty} \neq \overline{A}^m$ (unless all values of A_{ii} are equal over the energy shell $e^{(\alpha)}$). That is why we shall adopt the statistical point of view, by defining a uniform distribution of systems over the unit hypersphere in $2n$-dimensional space; this point of view (which we adopt in the whole of this chapter) will enable us to overcome the difficulties in relation to the stationary quantum states, thanks to the asymptotic geometrical properties of the unit hypersphere.

Since the macroscopic observation is incomplete, it does not permit us to describe the system by a well-determined wave function; a set of possible wave functions exists which is compatible with this observation and in which all wave functions have the form (IV.1) (in the case of an energy measure), with the values of the r_i^2 connected by the relation

$$\sum_{i=1}^{S(\alpha)} r_i^2 = 1, \tag{IV.5}$$

which can be rewritten as

$$\sum_i c_i(0)\, c_i^*(0) = 1, \qquad \text{(IV.5')}$$

where the values $c_i(0) = r_i e^{i\alpha_i}$ represent the initial components of the state vector Ψ.

The statistical method consists then in associating with such a macroscopic observation a statistical ensemble of systems corresponding to the set of wave functions defined by (IV.1) and (IV.5). This set is represented by all the points of a hypersphere of radius 1 in $2n$-dimensional space (in what follows we put $n \equiv S^{(\alpha)}$),† over which we must define an *a priori* distribution which is invariant under the equations of motion. In accordance with the fundamental hypothesis of quantum statistical mechanics we shall, for the probability law of the $c_i(0)$, take one which corresponds to the uniform distribution over the hypersphere: if $d\sigma_{2n}$ is an element of area of this hypersurface of measure σ_{2n}, the invariant probability density will be $d\sigma_{2n}/\sigma_{2n}$.

According to formula (27) of Appendix II, we have $\overline{r_i^2} = 1/n$, so that the chosen distribution corresponds just to the microcanonical ensemble (in future, we shall denote the averages taken over this ensemble by the superscript m);‡ thus, according to (IV.4) and (IV.2), we can write:

$$\overline{A(t)}^m = \overline{A(0)}^m = \sum_{i=1}^{n} \overline{r_i^2}^m A_{ii} = \frac{1}{n} \sum_{i=1}^{n} A_{ii} = \overline{A}^m, \qquad \text{(IV.6)}$$

since $\overline{e^{i(\alpha_i - \alpha_j)}}^m = 0$, if $i \neq j$. We note, likewise, that as in classical mechanics [cf. formula (I.24')] we have

$$\overline{\left[\overline{A(t)}^\infty\right]}^m = \overline{A}^m. \qquad \text{(IV.7)}$$

With these definitions the numbers $c_i(0)$ and, hence, the quantities $A(t)$ and $\overline{A(t)}^\infty$ are random variables defined over the hyper-

† Since n represents the number of states in an energy shell, we shall denote the number of particles in the system by N in this chapter.

‡ The averages are taken here over the initial conditions; however, because of the invariance property of the uniform ensemble, we can define these averages at any instant, as we can see by writing $\varrho_{ij}(t) = r_i r_j e^{i(\beta_i - \beta_j)}$, with $\beta_i = \alpha_i - (iE_i t/\hbar)$ and $\overline{e^{i(\beta_i - \beta_j)}}^m = \delta_{ij}$; it is also because of this invariance that the average $\overline{A(t)}^m$ is time-independent.

sphere (IV.5) or (IV.5′), with the distribution $d\sigma_{2n}/\sigma_{2n}$. In order to justify the methods of statistical mechanics, it is sufficient to prove that the second moment of $\overline{A(t)}^{\infty} - \overline{A}^{m}$ [i.e. the square of the standard deviation of the random variable $\overline{A(t)}^{\infty}$] can be made very small, when the number of degrees of freedom of the system is very large; we shall show that this second moment tends to zero as $1/n$, using the assumption that all the A_{ii} are bounded.

2. *Statement of the first probability ergodic theorem*

According to relations (IV.7) and (IV.4), we have

$$\overline{\left(\overline{A(t)}^{\infty} - \overline{A}^{m}\right)^2}^{m} = \overline{\left[\overline{A(t)}^{\infty}\right]^2}^{m} - \left(\overline{A}^{m}\right)^2, \tag{IV.8}$$

with

$$\left(\overline{A}^{m}\right)^2 = \frac{1}{n^2}\left(\sum_i A_{ii}^2 + \sum_{i \neq j} A_{ii}A_{jj}\right). \tag{IV.9}$$

By putting (IV.3) in (IV.8) we obtain finally

$$\overline{\left(\overline{A(t)}^{\infty} - \overline{A}^{m}\right)^2}^{m} = \sum_{i=1}^{n} A_{ii}^2 \left[\overline{r_i^4}^{m} - \left(\overline{r_i^2}^{m}\right)^2\right]$$
$$+ \sum_{i \neq j} A_{ii}A_{jj}\left[\overline{r_i^2 r_j^2}^{m} - \left(\overline{r_i^2}^{m}\right)\left(\overline{r_j^2}^{m}\right)\right]. \tag{IV.10}$$

However, according to formulae (27) and (30) of Appendix II, we have

$$\overline{r_i^2}^{m} = \frac{1}{n}, \quad \overline{r_i^4}^{m} = \frac{2}{n(n+1)}, \quad \overline{r_i^2 r_j^2}^{m} = \frac{1}{n(n+1)}, \tag{IV.11}$$

with the result that (IV.10) can be written as

$$\overline{\left(\overline{A(t)}^{\infty} - \overline{A}^{m}\right)^2}^{m} = \sum_{i=1}^{n} A_{ii}^2\left(\frac{2}{n(n+1)} - \frac{1}{n^2}\right)$$
$$+ \sum_{i \neq j} A_{ii}A_{jj}\left(\frac{1}{n(n+1)} - \frac{1}{n^2}\right)$$
$$= -\frac{1}{n+1}\,\frac{1}{n^2}\left(\sum_i A_{ii}^2 + \sum_{i \neq j} A_{ii}A_{jj}\right) + \frac{1}{n(n+1)}\sum_{i=1} A_{ii}^2, \tag{IV.12}$$

or, by virtue of (IV.9),

$$\overline{\left(\overline{A(t)}^{\infty} - \overline{A}^{m}\right)^2}^{m} = -\frac{1}{n+1}\left(\overline{A}^{m}\right)^2 + \frac{1}{n(n+1)}\sum_{i=1}^{n} A_{ii}^2. \tag{IV.13}$$

If we assume now that the matrix elements A_{ii} are all bounded by a number A_M, it is easy to evaluate (IV.13). We note first of all by (IV.9) that $(\overline{A}^{m})^2$ is, at the most, equal to A_M^2, which is obvious by the definition of $\overline{A}^{m}$: the first term on the right-hand side of (IV.13) is then of order $1/n$ and the same is true for the second term, since we have certainly

$$\sum_{i=1}^{n} A_{ii}^2 < nA_M^2.$$

Since $n = S^{(\alpha)}$ is always very large in macroscopic systems we can write

$$\overline{\left(\overline{A(t)}^{\infty} - \overline{A}^{m}\right)^2}^{m} = O\left(\frac{1}{S^{(\alpha)}}\right). \tag{IV.14}$$

Equation (IV.14) shows that it is legitimate to replace the time-average $\overline{A(t)}^{\infty}$ by the microcanonical average $\overline{A}^{m}$ for all macroscopic systems; thus, we have the first probability ergodic theorem: *the time-average $\overline{A(t)}^{\infty}$ is equal to the microcanonical average $\overline{A}^{m}$ except over a set of very small measure, corresponding to a set of exceptional initial conditions.*

We have had to accept the assumption of non-degeneracy of the Hamiltonian spectrum; this is essential for ensuring that $\overline{A(t)}^{\infty}$ is independent of the initial phases; we shall see later to what extent it is possible to dispense with this assumption. On the other hand, if the condition about the bound of the A_{ii} is not restrictive from the physical point of view, the fact that we consider A_M to be independent of n is an additional assumption which is not satisfied in general. We shall see, however, that for a simple model such as the one we shall study later (cf. Section IV) the increase of A_M with n is much slower than that of n, so that the result (IV.14) remains valid. In particular, (IV.14) can also be written

$$\text{Prob}\left\{\left|\overline{A(t)}^{\infty} - \overline{A}^{m}\right| > \varepsilon\right\} \to 0, \quad \text{as} \quad 1/S^{(\alpha)} \to 0. \tag{IV.15}$$

In conclusion, we are replacing the ergodic theorems of Chapter III by the probability theorems (IV.14) and (IV.15) which can serve as the foundation of quantum statistical mechanics, by depending essentially on the large number of degrees of freedom of the system and on the asymptotic geometrical properties of the hyperspheres in $2n$-dimensional space: this result makes obvious the principal role played by the law of large numbers in macroscopic systems. As in classical theory, we could dispense with the assumption of a uniform distribution over the unit hypersphere provided that the chosen distribution assigns a *very small probability* to a set of *very small measure*; this statement is valid for all the results in this chapter: we shall not return to it.

The foregoing developments have been obtained more easily than Khinchin's evaluation in classical mechanics and without making a special assumption concerning the observable $\hat{A}$: this is due to the fact that the formal integration of the equations of motion is always possible in quantum mechanics and that the time-dependence of the quantities $A(t)$ can be expressed in a simple way. We note also that the convergence towards zero of $\overline{A(t)}^{\infty} - \overline{A}^{m}$ is more rapid in quantum mechanics than in classical mechanics, the difference arising from the fact that we consider here averages taken in the functional space of $S^{(\alpha)}$ dimensions.

Actually, we can take for $S^{(\alpha)}$ either Fierz's approximation $S^{(\alpha)} \sim \exp(10^{20})$, or we can make a direct estimate of this same quantity by using a well-known result of quantum statistical mechanics: if we have a system with $3N$ degrees of freedom and if $\Delta\Phi$ is the classical extension in phase contained between the hypersurfaces E_1 and E_2, the number of quantum states contained between E_1 and E_2 is $\Delta\Phi/h^{3N}$. We can make simple use of this formula in the case of a perfect gas consisting of N indistinguishable point particles; then, for the extension in phase inside the hypersurface $E =$ constant, we have

$$\Phi = V^N \sigma_{3N} (2mE)^{3N/2},$$

whence

$$S^{(\alpha)} = \frac{\Delta\Phi}{h^{3N} N!} = \frac{V^N}{h^{3N}} \sigma_{3N} 3mN (2mE)^{(3N/2)-1}\, \delta E$$

$$\sim K \left[\left(\frac{4\pi m\bar{\varepsilon}}{3h^2} \right)^{3/2} v \right]^N \delta E, \tag{IV.16}$$

where V is the volume occupied by the gas, E is the energy of the system, m is the mass of a particle, σ_{3N} is the volume of the $3N$-

dimensional hypersphere, and δE is the energy range of the shell $e^{(\alpha)}$; $\bar{\varepsilon} = \frac{3}{2} kT$ denotes the average energy per particle and $v = V/N$ the specific volume of a particle.

Thus, we see (see also Appendix I, § 4) that $S^{(\alpha)}$ is an exponential function of N which provides, for the second moment (IV.8), a more rapid decrease than some power of $1/N$.

III. The Macroscopic Probability Ergodic Theorem

1. *Role of degeneracy of the spectrum of* $\hat{H}$

The results of the preceding section are only valid if the assumption of non-degeneracy of the Hamiltonian spectrum is satisfied—an assumption which is still too restrictive as we have seen. Before introducing macroscopic operators which will allow us to dispense with it, we shall analyse how these degenerate terms occur in the microscopic theory.

Let us consider an isolated quantum system whose wave function is of the form (IV.1); the quantum average of an observable $\hat{A}$ is then given by (IV.12), which can also be written as

$$A(t) = \sum_{i=1}^{n} r_i^2 A_{ii} + \sum_{i \neq j} r_i r_j \, e^{i(\alpha_i - \alpha_j)} \, e^{-i(E_i - E_j)t/\hbar} A_{ji}. \quad \text{(IV.17)}$$

If the Hamiltonian of the system is degenerate, the time-average $\overline{A(t)}^{\infty}$ is no longer given by (IV.3) and it is necessary to resolve the sum $\sum_{i \neq j}$ of (IV.17). If the n states i comprise n' non-degenerate states denoted by $k, l, \ldots$, and n'' degenerate states denoted by $\varkappa_p, \varkappa_q, \lambda_r, \ldots$, the suffix p varying from 1 to $g_\varkappa$ (where $g_\varkappa$ is the degree of degeneracy of the state $\varkappa$), we have by definition

$$\sum_{\varkappa} g_\varkappa = n'', \quad \text{(IV.18)}$$

and the $n(n-1)$ terms of $\sum_{i \neq j}$ are split into non-degenerate, degenerate and cross terms. We obtain

$$\begin{aligned}
\sum_{i \neq j} r_i r_j \, e^{i(\alpha_i - \alpha_j)} \, e^{-i(E_i - E_j)t/\hbar} A_{ji} &= \sum_{k \neq l} r_k r_l \, e^{i(\alpha_k - \alpha_l)} \, e^{-i(E_k - E_l)t/\hbar} A_{lk} \\
&+ \sum_{k, \varkappa_p} r_k r_{\varkappa_p} [e^{i(\alpha_k - \alpha_{\varkappa_p})} e^{-i(E_k - E_{\varkappa_p})t/\hbar} A_{\varkappa_p k} + e^{-i(\alpha_k - \alpha_{\varkappa_p})} e^{i(E_k - E_{\varkappa_p})t/\hbar} A_{k \varkappa_p}] \\
&+ \sum_{\varkappa \neq \lambda} \sum_{p, s} r_{\varkappa_p} r_{\lambda_s} \, e^{i(\alpha_{\varkappa_p} - \alpha_{\lambda_s})} \, e^{-i(E_{\varkappa_p} - E_{\lambda_s})t/\hbar} A_{\lambda_s \varkappa_p} \\
&+ \sum_{\varkappa} \sum_{p \neq q} r_{\varkappa_p} r_{\varkappa_q} \, e^{i(\alpha_{\varkappa_p} - \alpha_{\varkappa_q})} A_{\varkappa_q \varkappa_p}. \quad \text{(IV.19)}
\end{aligned}$$

The sum $\sum_{k \neq l}$ contains the $n'(n'-1)$ non-degenerate terms; the sum $\sum_{k, \varkappa_p}$ includes $2n'n''$ cross terms; the sum $\sum_{\varkappa \neq \lambda} \sum_{p, s}$ contains $\sum_{\varkappa \neq \lambda} g_\varkappa g_\lambda$ degenerate terms belonging to the different energy states; finally, the sum $\sum_{\varkappa} \sum_{p \neq q}$ contains the $\sum_{\varkappa} g_\varkappa(g_\varkappa - 1)$ terms grouped according to the state $E_\varkappa$. We can verify easily by (IV.18) that the sum of all these terms adds up to the $n(n-1)$ terms of $\sum_{i \neq j}$.

Having stated this, let us calculate the time-average $\overline{A(t)}^\infty$ of (IV.17): we can see from (IV.19) that only the terms in $\sum_i$ and $\sum_\varkappa \sum_{p \neq q}$ will make a contribution as $T \to \infty$; in fact, all the other terms contain an exponential factor of the form $\exp[-i(E_\alpha - E_\beta) t/\hbar]$ whose time-average is zero, since we have necessarily $E_\alpha \neq E_\beta$. Finally, if the Hamiltonian is degenerate, we have:

$$\overline{A(t)}^\infty = \sum_{i=1}^{n} r_i^2 A_{ii} + \sum_\varkappa \sum_{p \neq q} r_{\varkappa_p} r_{\varkappa_q} e^{i(\alpha_{\varkappa_p} - \alpha_{\varkappa_q})} A_{\varkappa_q \varkappa_p}, \tag{IV.20}$$

the sum $\sum_\varkappa$ containing $\sum_\varkappa g_\varkappa(g_\varkappa - 1)$ terms. As we have seen already in Chapter III, the time-average of $A(t)$ is no longer independent of the initial phases $\alpha_{\varkappa_p}$; on the other hand, one can prove again the relationship

$$\overline{\left[\overline{A(t)}^\infty\right]}^m = \frac{1}{n} \sum_{i=1}^{n} A_{ii} = \overline{A}^m. \tag{IV.21}$$

Let us try now to evaluate the standard deviation of the time-average $\overline{A(t)}^\infty$. According to (IV.21) we have

$$\overline{\left(\overline{A(t)}^\infty - \overline{A}^m\right)^2}^m = \overline{\left(\overline{A(t)}^\infty\right)^2}^m - \left(\overline{A}^m\right)^2, \tag{IV.22}$$

and for $(\overline{A(t)}^\infty)^2$ according to (IV.20)

$$\left(\overline{A(t)}^\infty\right)^2 = \sum_{i,j} r_i^2 r_j^2 A_{ii} A_{jj} + 2 \sum_i \sum_\varkappa \sum_{p \neq q} r_i^2 r_{\varkappa_p} r_{\varkappa_q} e^{i(\alpha_{\varkappa_p} - \alpha_{\varkappa_q})} A_{ii} A_{\varkappa_q \varkappa_p}$$
$$+ \sum_{\varkappa, \varkappa'} \sum_{p \neq q} \sum_{p' \neq q'} r_{\varkappa_p} r_{\varkappa_q} r_{\varkappa'_{p'}} r_{\varkappa'_{q'}} e^{i(\alpha_{\varkappa_p} - \alpha_{\varkappa_q})} e^{i(\alpha_{\varkappa'_{p'}} - \alpha_{\varkappa'_{q'}})} A_{\varkappa_q \varkappa_p} A_{\varkappa'_{q'} \varkappa'_{p'}}. \tag{IV.23}$$

When we form the microcanonical average $\overline{(\overline{A(t)}^{\infty})^2}^{m}$, only the terms independent of the phases α give contributions which are non-vanishing. Thus, the cross terms $\sum_i \sum_\varkappa \sum_{p \neq q}$ give a zero contribution and, in the sum $\sum_{\varkappa,\varkappa'} \sum_{p \neq q} \sum_{p' \neq q'}$, the only phase-independent terms are those for which $\varkappa'_{p'} = \varkappa_q$ and $\varkappa'_{q'} = \varkappa_p$. Because of the hermiticity of $\hat{A}$, these terms can be written as

$$\sum_\varkappa \sum_{p \neq q} |A_{\varkappa_p \varkappa_q}|^2 \, r^2_{\varkappa_p} r^2_{\varkappa_q}$$

and we have

$$\overline{\left(\overline{A(t)}^{\infty}\right)^2}^{m} = \sum_i \overline{r_i^4}^{m} A_{ii}^2 + \sum_{i \neq j} \overline{r_i^2 r_j^2}^{m} A_{ii} A_{jj} + \sum_\varkappa \sum_{p \neq q} \overline{r^2_{\varkappa_p} r^2_{\varkappa_q}}^{m} |A_{\varkappa_p \varkappa_q}|^2, \tag{IV.24a}$$

whence

$$\begin{aligned} \overline{\left(\overline{A(t)}^{\infty}\right)^2}^{m} - \left(\overline{A}^{m}\right)^2 = {} & \sum_{i=1}^{n} A_{ii}^2 \left[\overline{r_i^4}^{m} - \left(\overline{r_i^2}^{m}\right)^2\right] \\ & + \sum_{i \neq j} A_{ii} A_{jj} \left[\overline{r_i^2 r_j^2}^{m} - \left(\overline{r_i^2}^{m}\right)\left(\overline{r_j^2}^{m}\right)\right] \\ & + \sum_\varkappa \sum_{p \neq q} \overline{r^2_{\varkappa_p} r^2_{\varkappa_q}}^{m} |A_{\varkappa_p \varkappa_q}|^2 . \end{aligned} \tag{IV.24b}$$

By using the formulae of Appendix II for the moments $\overline{r_i^4}^{m}$ and $\overline{r_j^2 r_j^2}^{m}$, we obtain finally

$$\begin{aligned} & \overline{\left(\overline{A(t)}^{\infty}\right)^2}^{m} - \left(\overline{A}^{m}\right)^2 \\ & = \left(\frac{2}{n(n+1)} - \frac{1}{n^2}\right) \sum_{i=1}^{n} A_{ii}^2 + \left(\frac{1}{n(n+1)} - \frac{1}{n^2}\right) \sum_{i \neq j} A_{ii} A_{jj} \\ & \quad + \frac{1}{n(n+1)} \sum_\varkappa \sum_{p \neq q} |A_{\varkappa_p \varkappa_q}|^2 \\ & = -\frac{1}{n+1} \left(\overline{A}^{m}\right)^2 + \frac{1}{n(n+1)} \left(\sum_i A_{ii}^2 + \sum_\varkappa \sum_{p \neq q} |A_{\varkappa_p \varkappa_q}|^2\right). \end{aligned} \tag{IV.25}$$

If, as in the previous section, we suppose that the matrix elements are all bounded by a number A_M, we can see that we can write

$$\overline{\left(\overline{A(t)}^{\infty} - \overline{A}^{m}\right)^2}^{m} = O\left(\frac{1}{n}\right) + \frac{\sum_{\varkappa} g_{\varkappa}(g_{\varkappa} - 1)}{n(n+1)} A_M^2, \quad \text{(IV.26)}$$

so that the standard deviation of $\overline{A(t)}^{\infty}$ does not necessarily decrease with $1/n^{\frac{1}{2}}$, but depends on the numbers $g_{\varkappa}$ corresponding to the degrees of degeneracy of the various states $\varkappa$. In a general way, the standard deviation depends essentially on the value of the term

$$\frac{1}{n(n+1)} \sum_{\varkappa} \sum_{p \neq q} |A_{\varkappa_p \varkappa_q}|^2, \quad \text{(IV.27)}$$

and only an evaluation of the off-diagonal matrix elements $A_{\varkappa_p \varkappa_q}$ for macroscopic systems can allow the behaviour of (IV.25) to be defined more precisely as $n \to \infty$. We shall see now how the introduction of macroscopic observables enables us to overcome effectively this difficulty and we shall show in Section IV another assumption concerning the nature of the macroscopic system (an assumption similar to that of Khinchin's "sum" functions) which makes it possible to evaluate the term (IV.27).

2. *The macroscopic probability ergodic theorem*

Let us introduce now the macroscopic description of the system defined by the orthonormal base $\Omega_k^{(\alpha)}$ [cf. (II.55)]. The wave function (IV.1) can then be written as

$$\Psi(t) = \sum_{i=1}^{n} r_i \, e^{i\alpha_i} \, e^{-iE_i t/\hbar} \sum_{k=1}^{n} C_{ki}^{(\alpha)*} \Omega_k^{(\alpha)} \quad \text{(IV.28)}$$

$$= \sum_{k=1}^{n} \Omega_k^{(\alpha)} \sum_{i=1}^{n} r_i \, e^{i\alpha_i} \, e^{-iE_i t/\hbar} C_{ki}^{(\alpha)*} = \sum_{k=1}^{n} \bar{\omega}_k^{(\alpha)}(t) \, \Omega_k^{(\alpha)},$$

with

$$\bar{\omega}_k^{(\alpha)}(t) = \sum_{i=1}^{n} r_i \, e^{i\alpha_i} \, e^{-iE_i t/\hbar} C_{ki}^{(\alpha)*}. \quad \text{(IV.29)}$$

Let $\hat{\mathscr{A}}$ be the macroscopic operator associated with the microscopic observable $\hat{A}$; according to (II.54) its eigenvalues $a_\nu^{(\alpha)}$ are $s_\nu^{(\alpha)}$ times degenerate and its quantum average can be written, by virtue of (III.46′), (III.47) and (III.37), as

$$\mathscr{A}(t) = \sum_{\nu=1}^{N(\alpha)} a_\nu^{(\alpha)} \sum_{k=1}^{s_\nu(\alpha)} |\bar{\omega}_k^{(\alpha)}(t)|^2 = \sum_{\nu=1}^{N(\alpha)} a_\nu^{(\alpha)} w_\nu^{(\alpha)}(t), \quad \text{(IV.30)}$$

where $w_\nu^{(\alpha)}(t)$ is the statistical weight at t of the cell ν given by

$$w_\nu^{(\alpha)}(t) = \sum_{i,j} r_i r_j \, e^{i(\alpha_i - \alpha_j)} \, e^{-i(E_i - E_j)t/\hbar} D_{ij}^{(\nu)}, \qquad \text{(IV.31)}$$

$D_{ij}^{(\nu)}$ being the quantity defined by (II.64). As in the microscopic ergodic theory explained in the preceding section, we consider a microcanonical ensemble of systems having (IV.1) for their wave functions; the numbers $c_i = r_i e^{i\alpha_i}$ then become random variables, satisfying the relationship $\sum_i c_i^* c_i = \sum_i r_i^2 = 1$. In order to have a microcanonical ensemble, we must select for the c_i a uniform distribution over the complex hypersphere σ_{2n} with $n (= S^{(\alpha)})$ dimensions.

By taking the average of $\mathscr{A}(t)$ over the hypersphere σ_{2n} (or the phase average), we find the microcanonical average of the observable $\hat{\mathscr{A}}$; actually, whatever t may be, by (II.66) and the formulae from Appendix II, we have

$$\overline{\mathscr{A}(t)}^m = \sum_\nu a_\nu^{(\alpha)} \overline{w_\nu^{(\alpha)}(t)}^m = \sum_\nu a_\nu^{(\alpha)} \sum_{i,j} \overline{r_i r_j e^{i(\alpha_i - \alpha_j)}}^m \, e^{-i(E_i - E_j)t/\hbar} D_{ij}^{(\nu)}$$

$$= \sum_\nu a_\nu^{(\alpha)} \sum_i \overline{r_i^2}^m D_{ii}^{(\nu)} = \frac{1}{n} \sum_\nu a_\nu^{(\alpha)} \sum_i D_{ii}^{(\nu)} = \frac{1}{n} \sum_{\nu=1}^{N(\alpha)} a_\nu^{(\alpha)} s_\nu^{(\alpha)}. \qquad \text{(IV.32)}$$

On the other hand, we define likewise, starting from (IV.30), the time-average $\overline{\mathscr{A}(t)}^\infty$. We must discriminate, as in the preceding section, between the degenerate states (denoted by $\varkappa_p, \varkappa_q$) and the non-degenerate states (denoted by $k, l, \ldots$). Obviously, we have

$$\overline{\mathscr{A}(t)}^\infty = \sum_\nu a_\nu^{(\alpha)} \overline{w_\nu^{(\alpha)}(t)}^\infty \qquad \text{(IV.33)}$$

and, from (IV.31), we obtain for $\overline{w_\nu^{(\alpha)}(t)}^\infty$ an expression which is similar to (IV.20):

$$\overline{w_\nu^{(\alpha)}(t)}^\infty = \sum_{i=1}^{n} r_i^2 D_{ii}^{(\nu)} + \sum_\varkappa \sum_{p \neq q} r_{\varkappa_p} r_{\varkappa_q} \, e^{i(\alpha_{\varkappa_p} - \alpha_{\varkappa_q})} D_{\varkappa_p \varkappa_q}^{(\nu)}. \qquad \text{(IV.34)}$$

It is easy then to prove the relation

$$\left[\overline{\mathscr{A}(t)}^\infty\right]^m = \overline{\mathscr{A}}^m \qquad \text{(IV.35)}$$

which corresponds to (IV.21); actually, by (IV.34) we have

$$\overline{\left[\overline{w_\nu^{(\alpha)}(t)}^{\infty}\right]}^{m} = \sum_i \overline{r_i^2}^{m} D_{ii}^{(\nu)} = \frac{1}{n}\sum_i D_{ii}^{(\nu)} = \frac{s_\nu^{(\alpha)}}{n}. \tag{IV.36}$$

This being so, we must consider the expression $\overline{[\overline{\mathscr{A}(t)}^{\infty} - \overline{\mathscr{A}}^{m}]^2}^{m}$ in order to prove an ergodic theorem in quadratic mean and which can according to (IV.35) be written as:

$$\overline{\left[\overline{\mathscr{A}(t)}^{\infty} - \overline{\mathscr{A}}^{m}\right]^2}^{m} = \overline{\left[\overline{\mathscr{A}(t)}^{\infty}\right]^2}^{m} - \left(\overline{\mathscr{A}}^{m}\right)^2. \tag{IV.37}$$

First of all, by (IV.33) and (IV.34) we have

$$\left[\overline{\mathscr{A}(t)}^{\infty}\right]^2 = \sum_\nu (a_\nu^{(\alpha)})^2 \left[\overline{w_\nu^{(\alpha)}(t)}^{\infty}\right]^2 + \sum_{\nu \neq \nu'} a_\nu^{(\alpha)} a_{\nu'}^{(\alpha)} \left[\overline{w_\nu^{(\alpha)}(t)}^{\infty}\right]\left[\overline{w_{\nu'}^{(\alpha)}(t)}^{\infty}\right], \tag{IV.38}$$

with

$$\begin{aligned}\left[\overline{w_\nu^{(\alpha)}(t)}^{\infty}\right]^2 = {} & \sum_{i,j} r_i^2 r_j^2 D_{ii}^{(\nu)} D_{jj}^{(\nu)} \\ & + 2\sum_i \sum_\varkappa \sum_{p\neq q} r_i^2 r_{\varkappa_p} r_{\varkappa_q}\, e^{i(\alpha_{\varkappa_p}-\alpha_{\varkappa_q})} D_{ii}^{(\nu)} D_{\varkappa_p\varkappa_q}^{(\nu)} \\ & + \sum_{\varkappa,\varkappa'} \sum_{p\neq q} \sum_{p'\neq q'} r_{\varkappa_p} r_{\varkappa_q} r_{\varkappa'_{p'}} r_{\varkappa'_{q'}} \times \\ & \times e^{i(\alpha_{\varkappa_p}-\alpha_{\varkappa_q})} e^{i(\alpha_{\varkappa'_{p'}}-\alpha_{\varkappa'_{q'}})} D_{\varkappa_p\varkappa_q}^{(\nu)} D_{\varkappa'_{p'}\varkappa'_{q'}}^{(\nu)},\end{aligned} \tag{IV.39}$$

and

$$\begin{aligned}\left[\overline{w_\nu^{(\alpha)}(t)}^{\infty}\right]\left[\overline{w_{\nu'}^{(\alpha)}(t)}^{\infty}\right] = {} & \sum_{i,j} r_i^2 r_j^2 D_{ii}^{(\nu)} D_{jj}^{(\nu')} \\ & + \sum_i \sum_\varkappa \sum_{p\neq q} r_i^2 r_{\varkappa_p} r_{\varkappa_q}\, e^{i(\alpha_{\varkappa_p}-\alpha_{\varkappa_q})} (D_{ii}^{(\nu)} D_{\varkappa_p\varkappa_q}^{(\nu')} + D_{ii}^{(\nu')} D_{\varkappa_p\varkappa_q}^{(\nu)}) \\ & + \sum_{\varkappa,\varkappa'} \sum_{p\neq q} \sum_{p'\neq q'} r_{\varkappa_p} r_{\varkappa_q} r_{\varkappa'_{p'}} r_{\varkappa'_{q'}}\, e^{i(\alpha_{\varkappa_p}-\alpha_{\varkappa_q})} e^{i(\alpha_{\varkappa'_{p'}}-\alpha_{\varkappa'_{q'}})} D_{\varkappa_p\varkappa_q}^{(\nu)} D_{\varkappa'_{p'}\varkappa'_{q'}}^{(\nu')}.\end{aligned} \tag{IV.40}$$

Next, we must take the microcanonical average of (IV.38): the only terms making non-zero contributions are those which are independent of the phases α_i; as before these are, apart from the terms of the sum $\sum_{i,j}$, those terms of the sum $\sum_{\varkappa,\varkappa'}$ for which $\varkappa_q = \varkappa'_{p'}$ and $\varkappa_p = \varkappa'_{q'}$. We have thus

$$\overline{\left[\overline{w_\nu^{(\alpha)}(t)}^{\infty}\right]^2}^{m} = \sum_i \overline{r_i^4}^{m} (D_{ii}^{(\nu)})^2 + \sum_{i \neq j} \overline{r_i^2 r_j^2}^{m} D_{ii}^{(\nu)} D_{jj}^{(\nu)} + \sum_{\varkappa} \sum_{p \neq q} \overline{r_{\varkappa_p}^2 r_{\varkappa_q}^2}^{m} |D_{\varkappa_p \varkappa_q}^{(\nu)}|^2, \tag{IV.41}$$

$$\overline{\left[\overline{w_\nu^{(\alpha)}(t)}^{\infty}\right]\left[\overline{w_{\nu'}^{(\alpha)}(t)}^{\infty}\right]}^{m} = \sum_i \overline{r_i^4}^{m} D_{ii}^{(\nu)} D_{ii}^{(\nu')} + \sum_{i \neq j} \overline{r_i^2 r_j^2}^{m} D_{ii}^{(\nu)} D_{jj}^{(\nu')} + \sum_{\varkappa} \sum_{p \neq q} \overline{r_{\varkappa_p}^2 r_{\varkappa_q}^2}^{m} D_{\varkappa_p \varkappa_q}^{(\nu)} D_{\varkappa_q \varkappa_p}^{(\nu')}, \tag{IV.42}$$

whence

$$\overline{\left[\overline{\mathscr{A}(t)}^{\infty}\right]^2}^{m} = \frac{1}{n(n+1)} \times$$

$$\times \Big\{ \sum_\nu (a_\nu^{(\alpha)})^2 \Big[2 \sum_i (D_{ii}^{(\nu)})^2 + \sum_{i \neq j} D_{ii}^{(\nu)} D_{jj}^{(\nu)} + \sum_{\varkappa} \sum_{p \neq q} |D_{\varkappa_p \varkappa_q}^{(\nu)}|^2 \Big] + \sum_{\nu \neq \nu'} a_\nu^{(\alpha)} a_{\nu'}^{(\alpha)} \Big[2 \sum_i D_{ii}^{(\nu)} D_{ii}^{(\nu')} + \sum_{i \neq j} D_{ii}^{(\nu)} D_{jj}^{(\nu')} + \sum_{\varkappa} \sum_{p \neq q} D_{\varkappa_p \varkappa_q}^{(\nu)} D_{\varkappa_q \varkappa_p}^{(\nu')} \Big] \Big\}. \tag{IV.43}$$

We then consider the terms in $\sum_{\varkappa} \sum_{p \neq q}$; they can be written

$$\sum_{\varkappa} \sum_{p \neq q} \Big[\sum_\nu (a_\nu^{(\alpha)})^2 |D_{\varkappa_p \varkappa_q}^{(\nu)}|^2 + \sum_{\nu \neq \nu'} a_\nu^{(\alpha)} a_{\nu'}^{(\alpha)} D_{\varkappa_p \varkappa_q}^{(\nu)} D_{\varkappa_p \varkappa_q}^{(\nu')*} \Big] = \sum_{\varkappa} \sum_{p \neq q} \Big| \sum_\nu a_\nu^{(\alpha)} D_{\varkappa_p \varkappa_q}^{(\nu)} \Big|^2 \tag{IV.44}$$

(since we have $D_{\varkappa_p \varkappa_q}^{(\nu)} = D_{\varkappa_q \varkappa_p}^{(\nu)*}$). Each term of the sum (IV.44) is positive and so we have necessarily

$$\sum_{\varkappa} \sum_{p \neq q} \Big| \sum_\nu a_\nu^{(\alpha)} D_{\varkappa_p \varkappa_q}^{(\nu)} \Big|^2 < \sum_{i \neq j} \Big| \sum_\nu a_\nu^{(\alpha)} D_{ij}^{(\nu)} \Big|^2 = \sum_{i \neq j} \Big[\sum_\nu (a_\nu^{(\alpha)})^2 |D_{ij}^{(\nu)}|^2 + \sum_{\nu \neq \nu'} a_\nu^{(\alpha)} a_{\nu'}^{(\alpha)} D_{ij}^{(\nu)} D_{ij}^{(\nu')*} \Big] \tag{IV.45}$$

and, by substituting into (IV.43),

$$\overline{\left[\overline{\mathscr{A}(t)}^{\infty}\right]^2}^{m} < \frac{1}{n(n+1)} \times$$

$$\times \Big\{\sum_\nu (a_\nu^{(\alpha)})^2 \Big[2\sum_i (D_{ii}^{(\nu)})^2 + \sum_{i\neq j} (D_{ii}^{(\nu)} D_{jj}^{(\nu)} + |D_{ij}^{(\nu)}|^2)\Big]$$

$$+ \sum_{\nu\neq\nu'} a_\nu^{(\alpha)} a_{\nu'}^{(\alpha)} \Big[2\sum_i D_{ii}^{(\nu)} D_{ii}^{(\nu')} + \sum_{i\neq j} (D_{ii}^{(\nu)} D_{jj}^{(\nu')} + D_{ij}^{(\nu)} D_{ij}^{(\nu')*})\Big]\Big\}. \tag{IV.46}$$

Moreover, according to (II.64) we have

$$2\sum_i (D_{ii}^{(\nu)})^2 + \sum_{i\neq j} (D_{ii}^{(\nu)} D_{jj}^{(\nu)} + |D_{ij}^{(\nu)}|^2)$$

$$= 2\Big[\sum_k^{(\nu)} \sum_i |C_{ki}^{(\alpha)}|^4 + \sum_{k\neq k'}^{(\nu)} \sum_i |C_{ki}^{(\alpha)}|^2 \, |C_{k'i}^{(\alpha)}|^2\Big]$$

$$+ \Big[\sum_k^{(\nu)} \sum_{i\neq j} |C_{ki}^{(\alpha)}|^2 \, |C_{kj}^{(\alpha)}|^2 + \sum_{k\neq k'}^{(\nu)} \sum_{i\neq j} |C_{ki}^{(\alpha)}|^2 \, |C_{k'j}^{(\alpha)}|^2\Big]$$

$$+ \Big[\sum_k^{(\nu)} \sum_{i\neq j} |C_{ki}^{(\alpha)}|^2 \, |C_{kj}^{(\alpha)}|^2 + \sum_{k\neq k'}^{(\nu)} \sum_{i\neq j} C_{ki}^{(\alpha)*} C_{kj}^{(\alpha)} C_{k'i}^{(\alpha)} C_{k'j}^{(\alpha)*}\Big]; \tag{IV.47}$$

however, by virtue of formulae (II.61), (II.62) and (II.63), we have

$$2\sum_k^{(\nu)} \Big[\sum_i |C_{ki}^{(\alpha)}|^4 + \sum_{i\neq j} |C_{ki}^{(\alpha)}|^2 \, |C_{kj}^{(\alpha)}|^2\Big] = 2\sum_k^{(\nu)} \delta_{kk}^{(\nu)} = 2s_\nu^{(\alpha)}, \tag{IV.48}$$

$$\sum_{k\neq k'}^{(\nu)} \Big[2\sum_i |C_{ki}^{(\alpha)}|^2 \, |C_{k'i}^{(\alpha)}|^2 + \sum_{i\neq j} (|C_{ki}^{(\alpha)}|^2 \, |C_{k'j}^{(\alpha)}|^2 + C_{ki}^{(\alpha)*} C_{kj}^{(\alpha)} C_{k'i}^{(\alpha)} C_{k'j}^{(\alpha)*})\Big]$$

$$= \sum_{k\neq k'}^{(\nu)} \delta_{kk}\delta_{k'k'} = s_\nu^{(\alpha)}(s_\nu^{(\alpha)} - 1), \tag{IV.49}$$

whence

$$2\sum_i (D_{ii}^{(\nu)})^2 + \sum_{i\neq j} (D_{ii}^{(\nu)} \, D_{ij}^{(\nu)} + |D_{ij}^{(\nu)}|^2) = 2s_\nu^{(\alpha)} + s_\nu^{(\alpha)} \, (s_\nu^{(\alpha)} - 1)$$

$$= s_\nu^{(\alpha)} \, (s_\nu^{(\alpha)} + 1). \tag{IV.50}$$

Similarly, we can show that

$$2\sum_i D_{ii}^{(\nu)} D_{ii}^{(\nu')} + \sum_{i\neq j} (D_{ii}^{(\nu)} D_{jj}^{(\nu')} + D_{ij}^{(\nu)} D_{ij}^{(\nu')*}) = \sum_{k,k'}^{(\nu,\nu')} \delta_{kk}\delta_{k'k'} = s_{\nu j}^{(\alpha)} s_{\nu'}^{(\alpha)}. \tag{IV.51}$$

The result is then, using (IV.46),

$$\overline{\left[\overline{\mathscr{A}(t)}^{\infty}\right]^2}^{m} < \frac{1}{n(n+1)} \Big\{\sum_\nu (a_\nu^{(\alpha)})^2 s_\nu^{(\alpha)} (s_\nu^{(\alpha)} + 1) + \sum_{\nu\neq\nu'} a_\nu^{(\alpha)} a_{\nu'}^{(\alpha)} s_\nu^{(\alpha)} s_{\nu'}^{(\alpha)}\Big\} \tag{IV.52}$$

and according to (IV.37) and (IV.32),

$$\overline{\left[\overline{\mathscr{A}(t)}^{\infty}\right]^2}^{m} - \left(\overline{\mathscr{A}}^{m}\right)^2 < \sum_{\nu} (a_\nu^{(\alpha)})^2 \left(\frac{s_\nu^{(\alpha)2} + s_\nu^{(\alpha)}}{n(n+1)} - \frac{s_\nu^{(\alpha)2}}{n^2}\right)$$

$$+ \sum_{\nu \neq \nu'} a_\nu^{(\alpha)} a_{\nu'}^{(\alpha)} s_\nu^{(\alpha)} s_{\nu'}^{(\alpha)} \left(\frac{1}{n(n+1)} - \frac{1}{n^2}\right)$$

$$= -\frac{1}{n+1}\,\frac{1}{n^2}\left(\sum_{\nu} (a_\nu^{(\alpha)})^2 s_\nu^{(\alpha)2} + \sum_{\nu \neq \nu'} a_\nu^{(\alpha)} a_{\nu'}^{(\alpha)} s_\nu^{(\alpha)} s_{\nu'}^{(\alpha)}\right)$$

$$+ \frac{1}{n(n+1)} \sum_{\nu} (a_\nu^{(\alpha)})^2 s_\nu^{(\alpha)}, \tag{IV.53}$$

or

$$\overline{\left[\overline{\mathscr{A}(t)}^{\infty}\right]^2}^{m} - \left(\overline{\mathscr{A}}^{m}\right)^2 = -\frac{1}{n+1}\left(\overline{\mathscr{A}}^{m}\right)^2 + \frac{1}{n(n+1)} \sum_{\nu} (a_\nu^{(\alpha)})^2 s_\nu^{(\alpha)}. \tag{IV.54}$$

This last result will enable us to show that the quantity $\left\{\overline{\left[\overline{\mathscr{A}(t)}^{\infty}\right]^2}^{m} - \left(\overline{\mathscr{A}}^{m}\right)^2\right\} \Big/ \left(\overline{\mathscr{A}}^{m}\right)^2$ is very small for a macroscopic system. In fact, if we make no assumption concerning the eigenvalues $a_\nu^{(\alpha)}$, we can make $\left(\overline{\mathscr{A}}^{m}\right)^2$ appear in the second term on the right-hand side of (IV.54); if we denote the smallest value of $s_\nu^{(\alpha)}$ by s_m and if we put $\varepsilon = 1/s_m$, we have

$$\frac{1}{n(n+1)} \sum_{\nu} (a_\nu^{(\alpha)})^2 s_\nu^{(\alpha)} < \frac{1}{n^2} \sum_{\nu} (a_\nu^{(\alpha)})^2 s_\nu^{(\alpha)2} \frac{1}{s_\nu^{(\alpha)}}$$

$$< \frac{\varepsilon}{n^2} \sum_{\nu} (a_\nu^{(\alpha)})^2 s_\nu^{(\alpha)2} < \varepsilon\left(\overline{\mathscr{A}}^{m}\right)^2, \tag{IV.55}$$

whence, by substituting in (IV.37), we obtain easily

$$\frac{\overline{\left(\overline{\mathscr{A}(t)}^{\infty} - \overline{\mathscr{A}}^{m}\right)^2}^{m}}{\left(\overline{\mathscr{A}}^{m}\right)^2} = O\left(\frac{1}{s_m}\right). \tag{IV.56}$$

Since we have accepted that in a macroscopic system all the values $s_\nu^{(\alpha)}$ are large, it follows that *the time-average* $\overline{\mathscr{A}(t)}^{\infty}$ *can be considered as equal to the phase average* $\overline{\mathscr{A}}^m$, *except over a set of very small measure:* this is a probability ergodic theorem satisfied by macroscopic observables.

If, in addition, we suppose that the eigenvalues $a_\nu^{(\alpha)}$ are all bounded by a number a_M, we can even improve the preceding evaluation, since in this case we have

$$\frac{1}{n(n+1)}\sum_\nu (a_\nu^{(\alpha)})^2 s_\nu^{(\alpha)} < \frac{a_M^2}{n(n+1)}\sum_\nu s_\nu^{(\alpha)} = \frac{a_M^2}{n+1} = O\left(\frac{1}{n}\right), \tag{IV.57}$$

whence the result

$$\frac{\overline{\left(\overline{\mathscr{A}(t)}^\infty - \overline{\mathscr{A}}^m\right)^2}^m}{\left(\overline{\mathscr{A}}^m\right)^2} = O\left(\frac{1}{S^{(\alpha)}}\right), \tag{IV.58}$$

if $a_\nu^{(\alpha)} < a_M$, which naturally has the same consequences. Thus, we have established a probability ergodic theorem for the macroscopic observables $\mathscr{A}(t)$ similar to the one we obtained in classical mechanics, starting from Khinchin's results relative to sum functions (cf. Chapter I); we do not make any special assumption here concerning the structure of the system, except that the time evolution is governed by the unitary operator (II.6) and that the quantities accessible to experiment are well represented by so-called "macroscopic" observables when the system considered has a very large number of degrees of freedom.

It is the introduction of the macroscopic observables associated with the averages over the unit hypersphere with $2n$ dimensions which has permitted us to break away from the hypothesis on non-degeneracy of the spectrum of the Hamiltonian; we shall analyse in a little more detail (see Section IV) the exact role played by these macroscopic observables in the elimination of degenerate terms. We shall show also in that section that formula (IV.56) can be extended to the averages $\mathscr{A}(t)$ themselves, at any time t; in this way we shall rederive the results of recent works (Farquhar and Landsberg, 1957; Bocchieri and Loinger, 1958, 1959), which we shall compare with von Neumann's quantum ergodic theory, making its meaning more exact. Finally we shall establish the analogy between the quantum ergodic theory formulated in this way and the classical ergodic theory applied to Khinchin's sum functions;

by assuming a similar structure for macroscopic systems as that proposed by Khinchin for classical systems, we shall see that the microscopic observables themselves satisfy a relation of the type (IV.56), which completes the analogy between the classical and quantum aspects of the ergodic theory.

Before commencing this study, we shall conclude this section by showing that equivalent results are obtained by taking the average over a microcanonical ensemble of systems, or by taking an average over all possible macroscopic observers. In fact, in order to obtain another macroscopic observer, starting from the observer characterised by the vector $\Omega\{\Omega_k^{(\alpha)}\}$, we apply a unitary transformation $\hat{U}$ to Ω which leads to a new vector $\Omega'\{\Omega_l^{(\alpha)'}\}$; by definition we have

$$\Omega' = \hat{U}\Omega, \quad \Omega_l^{(\alpha)'} = \sum_k U_{lk}\Omega_k^{(\alpha)}, \quad \Omega_k^{(\alpha)} = \sum_l U_{lk}^{*}\Omega_l^{(\alpha)'}, \quad \text{(IV.59)}$$

whence, according to (IV.28)

$$\Psi(t) = \sum_l \bar{\omega}_l^{(\alpha)'}(t)\,\Omega_l^{(\alpha)'} = \sum_k \bar{\omega}_k^{(\alpha)}(t)\,\Omega_k^{(\alpha)} = \sum_l \Omega_l^{(\alpha)'} \sum_k \bar{\omega}_k^{(\alpha)}(t)\;U_{lk}^{*}, \quad \text{(IV.60)}$$

with

$$\bar{\omega}_k^{(\alpha)}(t) = \sum_l \bar{\omega}_l^{(\alpha)'}(t)\;U_{lk}, \quad \bar{\omega}_l^{(\alpha)'}(t) = \sum_k \bar{\omega}_k^{(\alpha)}(t) U_{lk}^{*}. \quad \text{(IV.61)}$$

In order to obtain all possible macroscopic observers, we must take all systems of U_{lk} satisfying the relationship

$$\sum_l |U_{lk}|^2 = 1, \quad \text{(IV.62)}$$

i.e. to each system of U_{lk} there corresponds one point of the complex unit sphere of n dimensions. Since, according to (IV.61), to each system of values U_{lk} there corresponds a system of $\bar{\omega}_l^{(\alpha)'}(t)$, it follows that the end-point of the vector $\{\bar{\omega}_l^{(\alpha)'}(t)\}$ traverses the entire unit sphere σ_{2n} when the U_{lk} take all values compatible with (IV.62); in fact, because of the unitarity of $\hat{U}$, we have

$$\begin{aligned}\sum_l |\bar{\omega}_l^{(\alpha)'}|^2 &= \sum_l \sum_{k,k'} \bar{\omega}_k^{(\alpha)}\bar{\omega}_{k'}^{(\alpha)*}U_{lk}^{*}U_{lk'} \\ &= \sum_l \Big\{ \sum_k |\bar{\omega}_k^{(\alpha)}|^2\,|U_{lk}|^2 + \sum_{k\neq k'} \bar{\omega}_k^{(\alpha)}\bar{\omega}_{k'}^{(\alpha)*}U_{lk}^{*}U_{lk'}\Big\} \\ &= \sum_k |\bar{\omega}_k^{(\alpha)}|^2 + \sum_{k\neq k'} \bar{\omega}_k^{(\alpha)}\omega_{k'}^{(\alpha)*}\delta_{kk'} = \sum_k |\bar{\omega}_k^{(\alpha)}|^2 = 1.\end{aligned} \quad \text{(IV.63)}$$

We can then make the set of systems satisfying (IV.63) correspond to the set of macroscopic observers characterised by (IV.62). In addition, we can say that the relation $(\Psi, \hat{U}\Omega_k^{(\alpha)}) = (\hat{U}^{-1}\Psi, \Omega_k^{(\alpha)})$ shows that rotation of the eigenvector $\Omega^{(\alpha)}$ of a macroscopic observer by applying $\hat{U}$ to it, is equivalent to rotating the representative vector Ψ of the system by applying the operator $\hat{U}^{-1}$ to it; the set of vectors Ψ thus obtained is represented completely by all the points of the unit hypersphere σ_{2n} with $2n$ dimensions: the averages taken over a microcanonical ensemble of systems or over the ensemble of macroscopic observers thus provide equivalent results. We shall see in the next section, however, that there are weighty reasons for preferring the methods based on averages of the microcanonical type.

IV. Statistical Properties of Macroscopic Observables. Comparison with Classical Theory

We shall now discuss the results obtained in the preceding sections and we shall examine their meaning. We shall commence by studying the standard deviation of the variables $\mathscr{A}(t)$ over the hypersphere σ_{2n} which will enable us to compare the approximated ergodic theorem (IV.56) with von Neumann's quantum ergodic theory. Then, having analysed the role of macroscopic observables in the theory, we shall establish a relation between the classical and quantum forms of the ergodic theory.

1. *Statistical properties of macroscopic observables*

In order to establish the macroscopic ergodic theorem, we have relied on the inequality (IV.46) in which a sum $\sum_{i \neq j}$ occurs as the second term, taken over all states i and j, degenerate or non-degenerate. The form of the second term of (IV.46) in this way leads us to study the second moment of the random variable $\mathscr{A}(t) - \overline{\mathscr{A}}^m$ over the hypersphere σ_{2n} at any time t (Jancel, 1960a); according to (IV.30) and (IV.31) we have

$$[\mathscr{A}(t)]^2 = \sum_{\nu} a_{\nu}^{(\alpha)2}[w_{\nu}^{(\alpha)}(t)]^2 + \sum_{\nu \neq \nu'} a_{\nu}^{(\alpha)} a_{\nu'}^{(\alpha)} w_{\nu}^{(\alpha)}(t) w_{\nu'}^{(\alpha)}(t), \quad \text{(IV.64)}$$

with

$$[w_\nu^{(\alpha)}(t)]^2 = \sum_{i,j} r_i^2 r_j^2 D_{ii}^{(\nu)} D_{jj}^{(\nu)}$$
$$+ 2\sum_i \sum_{k\neq l} r_i^2 r_k r_l \, e^{i(\alpha_k-\alpha_l)} \, e^{-i(E_k-E_l)t/\hbar} D_{ii}^{(\nu)} D_{kl}^{(\nu)}$$
$$+ \sum_{i\neq j} \sum_{k\neq l} r_i r_j r_k r_l \, e^{i(\alpha_i-\alpha_j)} \, e^{i(\alpha_k-\alpha_l)} \times$$
$$\times \, e^{-i(E_i-E_j)t/\hbar} \, e^{-i(E_k-E_l)t/\hbar} D_{ij}^{(\nu)} D_{kl}^{(\nu)}, \qquad \text{(IV.65)}$$

$$[w_\nu^{(\alpha)}(t) w_{\nu'}^{(\alpha)}(t)] = \sum_{i,j} r_i^2 r_j^2 D_{ii}^{(\nu)} D_{jj}^{(\nu')}$$
$$+ \sum_i \sum_{k\neq l} r_i^2 r_k r_l e^{i(\alpha_k-\alpha_l)} \times$$
$$\times \, e^{-i(E_k-E_l)t/\hbar} (D_{ii}^{(\nu')} D_{kl}^{(\nu)} + D_{ii}^{(\nu)} D_{kl}^{(\nu')})$$
$$+ \sum_{i\neq j} \sum_{k\neq l} r_i r_j r_k r_l e^{i(\alpha_i-\alpha_j)} e^{i(\alpha_k-\alpha_l)} \times$$
$$\times \, e^{-i(E_i-E_j)t/\hbar} e^{-i(E_k-E_l)t/\hbar} D_{ij}^{(\nu)} D_{kl}^{(\nu')}. \qquad \text{(IV.66)}$$

Since $\overline{\mathscr{A}(t)}^m = \bar{\mathscr{A}}^m$, the second moment of $\mathscr{A}(t) - \bar{\mathscr{A}}^m$ can be written as

$$\overline{\left(\mathscr{A}(t) - \bar{\mathscr{A}}^m\right)^2}^m = \overline{[\mathscr{A}(t)]^2}^m - \left(\bar{\mathscr{A}}^m\right)^2; \qquad \text{(IV.67)}$$

thus, we must calculate the microcanonical average of (IV.64) or the averages $\overline{[w_\nu^{(\alpha)}(t)]^2}^m$ and $\overline{[w_\nu^{(\alpha)}(t)\, w_{\nu'}^{(\alpha)}(t)]}^m$. As we have mentioned already in calculating $\overline{\left[\overline{\mathscr{A}(t)}^\infty\right]^2}^m$, the only terms of (IV.65) and (IV.66) which make non-vanishing contributions are those which are independent of the phases α_i; thus, the average of the sum $\sum_i \sum_{k\neq l}$ is zero and we obtain the phase-independent terms of $\sum_{i\neq j} \sum_{k\neq l}$ by putting $i = l$ and $j = k$. Then, according to (IV.50), (IV.51), (IV.65) and (IV.66) we have:

$$\overline{[w_\nu^{(\alpha)}(t)]^2}^m = \sum_{i=1}^n \overline{r_i^4}^m (D_{ii}^{(\nu)})^2 \qquad \text{(IV.68)}$$
$$+ \sum_{i\neq j} \overline{r_i^2 r_j^2}^m (D_{ii}^{(\nu)} D_{jj}^{(\nu)} + |D_{ij}^{(\nu)}|^2) = \frac{s_\nu^{(\alpha)}(s_\nu^{(\alpha)}+1)}{n(n+1)},$$

$$\overline{[w_\nu^{(\alpha)}(t)\, w_{\nu'}^{(\alpha)}(t)]}^m = \sum_{i=1}^n \overline{r_i^4}^m D_{ii}^{(\nu)} D_{ii}^{(\nu')} \qquad \text{(IV.69)}$$
$$+ \sum_{i\neq j} \overline{r_i^2 r_j^2}^m (D_{ii}^{(\nu)} D_{jj}^{(\nu')} + D_{ij}^{(\nu)} D_{ji}^{(\nu')}) = \frac{s_\nu^{(\alpha)} s_{\nu'}^{(\alpha)}}{n(n+1)}.$$

In this way we see that the right-hand side of (IV.46) is exactly equal to $\overline{[\mathscr{A}(t)]^2}^m$ and that equation (IV.53) can then be written as

$$\overline{\left[\overline{\mathscr{A}(t)}^{\infty}\right]^2}^m - \left(\overline{\mathscr{A}}^m\right)^2 \leqq \overline{[\mathscr{A}(t)]^2}^m - \left(\overline{\mathscr{A}}^m\right)^2, \qquad \text{(IV.70)}$$

with

$$\overline{[\mathscr{A}(t)]^2}^m - \left(\overline{\mathscr{A}}^m\right)^2 = \sum_{\nu} a_{\nu}^{(\alpha)2}\left[\frac{s_{\nu}^{(\alpha)}(s_{\nu}^{(\alpha)}+1)}{n(n+1)} - \frac{s_{\nu}^{(\alpha)2}}{n^2}\right]$$
$$+ \sum_{\nu \neq \nu'} a_{\nu}^{(\alpha)} a_{\nu'}^{(\alpha)} s_{\nu}^{(\alpha)} s_{\nu'}^{(\alpha)} \left[\frac{1}{n(n+1)} - \frac{1}{n^2}\right]$$
$$= -\frac{1}{n+1}\left(\overline{\mathscr{A}}^m\right)^2 + \frac{1}{n(n+1)}\sum_{\nu} a_{\nu}^{(\alpha)2} s_{\nu}^{(\alpha)}. \qquad \text{(IV.71)}$$

We can deduce from it, by reasoning similar to that of section III, that

$$\frac{\overline{\left(\mathscr{A}(t) - \overline{\mathscr{A}}^m\right)^2}^m}{\left(\overline{\mathscr{A}}^m\right)^2} = \begin{cases} O\left(\dfrac{1}{s_m}\right) & (\text{arbitrary } a_{\nu}^{(\alpha)}), \\[2ex] O\left(\dfrac{1}{S^{(\alpha)}}\right) & (a_{\nu}^{(\alpha)} < a_M). \end{cases} \qquad \text{(IV.72)}$$

Thus, *the standard deviation of* $\mathscr{A}(t)$, *compared with* $\overline{\mathscr{A}}^m$, *is independent of the time and is always at least of order* $1/s_m^{\frac{1}{2}}$. This result could, however, have been predicted: in fact, the time evolution causes only a variation of the phases $(\alpha_i - iE_i t/\hbar)$ which induces a displacement of the vector Ψ over the hypersurface σ_{2n}; however, the microcanonical ensemble represented by a uniform distribution over σ_{2n} is time-invariant, with the result that the average over σ_{2n} of a function of the time like $[\mathscr{A}(t)]^2$ must itself be independent of t.

This important statistical property of macroscopic observables depends only on the breakdown of Hilbert space into cells and on the unitarity of the operator of evolution; if the numbers $s_{\nu}^{(\alpha)}$ are all very large, the quantities $\mathscr{A}(t)$ are almost equal to the microcanonical averages $\overline{\mathscr{A}}^m$, except over sets of very small measure corresponding to exceptional initial conditions. The approximated ergodic theorem (IV.56) can be derived from it as well as von Neumann's ergodic theorem, as we shall see now.

2. *Return to von Neumann's quantum ergodic theorem*

In order to do this, we shall consider the expression

$$\overline{\left[\lim_{T\to\infty}\frac{1}{T}\int_0^T\left(\mathscr{A}(t)-\overline{\mathscr{A}}^m\right)^2 dt\right]}^m ; \qquad \text{(IV.73)}$$

it is easy to show that we can interchange the two integrations over the time and over the hypersphere σ_{2n}. In fact, if we evaluate $\frac{1}{T}\int_0^T [\mathscr{A}(t)]^2\,dt$, we can see by (IV.64), (IV.65) and (IV.66) that the terms of $\sum_{i\neq j}\sum_{k\neq l}$ which remain non-vanishing as $T\to\infty$ are of three kinds: (a) terms involving products of degenerate factors; (b) terms corresponding to resonance frequencies $E_i+E_k=E_j+E_l$ and, finally, (c) phase-independent terms. If we take the microcanonical average, terms (a) and (b) make a zero contribution and we have, finally,

$$\overline{\left[\lim_{T\to\infty}\frac{1}{T}\int_0^T[\mathscr{A}(t)]^2\,dt\right]}^m$$

$$= \sum_{\nu} a_\nu^{(\alpha)2}\left[\sum_{i=1}^{n}\overline{r_i^4}^m\,(D_{ii}^{(\nu)})^2+\sum_{i\neq j}\overline{r_i^2r_j^2}^m\,(D_{ii}^{(\nu)}D_{jj}^{(\nu)}+|D_{ij}^{(\nu)}|^2)\right]$$

$$+\sum_{\nu\neq\nu'}a_\nu^{(\alpha)}a_{\nu'}^{(\alpha)}\left[\sum_{i=1}^{n}\overline{r_i^4}^m\,D_{ii}^{(\nu)}D_{ii}^{(\nu')}+\sum_{i\neq j}\overline{r_i^2r_j^2}^m\,(D_{ii}^{(\nu)}D_{jj}^{(\nu')}+D_{ij}^{(\nu)}D_{ji}^{(\nu')}\right]$$

$$=\overline{[\mathscr{A}(t)]^2}^m=\lim_{T\to\infty}\frac{1}{T}\int_0^T\overline{[\mathscr{A}(t)]^2}^m\,dt, \qquad \text{(IV.74)}$$

according to (IV.68) and (IV.69); we can then interchange the two averaging processes. In particular, we shall have

$$\frac{\overline{\left[\lim_{T\to\infty}\frac{1}{T}\int_0^T\left(\mathscr{A}(t)-\overline{\mathscr{A}}^m\right)^2 dt\right]}^m}{\left(\overline{\mathscr{A}}^m\right)^2} =$$

$$= \frac{\overline{\lim_{T\to\infty} \frac{1}{T}\int_0^T \left(\mathscr{A}(t) - \overline{\mathscr{A}}^m\right)^2 dt}^m}{\left(\overline{\mathscr{A}}^m\right)^2} = \frac{\overline{\left(\mathscr{A}(t) - \overline{\mathscr{A}}^m\right)^2}^m}{\left(\overline{\mathscr{A}}^m\right)^2} = \begin{cases} O\left(\dfrac{1}{s_m}\right), \\ O\left(\dfrac{1}{S^{(\alpha)}}\right). \end{cases} \tag{IV.75}$$

We have here a result similar to von Neumann's quantum ergodic theorem and it can be seen that we have not made any assumption concerning the nature of the spectrum of H; actually, terms corresponding to degeneracies and to resonance frequencies of H are eliminated if we take the microcanonical average of $(\mathscr{A}(t) - \overline{\mathscr{A}}^m)^2$ Thus, it is the introduction of the statistical point of view *from the beginning of the proof* which has allowed us to break away from any assumption concerning the spectrum of the Hamiltonian. Obviously, it follows that the result (IV.75) is a theorem of general statistical mechanics and is valid only for an ensemble of systems; it belongs to the category of probability ergodic theorems.

In von Neumann's proof, we try to found statistical mechanics on a purely dynamical basis, by trying to obtain a true ergodic theorem which refers only to a single system; it is this which leads us to make the assumptions already mentioned concerning the nature of the spectrum of $\hat{H}$. These, however, are insufficient and we must return to statistical considerations about the ensemble of all possible macroscopic observers; as we have noted in the preceding section the identity of these averages and of the microcanonical averages, it follows that we can interchange the time-average operation and that of the average over the ensemble of macroscopic observers. According to the foregoing, it is obvious that von Neumann's quantum theorem is simply a consequence of the average over these observers; the assumptions concerning the spectrum of $\hat{H}$ become redundant, with the result that von Neumann's quantum ergodic theorem has only a statistical meaning, as we have mentioned already (see Chapter III), and that it falls within the category of probability ergodic theorems.

Moreover, if we actually calculate the averages of (IV.65) and (IV.66) over the ensemble of macroscopic observers with the help of formulae (36)–(43′) of Appendix II, we can confirm that we find the results (IV.68) and (IV.69), *independently of the nature of the time evolution of the system*: it suffices that the relationship $\sum_i c_i(0)^2 = 1$ is satisfied at each instant, i.e. that the norm of the vector $\Psi(t)$ is conserved. Such is not the case for the proof founded on microcanonical averages (equivalent to the averages over the initial conditions, see § IV.II.1), since the invariance of the microcanonical ensemble relative to the unitary operator of evolution plays an essential role here. Because of this difficulty, added to which is that of justifying the *a priori* assumption of equiprobability of the macroscopic observers, it is preferable to place ourselves at the onset within the framework of ensemble theory; this leads us, without special assumption, to the probability ergodic theorems sufficient for justifying the use of the microcanonical ensemble.

We may mention also that in von Neumann's method we compare $(\mathscr{A}(t) - \overline{\mathscr{A}}^m)^2$ with the microcanonical average of the observable $\mathscr{A}^2$; thus, we must show that the time-average of the factor

$$\lambda = \sum_\nu \frac{S^{(\alpha)}}{s_\nu^{(\alpha)}} \left(w_\nu^{(\alpha)}(t) - \frac{s_\nu^{(\alpha)}}{S^{(\alpha)}} \right)^2 \tag{IV.76}$$

is as small as we wish [cf. formula (III.44) applied to a single energy shell]. If we form the microcanonical average of λ, we obtain, according to (IV.68)

$$\begin{aligned} \overline{\lambda}^m = \overline{\left(\overline{\lambda}^\infty\right)}^m &= \sum_\nu \frac{S^{(\alpha)}}{s_\nu^{(\alpha)}} \left[\frac{s_\nu^{(\alpha)}\,(s_\nu^{(\alpha)}+1)}{S^{(\alpha)}(S^{(\alpha)}+1)} - \frac{s_\nu^{(\alpha)2}}{S^{(\alpha)2}} \right] \\ &= \sum_\nu \frac{S^{(\alpha)}}{s_\nu^{(\alpha)}} \left[\frac{s_\nu^{(\alpha)}(S^{(\alpha)} - s_\nu^{(\alpha)})}{S^{(\alpha)2}(S^{(\alpha)}+1)} \right] = \frac{N^{(\alpha)} - 1}{S^{(\alpha)} + 1} \\ &= \left\langle \overline{\lambda}^\infty \right\rangle. \end{aligned} \tag{IV.77}$$

Indeed, we find again equation (III.57′)—as would be expected—by virtue of the identity between the microcanonical averages and the averages over the ensemble of macroscopic observers, proved by equations (IV.62) and (IV.63). We point out, finally, that we can prove equally well an ergodic theorem by starting from the entropy of the system defined by formulae (III.64) and (III.65): it suffices to calculate the quadratic mean of the variable $S(t) - \mathfrak{S}$.

Summarising, the quantum ergodic theory—as we have emphasised already—is unable to prove the equivalence of the time and phase averages and it encounters difficulties at least as great as classical theory. Only the introduction of a statistical point of view allows us to prove probability ergodic theorems which constitute in some measure a generalisation of the normal ergodic theorems and suffice to justify the methods of statistical mechanics. From this point of view, von Neumann's quantum ergodic theorem is a special case of these probability theorems; if we wish to compare it with theorem (IV.56) we have, by applying Schwartz's inequality:

$$\left[\lim_{T\to\infty}\frac{1}{T}\int_0^T \mathscr{A}(t)\,dt\right]^2 \leqq \lim_{T\to\infty}\frac{1}{T}\int_0^T [\mathscr{A}(t)]^2\,dt, \tag{IV.78}$$

whence, since both sides are positive

$$\overline{\left[\lim_{T\to\infty}\frac{1}{T}\int_0^T \mathscr{A}(t)\,dt\right]^2}^{m} = \overline{\left[\overline{\mathscr{A}(t)}^{\infty}\right]^2}^{m} \leqq \overline{\left[\lim_{T\to\infty}\frac{1}{T}\int_0^T [\mathscr{A}(t)]^2\,dt\right]}^{m}, \tag{IV.79}$$

which can also be written as

$$\overline{\left(\overline{\mathscr{A}(t)}^{\infty} - \overline{\mathscr{A}}^{m}\right)^2}^{m} \leqq \overline{\left[\lim_{T\to\infty}\frac{1}{T}\int_0^T \left(\mathscr{A}(t) - \overline{\mathscr{A}}^{m}\right)^2 dt\right]}^{m}$$

$$= \overline{\left(\mathscr{A}(t) - \overline{\mathscr{A}}^{m}\right)^2}^{m}, \tag{IV.80}$$

this result being already contained in the inequalities (IV.46) and (IV.70). Before comparing these results as a whole with those of classical theory, we must analyse in detail the role played by macroscopic observables in the proofs.

3. *The role of macroscopic observables*

The interesting point in these observables is obviously the introduction of an additional "macroscopic" degeneracy, implied by the division of Hilbert space into "cells" with $s_\nu^{(\alpha)}$ dimensions.

We can say that this division into cells is involved in three different ways in the foregoing proofs.

(a) It permits us to give an upper limit to expressions such as (IV.54) without making an assumption about $a_\nu^{(\alpha)}$; actually, by returning to the corresponding microscopic expression (IV.13), we see that it is necessary to make certain assumptions about the numbers A_{ii} if we wish to find an upper bound for the quantity $\frac{1}{n(n+1)} \sum_{i=1}^{n} A_{ii}^2$ or to compare it with another quantity, such as $(\overline{A}^m)^2$.

(b) It allows us also to evaluate terms arising from degeneracies of the Hamiltonian; in macroscopic theory, these are of the form $\sum_{\varkappa} \sum_{p \neq q} \left| \sum_{\nu} a_\nu^{(\alpha)} D_{\varkappa_p \varkappa_q}^{(\nu)} \right|^2$ and an upper bound is, according to (IV.45), determined by the terms

$$\sum_{i \neq j} \left| \sum_{\nu} a_\nu^{(\alpha)} D_{ij}^{(\nu)} \right|^2 \tag{IV.81}$$

which originate from $\overline{[\mathscr{A}(t)]^2}^m$. In the microscopic theory, the sum $\sum_{\varkappa} \sum_{p \neq q} |A_{\varkappa_p \varkappa_q}|^2$ corresponds to them and this sum has as upper bound $\sum_{i \neq j} |A_{ij}|^2$; this latter expression originates from the average $\overline{[A(t)]^2}^m$; in fact, by (IV.17), we have

$$\overline{[\mathscr{A}(t)]^2}^m = \sum_i \overline{r_i^4}^m A_{ii}^2 + \sum_{i \neq j} \overline{r_i^2 r_j^2}^m (A_{ii} A_{jj} + |A_{ij}|^2). \tag{IV.82}$$

By comparing this with $\overline{[\mathscr{A}(t)]^2}^m$, we can verify that the transition from microscopic to macroscopic observables is expressed by the relation

$$A_{ii} \to \sum_{\nu} a_\nu^{(\alpha)} D_{ii}^{(\nu)}, \quad A_{ij} \to \sum_{\nu} a_\nu^{(\alpha)} D_{ji}^{(\nu)}. \tag{IV.83}$$

These formulae can be obtained directly by using the change of variables (II.55) and assuming the operator $\hat{A}$ to be diagonal and with degeneracy $s_\nu^{(\alpha)}$ in the representation $\Omega_k^{(\alpha)}$, which reduces to identifying it with the macroscopic operator.

Given this, let us consider the sum $\sum_{i \neq j} |A_{ij}|^2$; if we assume, for a rough approximation, that $|A_{ij}| < A_M$, it can be seen that this sum is of order $n(n-1) A_M^2$ and that

$$\frac{\sum_{i \neq j} |A_{ij}|^2}{n(n+1)} \simeq \frac{n(n-1)}{n(n+1)} A_M^2. \tag{IV.84}$$

Obviously, this expression does not tend to zero with $1/n$, which makes it difficult to eliminate the degenerate terms given by (IV.27). If, on the contrary, we use the relation (IV.83), we have immediately by formulae (II.61) and (II.63):

$$\sum_{i\neq j} |A_{ij}|^2 \to \sum_{i\neq j}\left|\sum_{\nu} a_\nu^{(\alpha)} D_{ji}^{(\nu)}\right|^2 < \sum_{i,j}\left|\sum_{\nu} a_\nu^{(\alpha)} D_{ji}^{(\nu)}\right|^2$$

$$= \sum_{\nu} (a_\nu^{(\alpha)})^2 \sum_{i,j} |D_{ij}^{(\nu)}|^2 = \sum_{\nu} s_\nu^{(\alpha)} (a_\nu^{(\alpha)})^2, \tag{IV.85}$$

whence, by assuming that the $a_\nu^{(\alpha)}$ are bounded by a_M,

$$\sum_{i\neq j}\left|\sum_{\nu} a_\nu^{(\alpha)} D_{ji}^{(\nu)}\right|^2 < a_M^2 \sum_{\nu} s_\nu^{(\alpha)} = n a_M^2. \tag{IV.86}$$

Thus, the transition from the microscopic to the macroscopic representation enables us to find an upper bound for the sum $\sum_{i\neq j} |A_{ij}|^2$ by an expression which is proportional to n; it is this which enables us to eliminate the degeneracies of the Hamiltonian and to arrive at the general result (IV.72). Once we have arrived at this stage of the reasoning, it is natural to ask ourselves the following question: is it possible to evaluate the matrix elements A_{ij} by taking into account the macroscopic nature of the system, in such a way that an upper limit for the sum $\frac{1}{n(n+1)} \sum_{i\neq j} |A_{ij}|^2$ can be found in the form of a quantity which approaches zero when n increases indefinitely? We shall try to give certain elements of an answer to this question in the next section.

(c) This division into "phase cells" enables us also to show that the statistical weight $w_\nu^{(\alpha)}(t)$ of each cell is, at every instant, practically equal to its relative weight in the microcanonical ensemble, i.e. $s_\nu^{(\alpha)}/S^{(\alpha)}$. Actually, according to (IV.68) we have:

$$\frac{\overline{\left(w_\nu^{(\alpha)}(t) - \overline{w_\nu^{(\alpha)}}^{m}\right)^2}^{m}}{\left(\overline{w_\nu^{(\alpha)}}^{m}\right)^2} = \frac{\overline{\left(w_\nu^{(\alpha)}(t) - \frac{s_\nu^{(\alpha)}}{S^{(\alpha)}}\right)^2}^{m}}{\frac{s_\nu^{(\alpha)2}}{S^{(\alpha)2}}} \tag{IV.87}$$

$$= \frac{S^{(\alpha)} - s_\nu^{(\alpha)}}{s_\nu^{(\alpha)}(S^{(\alpha)}+1)} \approx \frac{1}{s_\nu^{(\alpha)}} - \frac{1}{S^{(\alpha)}+1} \sim O\left(\frac{1}{s_\nu^{(\alpha)}}\right),$$

which proves the stated theorem provided that $s_\nu^{(\alpha)}$ is sufficiently large, whatever the value of ν. We could establish a similar formula by using the time-average $\overline{w_\nu^{(\alpha)}(t)}^{\infty}$ in place of $w_\nu^{(\alpha)}(t)$ and we would find in this way the results obtained by various authors (Bocchieri and Loinger, 1958, 1959; Farquhar and Landsberg, 1957) in recent papers. In addition, we note that in the first and third references the averages are taken with respect to the ensemble of macroscopic observers, which changes nothing in the final conclusions because of formulae (39) to (43) of Appendix II. Moreover, by summing (IV.87) over the $N^{(\alpha)}$ cells of the energy shell we obtain the results of Bocchieri and Loinger (1959):

$$\overline{\sum_\nu \left(\frac{w_\nu^{(\alpha)}(t) - \dfrac{s_\nu^{(\alpha)}}{S^{(\alpha)}}}{\dfrac{s_\nu^{(\alpha)}}{S^{(\alpha)}}} \right)^2}^{m} < \sum \frac{1}{s_\nu^{(\alpha)}} \qquad \text{(IV.87')}$$

and, by taking the time-average,

$$\overline{\left\{ \lim_{T\to\infty} \frac{1}{T} \int_0^T \left[\sum_\nu \left(\frac{w_\nu^{(\alpha)}(t) - \dfrac{s_\nu^{(\alpha)}}{S^{(\alpha)}}}{\dfrac{s_\nu^{(\alpha)}}{S^{(\alpha)}}} \right)^2 \right] dt \right\}}^{m} < \sum_\nu \frac{1}{s_\nu^{(\alpha)}}. \qquad \text{(IV.87'')}$$

We point out also that Prosperi and Scotti (1959) have recently evaluated the probability that one of the quantities (IV.87) is larger than a given value during a fraction of time exceeding a specified duration.

In order to emphasise the role played by dividing Hilbert space into cells, we must now calculate the expression corresponding to (IV.87) for a single quantum state; we have

$$\frac{\overline{\left(r_i^2 - \overline{r_i^2}^{m}\right)^2}^{m}}{\left(\overline{r_i^2}^{m}\right)^2} = \frac{\overline{r_i^4}^{m} - \left(\overline{r_i^2}^{m}\right)^2}{\left(\overline{r_i^2}^{m}\right)^2} = \frac{n-1}{n+1} \to 1. \qquad \text{(IV.88)}$$

Thus, we prove—as we could have predicted—that a property such as (IV.87) is only satisfied in the case when a large number of states are grouped in the same cell: the role of macroscopic degeneracy is thus found to be clearly defined.

4. *Comparisons with the classical ergodic theory*

We have seen that it is impossible in classical mechanics to prove the equality of time-averages (whose existence is assured by the ergodic theorems) and phase-averages without involving unverifiable assumptions (metric transitivity) concerning the nature of the system; even then, the ergodic theorems mentioned are valid apart from a set of zero measure. Thus, we have seen how a statistical element is introduced into the theory and we have shown, by following Khinchin's reasoning, that the methods of statistical mechanics could be justified if we took account of the large number of degrees of freedom of a macroscopic system. This result is essentially based on the following statistical property of the "sum" functions of a system of N particles:

$$\frac{\overline{\left(f-\overline{f}^{m}\right)^{2}}^{m}}{\left(\overline{f}^{m}\right)^{2}}=O\left(\frac{1}{N}\right), \qquad \text{(IV.89)}$$

where the microcanonical average denotes the average over a hypersphere of constant energy. The analogy between this relationship and (IV.72), which is valid for quantum macroscopic observables, is obvious and they are both fundamental for proving the probability ergodic theorems; actually, (I.52) corresponds to relation (IV.89), which can be written as

$$\frac{\overline{\left(\overline{f}^{\infty}-\overline{f}^{m}\right)^{2}}^{m}}{\left(\overline{f}^{m}\right)^{2}} \leqq \frac{\overline{\left(f-\overline{f}^{m}\right)^{2}}^{m}}{\left(\overline{f}^{m}\right)^{2}}=O\left(\frac{1}{N}\right), \qquad \text{(IV.90)}$$

so that the state of the ergodic theory is the same in its quantum and classical aspects; however, two important differences must be pointed out.

The first arises from the differences existing between the order of magnitude of $1/N$ in (IV.89) and $1/S^{(\alpha)}$ in (IV.72); if, in fact, we take for $S^{(\alpha)}$ the estimate (IV.16), we see that $1/S^{(\alpha)} \ll 1/N$ since $S^{(\alpha)}$ increases exponentially with N. This must be attributed to the functional representation of particles in wave mechanics; actually, in classical mechanics, the representative point of the system is a point in $6N$-dimensional space and the microcanonical average is taken over a hypersurface of $(6N - 1)$ dimensions; in quantum mechanics the vector Ψ is defined in a functional space of $n \equiv S^{(\alpha)}$ dimensions and the averages are taken over a hypersphere whose number of dimensions is proportional to the measure of the hypersurface with $(6N - 1)$ dimensions. It follows that the sets of possible states of quantum statistical mechanics are much richer than the corresponding classical sets.†

The second difference lies in the use of macroscopic observables in quantum statistical mechanics, whilst the classical theorem (IV.90) is valid only for sum phase functions. This brings us back to the question we raised at the end of subsection (b) of the preceding section and which we can reformulate thus: is it possible to justify, for a system with a large number of degrees of freedom, the use of macroscopic observables possessing the required properties? We may note first of all, by referring to the intuitive definition of these observables in Chapter II, that we have made use largely of the macroscopic nature of the observed system and that in consequence the definition of macroscopic observables seems to be quite natural. Obviously, however, this is only a qualitative argument which does not eliminate the external interference of the macroscopic observer and which can not prevent us from considering the introduction of these observables as a supplementary assumption, inherent in quantum theory.

In order to try to answer this objection, we shall conclude this section by constructing a quantum model similar to Khinchin's model and for which the ergodic theorems can be written directly in terms of microscopic observables (Jancel, 1960b); in this way we shall have taken an additional step in reconciling the classical and quantum ergodic theorems by breaking away from the concept of macroscopic observables. We note first of all that a property of the form (IV.80) can be extended to microscopic observables; in fact, by applying Schwartz's inequality we have

$$\overline{\left[\overline{A(t)}^{\infty} - \left(\overline{A}^{m}\right)^{2}\right]}^{m} \leq \overline{\left[\lim_{T\to\infty} \frac{1}{T}\int_{0}^{T}\left(A(t) - \overline{A}^{m}\right)^{2} dt\right]}^{m} = \overline{\left(A(t) - \overline{A}^{m}\right)^{2}}^{m}, \tag{IV.91}$$

† For an introduction of functional space in the classical ergodic theorem, see especially the paper by Albertoni, Bocchieri, and Loinger (1960).

where, according to (IV.17) and (IV.82),

$$\overline{\left[A(t) - \overline{A}^{m}\right]^2}^{m} = \sum_i A_{ii}^2 \left[\overline{r_i^4}^{m} - \left(\overline{r_i^2}^{m}\right)^2\right]$$
$$+ \sum_{i \neq j} A_{ii}A_{jj} \left[\overline{r_i^2 r_j^2}^{m} - \left(\overline{r_i^2}^{m}\right)\left(\overline{r_j^2}^{m}\right)\right] + \sum_{i \neq j} \overline{r_i^2 r_j^2}^{m} |A_{ij}|^2$$
$$= -\frac{1}{n+1}\left(\overline{A}^{m}\right)^2 + \frac{1}{n(n+1)}\left(\sum_i A_{ii}^2 + \sum_{i \neq j} |A_{ij}|^2\right). \tag{IV.92}$$

As we have pointed out already in § 3(b), the difficulties concerning the degenerate terms originate from the sum $\sum_{i \neq j} |A_{ij}|^2$ depending on the off-diagonal terms [see also formula (IV.25)].

We shall now evaluate the terms $\frac{1}{n(n+1)}\left(\sum_i A_{ii}^2 + \sum_{i \neq j} |A_{ij}|^2\right)$ by using a method which is similar to that of Khinchin. Let us suppose first of all, that our system consisting of N particles is described by the Hamiltonian

$$\hat{H} = \sum_{\varkappa=1}^{N} \hat{H}^{(\varkappa)}, \tag{IV.93}$$

where $\hat{H}^{(\varkappa)}$ represents the Hamiltonian of the $\varkappa$th particle; this amounts to neglecting, to a first approximation, the supposedly small interaction between the particles themselves. If the particles are identical, the Hamiltonians are equal and their eigenfunctions are given by

$$\hat{H}^{(\varkappa)} \varphi_{\varkappa_i}(\varkappa) = E_{\varkappa_i} \varphi_{\varkappa_i}(\varkappa); \tag{IV.94}$$

the eigenfunctions of $\hat{H}$ can then be expressed as products of functions $\varphi_{\varkappa_i}$, or

$$\psi_i(1, 2, \ldots, N) = \prod_{\varkappa_i} \varphi_{\varkappa_i}(\varkappa) = \varphi_{\alpha_i}(1)\, \varphi_{\beta_i}(2) \ldots \varphi_{\varkappa_i}(\varkappa) \ldots \varphi_{\nu_i}(N). \tag{IV.95}$$

In order not to complicate the proceedings needlessly, we shall consider Maxwell–Boltzmann statistics, that is, we shall not take into account the indistinguishability of the particles; it does not play an essential role in the discussions which follow.

Having said this, let us consider the quantum microscopic observables of the "sum" type as Khinchin did in classical theory.

These observables are of the form

$$\hat{A} = \sum_{\varkappa=1}^{N} \hat{A}^{(\varkappa)}, \tag{IV.96}$$

where the operator $\hat{A}^{(\varkappa)}$ acts only on the coordinates of the $\varkappa$th particle; we get immediately

$$A_{ii} = \sum_{\varkappa=1}^{N} A_{ii}^{(\varkappa)}, \quad A_{ij} = \sum_{\varkappa=1}^{N} A_{ij}^{(\varkappa)}, \tag{IV.97}$$

where i and j denote eigenstates represented by the functions (IV.95), with energies E_i and E_j contained within the energy shell $(E, E + \delta E)$. Thus, since the $\varphi_{\varkappa_i}(\varkappa)$ form an orthonormal system, we have

$$\left.\begin{aligned}
A_{ii}^{(\varkappa)} &= (\varphi_{\alpha_i}(1)\,\varphi_{\beta_i}(2)\,\ldots\,\varphi_{\varkappa_i}(\varkappa)\,\ldots\,\varphi_{\nu_i}(N),\\
&\qquad \hat{A}^{(\varkappa)}\,\varphi_{\alpha_i}(1)\,\varphi_{\beta_i}(2)\,\ldots\,\varphi_{\varkappa_i}(\varkappa)\,\ldots\,\varphi_{\nu_i}(N))\\
&= (\varphi_{\varkappa_i}, \hat{A}^{(\varkappa)}\,\varphi_{\varkappa_i}) = A_{\varkappa_i\varkappa_i}^{(\varkappa)},\\
A_{ij}^{(\varkappa)} &= (\varphi_{\alpha_i}(1)\,\varphi_{\beta_i}(2)\,\ldots\,\varphi_{\varkappa_i}(\varkappa)\,\ldots\,\varphi_{\nu_i}(N),\\
&\qquad \hat{A}^{(\varkappa)}\varphi_{\alpha_j}(1)\,\varphi_{\beta_j}(2)\,\ldots\,\varphi_{\varkappa_j}(\varkappa)\,\ldots\,\varphi_{\nu_j}(N))\\
&= (\varphi_{\varkappa_i}, \hat{A}^{(\varkappa)}\varphi_{\varkappa_j})\,\delta_{\alpha_i\alpha_j}\,\ldots\,\delta_{\varkappa_i-1,\varkappa_j-1}\delta_{\varkappa_i+1,\varkappa_j+1}\,\ldots\,\delta_{\nu_i\nu_j}\\
&= A_{\varkappa_i\varkappa_j}^{(\varkappa)}\delta_{\alpha_i\alpha_j}\,\ldots\,\delta_{\varkappa_i-1,\varkappa_j-1}\delta_{\varkappa_i+1,\varkappa_j+1}\,\ldots\,\delta_{\nu_i\nu_j}.
\end{aligned}\right\} \tag{IV.98}$$

Thus, the off-diagonal terms $A_{ij}^{(\varkappa)}$ are non-vanishing only for states which differ between themselves only in the wave function of the $\varkappa$th particle. We note, in addition, that the eigenstates of a single particle are represented by all values of the spectrum of $\hat{H}^{(\varkappa)}$ below $E + \delta E$, which is the total maximum energy of our system.

In order to be able to evaluate the sums $\sum_i A_{ii}^2$ and $\sum_{i \neq j} |A_{ij}|^2$, we shall make the following assumption: we shall assume that the matrix elements *referring to a single particle* are bounded, i.e. we shall put $|A_{ii}^{(\varkappa)}| \leq A_M^{(1)}$ and $|A_{ij}^{(\varkappa)}| \leq A_M^{(1)}$, whatever the value of i and j. This assumption is physically natural and does not encounter the difficulty which we pointed out in section II of this chapter: in fact, the matrix elements depend neither on the number of particles nor on the number of states contained in the energy shell δE. With this assumption, we have immediately for A_{ii} and for the microcanonical average $\overline{A}^m$:

$$A_{ii} \leqq NA_M^{(1)} = O(N), \tag{IV.99}$$

$$\overline{A}^m = \frac{1}{n}\sum_{i=1}^{n} A_{ii} = \sum_{\varkappa=1}^{N}\left(\frac{1}{n}\sum_{i=1}^{n} A_{ii}^{(\varkappa)}\right) \leqq NA_M^{(1)} = O(N), \tag{IV.100}$$

whence

$$A_{ii}^2 \approx O(N^2), \quad \left(\overline{A}^m\right)^2 \approx O(N^2). \tag{IV.101}$$

If, then, we consider the sum $\sum_{i=1}^{n} A_{ii}^2$, we can certainly write

$$\sum_{i=1}^{n} A_{ii}^2 \leqq N^2 n A_M^{(1)2}, \tag{IV.102}$$

whence

$$\frac{1}{n(n+1)}\sum_{i=1}^{n} A_{ii}^2 \leqq A_M^{(1)2}\frac{N^2}{n+1}. \tag{IV.103}$$

Since, according to (IV.16), n is an exponential function of N of the form $K_N a^N\,\delta E$, it follows that N^2/n rapidly approaches zero when N becomes very large; we can then write

$$\frac{1}{n(n+1)}\sum_{i=1}^{n} A_{ii}^2 = O\left(\frac{N^2}{n}\right) \sim O\left(\frac{1}{n}\right), \tag{IV.104}$$

which justifies *a posteriori* the statement at the end of section II, according to which it was legitimate to neglect the dependence of the A_{ii} on n.

Let us consider now the sum of the off-diagonal terms $|A_{ij}|^2$. According to (IV.97) we have:

$$|A_{ij}|^2 = \sum_{\varkappa=1}^{N} |A_{ij}^{\varkappa}|^2 + \sum_{\varkappa \neq \varkappa'} A_{ij}^{\varkappa} A_{ij}^{(\varkappa')*} \tag{IV.105}$$

and, by virtue of (IV.98) we have

$$\begin{aligned} |A_{ij}^{(\varkappa)}|^2 &= |A_{\varkappa_i\varkappa_j}^{(\varkappa)}|^2\, \delta_{\alpha_i\alpha_j} \ldots \delta_{\varkappa_i-1,\varkappa_j-1}\,\delta_{\varkappa_i+1,\varkappa_j+1} \ldots \delta_{\nu_i\nu_j}, \\ A_{ij}^{(\varkappa)} A_{ij}^{(\varkappa')*} &= A_{\varkappa_i\varkappa_j}^{(\varkappa)} A_{\varkappa'_i\varkappa'_j}^{(\varkappa')*} (\delta_{\alpha_i\alpha_j} \ldots \delta_{\varkappa_i-1,\varkappa_j-1} \times \\ &\quad \times \delta_{\varkappa_i+1,\varkappa_j+1} \ldots \delta_{\varkappa'_i\varkappa'_j} \ldots \delta_{\nu_i\nu_j}) \times \\ &\quad \times (\delta_{\alpha_i\alpha_j} \ldots \delta_{\varkappa_i\varkappa_j} \ldots \delta_{\varkappa'_i-1,\varkappa'_j-1}\,\delta_{\varkappa'_i+1,\varkappa'_j+1} \ldots \delta_{\nu_i\nu_j}) = 0, \quad \varkappa \neq \varkappa'. \end{aligned} \tag{IV.106}$$

(The last equation is due to the fact that we have always $\varkappa \neq \varkappa'$, by assumption and $\varkappa_i \neq \varkappa_j$, $\varkappa'_i \neq \varkappa'_j$ as otherwise one of the δ's would necessarily be zero, since $i \neq j$ for the total state; but $\delta_{\varkappa_i\varkappa_j}$ and $\delta_{\varkappa'_i\varkappa'_j}$ occur in $A_{ij}^{(\varkappa')*}$ and $A_{ij}^{(\varkappa)}$ respectively, whence the vanishing of the product $A_{ij}^{(\varkappa)}A_{ij}^{(\varkappa')*}$.) By putting it in (IV.105), we have

$$|A_{ij}|^2 = \sum_{\varkappa=1}^{N} |A_{ij}^{(\varkappa)}|^2 = \delta_{\alpha_i\alpha_j} \dots \delta_{\varkappa_i-1,\varkappa_j-1}\, \delta_{\varkappa_i+1,\varkappa_j+1} \dots \delta_{\nu_i\nu_j} \sum_{\varkappa=1}^{N} |A_{\varkappa_i\varkappa_j}^{(\varkappa)}|^2. \tag{IV.107}$$

We consider now the sum

$$\sum_{i\neq j} |A_{ij}|^2 = \sum_{\varkappa=1}^{N} \sum_{i\neq j} \delta_{\alpha_i\alpha_j} \dots \delta_{\varkappa_i-1,\varkappa_j-1}\, \delta_{\varkappa_i+1,\varkappa_j+1} \dots \delta_{\nu_i\nu_j} |A_{\varkappa_i\varkappa_j}^{(\varkappa)}|^2, \tag{IV.108}$$

where the indices i and j correspond to all possible states represented by (IV.95). We begin by carrying out the summation over i with j fixed: the indices $\alpha_j \dots \nu_j$ are then completely defined (as well as their order) and the only states i giving non-zero contributions are those for which $\alpha_i = \alpha_j, \dots, \varkappa_i - 1 = \varkappa_j - 1, \varkappa_i + 1 = \varkappa_j + 1, \dots, \nu_i = \nu_j$. These states are then obtained by varying the index $\varkappa_i$, so that the total energy E_i remains inside the shell δE; thus, we have the relation

$$E_i - E_j = E_{\varkappa_i} - E_{\varkappa_j} \leqq \delta E \tag{IV.109}$$

which shows that the difference $E_{\varkappa_i} - E_{\varkappa_j}$ is at most equal to δE. The result is that the number of states i making a non-zero contribution to (IV.108) with i fixed, is equal to the number of states $E_{\varkappa_i}$ contained in an energy shell E', $E' + \delta E$, with $E' \sim E_{\varkappa_j}$. Since this number is given by $n_1 = K_1 E'^{\frac{1}{2}} \delta E$ (see Appendix I, § 4) and we have certainly $E' \leqq E$, n_1 has $K_1 E^{\frac{1}{2}} \delta E$ as upper limit. Evaluating now the summation in (IV.108) over j (from 1 to $S^{(\alpha)}$) and over $\varkappa$ (from 1 to N), we have (always with the assumption $|A_{\varkappa_i\varkappa_j}^{(\varkappa)}| < A_M^{(1)}$):

$$\sum_{i\neq j} |A_{ij}|^2 \leqq N n_1 n A_M^{(1)2}, \tag{IV.110}$$

whence

$$\frac{1}{n(n+1)} \sum_{i\neq j} |A_{ij}|^2 \leqq \frac{N n_1}{n+1} A_M^{(1)2}. \tag{IV.111}$$

Since n is always of order $K_N a^N \,\delta E$, the ratio $Nn_1/(n+1)$ rapidly approaches zero when N becomes very large. We can then write approximately

$$\frac{1}{n(n+1)} \sum_{i \neq j} |A_{ij}|^2 = O\left(\frac{Nn_1}{n+1}\right) \sim O\left(\frac{1}{n}\right), \quad \text{(IV.112)}$$

a relation which is valid when N is sufficiently large.

By putting the asymptotic formulae (IV.104) and (IV.112) in (IV.92), it can be seen that it is possible to write

$$\frac{\overline{\left(A(t) - \overline{A}^m\right)^2}^m}{\left(\overline{A}^m\right)^2} \sim O\left(\frac{1}{n}\right), \quad \text{(IV.113)}$$

for the second moment of the microscopic quantity $A(t) - \overline{A}^m$. This result is proved for a "sum" observable of the form (IV.96); according to (IV.91), we then obtain a probability ergodic theorem for the observable $\hat{A}$. Comparison of formulae (IV.89) and (IV.113) on the one hand, and (IV.90) and (IV.91) on the other hand, each obtained for "sum" functions, emphasises the straight analogy between the quantum and classical forms of the ergodic theory. In this way, the simple quantum model which we have just analysed enables us to establish an even more accurate agreement between the two theories, by showing that it is possible to break away from the use of macroscopic observables, at least in certain cases.

In conclusion, we have been able to develop the quantum ergodic theory in a manner similar to classical theory. This has been made possible by the important distinction between the probabilities of quantum mechanics and the probabilities in the classical sense which are used to define, in quantum statistics, ensembles of systems: this difference is apparent in the definition of the matrices $\hat{\varrho}$ and $\hat{\hat{\varrho}}$ of Chapter II; the matrix $\hat{\hat{\varrho}}$, which satisfies the quantum analogy of Liouville's theorem, thus plays the role of the phase probability density of classical mechanics.

It is important to emphasise that, even if we have justified the fundamental principles of statistical mechanics by proving the probability ergodic theorems, we have not been able to prove a true ergodic theorem establishing the equality of the time-averages and the microcanonical averages for *a single system*, without recourse to an assumption concerning the nature of the system. In this connection, the differences which we have encountered in Chapter III between the classical and quantum forms of the theory are due mainly, as we have mentioned already, to the special properties of quantum observables which are not measurable simultaneously and whose quantum averages are time-dependent only through the phase factors (stationary states). In the second ergodic theorem, this difficulty has led us to introduce macroscopic operators and to be satisfied with a convergence in probability. Thus, not only has it been impossible for us to base quantum statistical mechanics on a purely dynamic basis, but there is still no rigorous proof of the quantum parallel of Hopf's theorem; only the papers by Ludwig (1958a, b) already mentioned perhaps open up a path of research in this field.

Thus, we have been led quite naturally to accept the point of view of statistical ensembles of systems and to prove in the present chapter quantum probability ergodic theorems similar to the corresponding classical theorem (I.52). The result is that if we consider a microcanonical ensemble of virtual systems identical with the observed system, the time-average and the microcanonical average of a physical quantity are almost equal for all systems of the ensemble, except for an ensemble of systems (associated with a set of exceptional initial conditions) whose measure is negligible when the number of degrees of freedom of the system is sufficiently large. Even if such a statement were sufficient for the basis of statistical mechanics, nevertheless there exists the possibility of a macroscopic system being found in exceptional states which do not satisfy the properties of ergodicity, but whose observation is highly improbable; we recall, moreover, that the classical ergodic theorems of Birkhoff and von Neumann also involve the possibility of exceptional trajectories whose total measure was zero. Finally, the justification of the fundamental hypothesis of statistical mechanics appears, in the final analysis, as an application of the law of large numbers and of the geometrical properties of certain hypersurfaces in $2n$-dimensional space.

The foregoing results can be extended in various directions; we point out especially the recent paper by Prosperi and Scotti (1960; see also Chapter VI, section III), who, by replacing the microcanonical averages by averages over all states belonging to a single quantum cell, obtain in this way new conditions of ergodicity: we shall return to this point in the course of studying the generalised H-theorem. We point out also that Albertoni, Bocchieri, and Loinger (1960) have been able to prove a classical theorem which justifies the methods of classical statistical mechanics, without any condition of ergodicity, for systems possessing a large number of degrees of freedom; it is the classical analogue of the quantum probability ergodic theorems and a generalisation of Khinchin's results (valid only for "sum" phase functions).

Finally, we could consider as an additional difficulty of the quantum ergodic theorem, the necessity for calling on macroscopic observables: this theory would then have a "subjective" nature connected with the essential role of the macroscopic observer. We can object to this that we have proved a probability ergodic theorem, valid for microscopic observables of the "sum" type, which in this case eliminates the difficulties relating to the interference of the macroscopic observer. Moreover, it is probable that this type of proof could be extended to more general microscopic observables (for example, of the form $\hat{A} = \sum_{\varkappa} \hat{A}^{(\varkappa)} + \sum_{\varkappa,\lambda} \hat{A}^{(\varkappa,\lambda)}$), provided that we take into account explicitly the macroscopic nature of the system studied. Thus, the introduction of macroscopic observables would be, above all, a convenient way of expressing formally the special properties of quantum observables associated with a system having a large number of degrees of freedom. Be that as it may, the similarity between the classical and quantum formulation of the ergodic theory shows that classical and quantum statistical mechanics rest on the same fundamental principles, whose interpretation is connected with that of the classical concept of probability in physics.†

V. Relations between Microcanonical, Canonical and Grand-Canonical Ensembles

The proof of the ergodic theorems allows us to justify the use in statistical mechanics of the microcanonical ensemble: it is applicable to isolated systems whose energy is known either precisely or within a range δE. However, it is interesting from the point of view of physics, and particularly of thermodynamics, to study the

† In this connection, see the study by Bohm and Schützer (1955) and the epistemological papers of Costa de Beauregard (1958, 1960).

distribution of systems in thermal contact with their surroundings: it is known that this situation is properly represented by the canonical distribution. We shall conclude this chapter by showing how we can deduce it from a microcanonical distribution without assumptions. We shall carry out this study for quantum systems and we shall proceed according to a method which is similar in all points to that followed by Khinchin (1949; see also Rosenfeld, 1952, 1955) in classical theory; in addition, we shall see that this can be applied also to the derivation of the grand-canonical ensemble: in this way, we can justify the use of canonical and grand-canonical distributions starting from the results of the ergodic theory.

1. *Average values in coupled systems*

We consider a system s, which is in thermal equilibrium with the surroundings or with a thermostat T, of which we suppose that the heat capacity is so large that exchanges of heat with s do not change its temperature. (The number of degrees of freedom of s is thus very small in comparison with that of T.) The two systems s and T, by free exchange of energy, are at the same temperature (that of the thermostat) and form a total system S which can be considered as isolated: we shall suppose that the system S is described by a microcanonical ensemble; our problem then is to calculate the statistical distribution of s, starting from that of S.

We shall assume also that the magnitude of the interaction between S and T is negligible, so that the total energy E_S is the sum of the energies E_s and E_T of the systems s and T. Since we are interested in the system s, we shall calculate the quantities referring to this system, i.e. the statistical averages of the observables of the system s taken over the microcanonical ensemble corresponding to S. In order to put the problem mathematically, we consider two quantum systems (1) and (2) described by the Hamiltonians $\hat{H}^{(1)}$ and $\hat{H}^{(2)}$ and by the wave functions $\Psi^{(1)}$ and $\Psi^{(2)}$. The Hamiltonian of the total system is described by

$$\hat{H} = \hat{H}^{(1)} \times \hat{I}^{(2)} + \hat{H}^{(2)} \times \hat{I}^{(1)}, \qquad \text{(IV.114)}$$

where $\hat{I}^{(\alpha)}$ represents the unit operator in Hilbert space of the system α and where the multiplication signs denote the direct products of operators which operate in different Hilbert spaces.

The total wave function, which can be written as $\Psi^{(1)} \times \Psi^{(2)}$, is a vector in the Hilbert space which is the direct product of the two Hilbert spaces corresponding to systems (1) and (2). The eigen-energies of the total system are $E_i^{(1)} + E_j^{(2)}$, with the projection operators $P_{\psi_i(1)} \times P_{\psi_j(2)}$; in particular, we can associate the projection operator

$$\hat{U}(\lambda) = \sum_{E_i+E_j \leqq \lambda} \hat{P}_{\psi_i(1)} \times \hat{P}_{\psi_j(2)}. \qquad \text{(IV.115)}$$

with the total energy. We deduce from this the fundamental relation

$$d\hat{U}(\lambda) = \int_{-\infty}^{+\infty} d\hat{U}^{(1)}(\lambda')\, d\hat{U}^{(2)}(\lambda - \lambda'), \qquad \text{(IV.116)}$$

where the integration is carried out over λ'.

By applying this relation to the system s and to the thermostat T, when the interaction energy is weak we have

$$d\hat{U}(\lambda) = \int_{-\infty}^{+\infty} d\hat{U}_s(\lambda')\, d\hat{U}_T(\lambda - \lambda'). \qquad \text{(IV.117)}$$

Since the state of the total system is described by a microcanonical ensemble, the corresponding density matrix $\hat{\varrho}$ will be equal to the projection operator over the energy range in which the total system exists. Thus, by denoting this operator by $\Delta\hat{U}(\lambda)$, we have

$$\hat{\varrho} = \frac{\Delta\hat{U}(\lambda)}{\mathrm{Tr}\,\Delta\hat{U}(\lambda)}, \qquad \text{(IV.118)}$$

whence, by (IV.117),

$$\frac{\Delta\hat{U}(\lambda)}{\mathrm{Tr}\,\Delta\hat{U}(\lambda)} = \int_{-\infty}^{+\infty} d\hat{U}_s(\lambda') \frac{\Delta\hat{U}_T(\lambda - \lambda')}{\mathrm{Tr}\,\Delta\hat{U}(\lambda)}. \qquad \text{(IV.119)}$$

We can then find an expression for the average value of an observable $\hat{A}^{(s)}$ of the system s; actually, the operator

$$\hat{A} = \hat{A}^{(s)} \times \hat{I}^{(T)} \qquad \text{(IV.120)}$$

corresponds to the operator $\hat{A}^{(s)}$ in Hilbert space of the total system; its mean value is given, according to (IV.118) and (IV.119) by

$$\bar{A}^{(s)} = \mathrm{Tr}(\hat{\varrho}\hat{A}) = \int_{-\infty}^{+\infty} \mathrm{Tr}[\hat{A}^{(s)} d\hat{U}_s(\lambda')] \frac{\mathrm{Tr}\Delta\hat{U}_T(\lambda - \lambda')}{\mathrm{Tr}\Delta\hat{U}(\lambda)}. \qquad \text{(IV.121)}$$

As the integration is over λ', we see that the average values of the observables $\hat{A}^{(s)}$ are calculated with a distribution which depends only on the variable λ' of s and which is defined by (IV.121). We must evaluate therefore the expression $\dfrac{\operatorname{Tr} \Delta \hat{U}_T(\lambda - \lambda')}{\operatorname{Tr} \Delta \hat{U}(\lambda)}$ which defines the effect of the thermostatic contact on the distribution of the system s. We shall calculate now an asymptotic value for the preceding expression using the properties of the thermostat.

2. *Properties of the thermostat*

The thermostat only needs satisfy the condition of being a system with a very large number of degrees of freedom (this number is much higher than that of the observed system s); therefore, we can suppose that it comprises a very large number of parts $T_1, T_2, \ldots, T_i, \ldots, T_n$, the interactions between which can be neglected [system with weakly coupled components†]; by using formula (IV.116) we can then write:

$$d\hat{U}_T(\Lambda) = \int_{\Lambda_1} \cdots \int_{\Lambda_{n-1}} \left\{ \prod_{i=1}^{n-1} d\hat{U}_{T_i}(\Lambda_i) \right\} d\hat{U}_{T_n}\left(\Lambda - \sum_{i=1}^{n-1} \Lambda_i \right), \qquad \text{(IV.122)}$$

whence we obtain

$$\operatorname{Tr}[d\hat{U}_T(\Lambda)] = \int_{\Lambda_1} \cdots \int_{\Lambda_{n-1}} \left\{ \prod_{i=1}^{n-1} \operatorname{Tr}[d\hat{U}_{T_i}(\Lambda_i)] \right\} \operatorname{Tr}\left[d\hat{U}_{T_n}\left(\Lambda - \sum_{i=1}^{n-1} \Lambda_i \right) \right]. \qquad \text{(IV.123)}$$

These formulae are the quantum parallels of the law of compounding invariant measures of energy hypersurfaces in classical mechanics (Khinchin, 1949, pp. 41 and 81; see also Appendix I, § 4). The formal analogy of (IV.123) with the law of compounding of probabilities suggests that, with Khinchin, we use the central limit theorem for calculating the probabilities, in order to obtain an asymptotic expression for (IV.123):

If we consider n stochastic variables x_i with independent distributions $u_i(x_i)\, dx_i$, the distribution of the sum $x = \sum_{i=1}^{n} x_i$

† For the definition of such systems, see Appendix I, § 4.

obeys the law

$$u(x) = \int \dots \int \left\{ \prod_{i=1}^{n-1} u_i(x_i) dx_i \right\} u_n \left(x - \sum_{i=1}^{n-1} x_i \right). \qquad \text{(IV.124)}$$

If the distributions $u_i(x_i)$ fulfil certain conditions (the most important of which is that the dispersion of the variables x_i is finite), the distribution law for x has the asymptotic form

$$u(x)\, dx = \frac{1}{\sqrt{2\pi B}} \exp\left[-\frac{(x - \bar{x})^2}{2B} \right] dx, \qquad \text{(IV.125)}$$

where B is the sum of the dispersions of the variables x_i. Given in the form of (IV.123), we shall find an asymptotic formula for Tr $[d\hat{U}_T(\Lambda)]$ by using the previous theorem.

Before we can use this theorem, we assign to each Tr $[d\hat{U}_{T_i}(\Lambda_i)]$ a normalised probability distribution:

$$du^{\beta}_{T_i}(\Lambda_i) = \frac{e^{-\beta\Lambda_i}\, \mathrm{Tr}[d\hat{U}_{T_i}(\Lambda_i)]}{\Phi_{T_i}(\beta)} \qquad \text{(IV.126)}$$

which satisfies a law of compounding similar to (IV.123). For the normalisation function $\Phi_{T_i}(\beta)$, we have:

$$\Phi_{T_i}(\beta) = \int_{-\infty}^{+\infty} e^{-\beta\Lambda_i} \mathrm{Tr}[d\hat{U}_{T_i}(\Lambda_i)] = \mathrm{Tr}[\exp(-\beta \hat{H}_{T_i})], \qquad \text{(IV.127)}$$

so that the asymptotic law for the total system can be written as

$$du^{\beta}_T(\Lambda) = \frac{1}{\sqrt{2\pi B_T}} \exp\left[-\frac{(\Lambda - \bar{E}_T)^2}{2B_T} \right] d\Lambda. \qquad \text{(IV.128)}$$

where $\bar{E}_T$ represents the average value of the energy in this distribution and B_T represents the sum of the energy fluctuations of the various components. We obtain from (IV.128) the asymptotic law for Tr $[d\hat{U}_T(\Lambda)]$:

$$\mathrm{Tr}[d\hat{U}_T(\Lambda)] = \frac{\Phi_T(\beta)}{\sqrt{2\pi B_T}} \exp\left[\beta\Lambda - \frac{(\Lambda - \bar{E}_T)^2}{2B_T} \right] d\Lambda, \qquad \text{(IV.129)}$$

where $\Phi_T(\beta)$ is the product $\prod_{i=1}^{n} \Phi_{T_i}(\beta)$.

3. *The canonical and grand-canonical ensembles*

Let us apply this result to the argument $\lambda - \lambda' = \Lambda$; by observing that

$$(\Lambda - \bar{E}_T)^2 = (\lambda' - \bar{E}_s)^2$$

and that the fluctuations of the system studied are generally several orders of magnitude smaller than those of the thermostat whose average is B_T, we have

$$\mathrm{Tr}[d\hat{U}_T(\lambda - \lambda')] \simeq \frac{\Phi_T(\beta)}{\sqrt{2\pi B_T}}\, e^{\beta(\lambda-\lambda')} d\Lambda. \qquad \text{(IV.130)}$$

Likewise, for the total system we shall have

$$\mathrm{Tr}[d\hat{U}(\lambda)] \simeq \frac{\Phi_S(\beta)}{\sqrt{2\pi B_S}}\, e^{\beta\lambda} d\Lambda. \qquad \text{(IV.131)}$$

On the other hand, we have $\Phi_S(\beta) = \Phi_T(\beta)\,\Phi_s(\beta)$ (by definition and because the interaction between s and T is negligible) and $B_S \approx B_T$, since B_S differs from B_T only by the contribution of the small system s, which can be neglected. Thus, we have (by considering the interval of energy $\Delta\lambda$ as physically very small):

$$\frac{\mathrm{Tr}[\Delta\hat{U}_T(\lambda - \lambda')]}{\mathrm{Tr}[\Delta\hat{U}(\lambda)]} \simeq \frac{1}{\Phi_s(\beta)}\, e^{-\beta\lambda'}, \qquad \text{(IV.132)}$$

with the result that the distribution defined by (IV.121) can be written as

$$\hat{\bar{\varrho}}_s = \frac{1}{\Phi_s(\beta)} \int_{-\infty}^{+\infty} e^{-\beta\lambda'}\, d\hat{U}_s(\lambda') = \frac{1}{\Phi_s(\beta)}\, e^{-\beta\hat{H}_s} = \frac{e^{-\beta\hat{H}_s}}{\mathrm{Tr}(e^{-\beta\hat{H}_s})} \qquad \text{(IV.133)}$$

[by using formula (IV.127)].

This is the canonical distribution for the system s: the effect of the thermostat is found to be concentrated in the parameter β. Thus, we see that a canonical distribution can be derived which corresponds to a system in thermal equilibrium with a thermostat, by using the microcanonical ensemble for an isolated system.

In this way, we see that the whole of statistical mechanics rests, in the ultimate analysis, on the study of isolated systems and on the justification of the replacement of observed quantities concerning these systems by the averages taken over the microcanonical ensemble: this is the problem dealt with by the ergodic theory.

We can even interpret the foregoing result by saying that in statistical equilibrium the state of some small part of an isolated macroscopic system (whose interaction with the remainder of the system is weak) is properly represented by the canonical ensemble defined by formula (IV.133). This property is satisfied by any subsystem (even microscopic) provided that it is small compared with the total system; in particular, we can apply it to the case of an atom in a perfect gas:† we obtain in this way the Maxwell–Boltzmann distribution, which shows us that the analogy between this and the canonical distribution is not fortuitous.

On the other hand, it is easy to connect the distribution parameters with the observed macroscopic quantities; in fact, we have for the mean value of the energy

$$\overline{E}_s = \frac{\mathrm{Tr}\left(\hat{H}^s\, e^{-\beta \hat{H}^s}\right)}{\mathrm{Tr}(e^{-\beta \hat{H}^s})} = -\frac{d \ln \Phi_s(\beta)}{d\beta}, \qquad \text{(IV.134)}$$

where $\Phi_s(\beta)$ is *the partition function* corresponding to the canonical ensemble, starting from which we can calculate the thermodynamic quantities. Formula (IV.134) then allows us to establish:

(a) the principle of equipartition of energy;

(b) the relation $\beta = 1/kT$, where T is the absolute temperature [it is sufficient to apply (IV.134) to a monatomic perfect gas].

By comparing now (IV.133) and (II.33), we can see that the parameters of the canonical distribution (II.33) can be written as

$$\left.\begin{aligned} \theta &= \frac{1}{\beta} = kT, \\ e^{-\beta \Psi} &= \mathrm{Tr}\left(e^{-\beta \hat{H}^s}\right) = \Phi_s(\beta). \end{aligned}\right\} \qquad \text{(IV.135)}$$

Thermodynamical considerations lead, in addition, to identifying Ψ with the (Helmholtz) free energy F of the system. Without further emphasis on these applications, we can see that the canonical ensemble plays a principal role in the thermodynamic applications of statistical mechanics by enabling us to define the concept of thermal equilibrium; this is why it was fundamental to establish

† Cf. Appendix I, § 4.

the relationship between this ensemble and the microcanonical ensemble.

The same considerations can be developed for the grand-canonical ensemble. Actually, the proof of the canonical distribution rests on the use of the central limit theorem (IV.125): the use of this theorem is legitimate in the case where we are looking for the statistical distribution between different parts of a system, for a quantity which satisfies a conservation law and a law of additivity; this is the case with the energy according to formula (IV.114). It is the same for numbers of particles of different kinds which we introduce into the definition of grand ensembles; thus, we can apply the previous reasoning to a small part of a large system composed of h species of particles: the large system plays here, at the same time, the role of a thermostat and of a "particle reservoir" for the small sub-system considered. The law of distribution for the numbers of particles of each kind in the sub-system thus has a form comparable with the canonical law for energy and new macroscopic parameters μ_i make their appearance, which are related to the average numbers of particles $\bar{n}_i$ in the same way as the modulus of the canonical distribution is related to the average energy; thus, a system composed of a variable number of particles in statistical equilibrium with its surroundings is described adequately by a grand-canonical ensemble of the form (II.35).

As for the canonical ensemble, we can express the parameters Ω, θ and μ_i as functions of the average values of the energy and of the numbers of particles of the system considered. For this purpose, we calculate the following average values:

$$\left.\begin{aligned} \bar{E} &= \sum_{n_1 \ldots n_h, n} E_n(n_i) \exp \frac{\Omega + \sum_i \mu_i n_i - E_n(n_i)}{\theta} \\ \bar{n}_i &= \sum_{n_1 \ldots n_h, n} n_i \exp \frac{\Omega + \sum_i \mu_i n_i - E_n(n_i)}{\theta} \qquad (i = 1, 2, \ldots, h), \end{aligned}\right\} \quad \text{(IV.136)}$$

to which we add the relation

$$e^{-\Omega/\theta} = \sum_{n_1 \ldots n_h, n} \exp \frac{\sum_i \mu_i n_i - E_n(n_i)}{\theta} \quad \text{(IV.137)}$$

obtained by starting from (II.36′). We then show that, just as for the canonical ensemble, we have $\theta = kT$ and that the parameters μ_i

must be identified with the chemical potentials introduced by Gibbs.

We shall end this discussion by mentioning that in the case of systems with a very large number of degrees of freedom, we arrive at the same results for the average values of the sum functions whether we calculate them in a microcanonical ensemble or in canonical or grand-canonical ensembles. More precisely, we can say that the average values calculated in a microcanonical ensemble corresponding to a given energy E_0 are equal to those calculated in a canonical ensemble corresponding to an average energy $\bar{E} = E_0$; however, this is valid only if the quantities whose averages we are taking can be put into the form of sum functions in the sense of Khinchin† (see Chapter I). This result allows us to simplify the calculation of mean values, since we are dealing now with independent variables which are being restricted no longer by constraints such as the fixing of the energy or of the number of particles. Thus, we can justify the use in statistical mechanics of canonical and grand-canonical ensembles by understanding it as a purely mathematical procedure, suitable for simplifying the calculation of the average values of the sum functions: it is possible also to develop statistical mechanics by taking the canonical or grand-canonical distributions as a starting assumption without forgetting, however, that these are based on the microcanonical distribution.

Concerning the physical significance of these ensembles, it is made more precise by the analysis of the present section, according to which the canonical and grand-canonical ensembles representing a system in statistical equilibrium with its surroundings, are deduced from the microcanonical distribution associated with the total system (observed system + surroundings) considered as isolated. As the justification for the microcanonical ensemble itself depends on the existence of an ergodic theorem, we see that the ergodic theorem in its precise or "probabilistic" form actually contributes the indispensable foundation of statistical mechanics. Thus, the objections to the ergodic theorem based on a criticism of the concept of an isolated system appear to be without foundation; however, if such a concept is frequently an abstraction, from the point of view of physics, because of the small but numerous and

† If this condition were not fulfilled, we would risk introducing serious errors; thus, for example, the energy dispersion is zero for the microcanonical ensemble but has a positive value in the canonical ensemble.

almost uncontrollable interactions of a given system with its surroundings, it is conceivable nevertheless that the changes in energy of a macroscopic system with its surroundings are made sufficiently small to be able to neglect them on our scale. Moreover, if we accept that it is preferable, according to the degree of isolation of the system,† to represent it by a canonical ensemble, nothing remains except that the use of such representative ensembles rests on the fundamental postulate of statistical mechanics about *a priori* probabilities; this postulate relies necessarily for justification on either the ergodic theorem—at least in its approximate form—or on the H-theorems and on the kinetic method which we shall be studying in the second part. (We shall see, nevertheless, that the H-theorem in its generalised form can be proved strictly only by recourse to the results of Hopf's ergodic theorem; see Chapter V, § II.4 and Chapter VI, § III.2).

4. *System–thermostat coupling and irreversibility*

To conclude, we shall point out that the concept of interaction of a system with its surroundings can be used with a view to placing on a wider basis the problem of proving the foundations of statistical mechanics. In this class of ideas, we can visualise two research methods, according to the manner in which the coupling between the observed system and its environment is described. In the first method, the system being studied is considered as a small part of a larger system whose total Hamiltonian comprises three terms: the Hamiltonian of the observed system, the Hamiltonian of the "surroundings" forming the thermostat, and a generally very small interaction potential. We then examine under what conditions the time average of an observable of the system being studied can be equal to its phase average. We must classify under this line of research the work of Klein (1952) who finds the ergodicity conditions for the expansion coefficients of the wave vector $\Psi(t)$ of the total system. Although these conditions are in fact almost identical to those we encountered when studying the

† See, concerning this subject, the analysis of R. C. Tolman (1938), Chapter XII. We point out also that some authors, for instance Landau and Lifshitz, consider that the inevitable interaction between a quantum system and the measurement equipment being used by the observer could play an essential role in the irreversible evolution towards equilibrium. We shall postpone the discussion of this problem until Chapter VII.

earlier quantum ergodic theorem (cf. Chapter III, § II), they must be interpreted differently here, since they refer now not to the system itself but to its interaction with the surroundings; they remain, nevertheless, completely formal and come up against the same objections as in the case of an isolated system. Ekstein's attempt (1957) also belongs to this same method of research; the thermostat is considered as if formed from N identical systems interacting with the system being studied through a potential of the form λV. Ekstein then shows that the time-average and the phase average of an observable in the observed system are equal for "almost all" interactions λV, when N becomes very large and the coupling constant becomes very small; "almost all" denotes here that the averages are taken with a uniform probability distribution, as we have done in the present chapter for isolated systems. This method is therefore similar to that which leads to the probability ergodic theorems and, in consequence, suffers from the same disadvantages, namely, the incompatibility between a uniform probability distribution and the possible existence of primary integrals other than energy (see Chapter I, § III.4 and Chapter III, § IV.4). We shall remember, however, from this attempt by Ekstein the essential role played once more by transition to the limit as $N \to \infty$. We have already emphasised several times, especially in connection with the theorem of Golden and Longuet-Higgins (cf. Chapter III, § IV.4), the importance of this limiting process which does not commute with the transition to the limit as $T \to \infty$. It appears as though the systematic application of this limiting process to the situation envisaged here would open an interesting way of research, as seems likely from certain calculations on which we are engaged at present.

Another route consists in describing the interaction of the system with its environment by a perturbation of a statistical nature which is responsible for the irreversible evolution of the system. This line of research, which has been developed by various authors,† leads to the introduction of a stochastic term in the Liouville equation which replaces the usual perturbation potential λV. This term has the effect of producing sudden changes in the tra-

† From amongst the papers relating to the irreversible evolution of a system towards a stationary state, we cite those by Bergmann and Lebowitz (1955) and Lebowitz and Bergmann (1957). For the case of systems in equilibrium, we cite papers by Blatt (1959) and Mayer (1961).

jectory followed by the unperturbed system, and thus giving rise to a "scintillating" motion of the representative point P_t of the system which describes a discontinuous trajectory in phase space. In order to obtain the desired evolution, it is necessary to impose on this random perturbation certain supplementary conditions; in particular, it must be such that the point P_t finally passes through all the "accessible" points of the phase space of the system. By virtue of this latter condition, suggested earlier by Maxwell (1879; see also Truesdell, 1961), the integrals of motion of the unperturbed system are destroyed by the presence of the stochastic perturbation. Although the irreversibility of the evolution thus appears to be well established, we must not forget however that it is actually introduced *a priori* into the description of the interaction between the observed system and its surroundings, which is expressed as a "Zitterbewegung". Finally, the supplementary conditions which it is necessary to impose on this interaction contribute to reducing further the interest in this method of research: they are of an essentially statistical nature and completely comparable with the hypothesis of molecular chaos introduced right at the first appearance of the kinetic theory of gases; as we shall have occasion to see in later chapters, this kind of hypothesis is, in fact, just as restrictive as the ergodic hypothesis.

PART II

H-Theorems

Introduction

In the first part we have studied how the classical and quantum ergodic theorems enabled us to answer the fundamental question of statistical mechanics by identifying the limit of the time-averages of mechanical quantities with the statistical averages taken over the microcanonical ensemble. According to the properties of microcanonical ensembles (stationary ensembles describing an isolated system), we can assume that the ergodic theory, although justifying the methods of statistical mechanics for macroscopic systems, shows that an isolated system tends naturally after quite a long time (measured in terms of the microscopic evolution) towards statistical equilibrium states. This is particularly obvious in the case of Hopf's theorem since we determine by this theorem the conditions under which some initial distribution approaches a limit distribution. We have mentioned already, however, that in this connection we could not draw precise conclusions from the ergodic theory concerning the manner and rapidity with which the equilibrium state was achieved, without integrating the equations of motion; now, this is precisely what statistical mechanics aims to avoid.

In fact, by depending on time-averages over very long periods and on stationary ensembles, the ergodic theory enables us above all to establish the principles of statistical mechanics for systems in equilibrium; with regard to the study of the irreversible evolution of non-equilibrium systems, we must turn to other more suitable methods which we are going to discuss in the second part of this book. These must achieve a dual purpose; on the one hand, to take account of the irreversibility of evolution, which forms the objective of the H-theorems, and, on the other hand, to enable a quantitative study of irreversible events and of their tendency towards equilibrium, which is described by kinetic equations.

Historically, the first method used with success was that of the *kinetic theory of gases*, based on the concept of the velocity distribution function, which led to Boltzmann's equation relating the macroscopic evolution of a gas with its microscopic (atomistic) structure. We know that this equation depends on the dynamic study of binary collisions between atoms and on the postulate of *molecular chaos*, whence we deduce the irreversibility of evolution with the aid of Boltzmann's H-theorem. But the conclusions of this theorem appear to be contradicted by the objections of Loschmidt and Zermelo based on the dynamic properties of a mechanical system which led Boltzmann afterwards to draw attention to the statistical significance of the H-theorem; this significance still had to be examined thoroughly by the work of P. and T. Ehrenfest. Finally, Boltzmann's equation and the H-theorem are based on the postulate of molecular chaos, whose statistical nature is well-established but whose real content is far from being completely elucidated; the fundamental problem of statistical mechanics of non-equilibrium systems is thus to explain the necessary assumptions for establishing an H-theorem and for deducing the kinetic equations which generalise Boltzmann's equation. If we adopt the point of view of Gibbs' ensembles, this problem becomes one of analysing under which conditions a statistical ensemble of systems represented by a non-steady-state distribution tends towards a steady-state ensemble. We can then prove that the fine-grained densities of the first part are not suitable for a proof of H-theorems, since they obey Liouville's equation (or its quantum parallel) which is reversible like all the equations of classical or quantum mechanics. Thus, the important question is posed immediately of *the compatibility of the dynamic properties of a mechanical system and of its irreversible evolution from the macroscopic point of view.*

A first answer to this question is provided by the introduction of coarse-grained statistical densities in the sense of Ehrenfest, whose definition in classical mechanics is based on the division of Γ-space into cells of finite volume and, in wave mechanics on the quantum "cells" associated with the macroscopic operators of Chapter II. With the help of these definitions we can then establish completely analogous classical and quantal generalised $\overline{H}$-theorems, valid for ensembles of systems, by relying precisely upon the difference in evolution between the fine-grained densities ϱ and the coarse-grained densities P. We can see at once that the quantum proof of the $\bar{H}$-theorem includes, with Klein's lemma, an extra element of irreversibility inherent to the quantal density matrices $\hat{\varrho}$. We shall show also that the introduction of macroscopic "cells" in quantum theory enables us to avoid the difficulties concerned with the evolution of an isolated system towards a state of equilibrium (by breaking away from perturbation potentials external to the system).

Unfortunately, the generalised $\bar{H}$-theorem does not constitute a strict proof of the tendency of a system towards an equilibrium state, but it contributes only to the qualitative arguments which make such an evolution probable. Such an evolution can be determined effectively only by integrating the equations of motion and hence by looking for solutions of Liouville's equation; actually, in this model, *the probability is involved only in the initial distribution* $P(0)$, *the evolution of the distribution being governed by the deterministic equations of mechanics*. In order to get away from this difficulty, we can try to reject the mechanical description of Liouville's equation in order to make use of models inspired by the theory of stochastic processes.

Accepting that only binary collisions determine the mechanism of evolution, we can calculate first of all the probabilities for transition between the various cells of Γ-space, relying on the fundamental assumption of statistical mechanics, which is equivalent—in the case of a uniform gas and for a given instant—to the hypothesis of molecular chaos (Stosszahlansatz). If we consider a single system, we can then express the most probable value for the time derivative of H and thus obtain an H-theorem which gives rise to the well-known discussions of Ehrenfest concerning the statistical nature of this theorem. In doing this, we shall pay particular attention to examining thoroughly the fundamental assumptions which are essential to the kinetic theory and to establishing Boltzmann's equation; we shall see, in particular, that to use the transition probabilities calculated with the help of the fundamental principle of statistical mechanics is strictly valid only during a *quite short* interval of time after the instant of macroscopic observation which determines the initial statistical ensemble. It follows that if we wish to apply the results obtained previously to study the evolution of the system over a long period (for example, for a calculation of relaxation times) and to derive Boltzmann's equation, we must assume that the hypothesis of molecular chaos is valid at *every instant* of evolution. Compared with the fundamental assumption of equiprobability of statistical mechanics, this latter assumption is naturally much more restrictive, since it introduces the concept of probability into the evolution mechanism itself.

This is why we are led, with a view to analysing the real significance of this hypothesis, to attempt to describe the evolution of a uniform gas by a Markovian stochastic process in Γ-space. By choosing for the transition probabilities those provided by the "Stosszahlansatz", we find that the distribution functions of Γ-space must obey a linear equation called the "Master Equation", from which we can derive a non-linear equation of the Boltzmann type, by means of appropriate assumptions. Moreover, we can show that these assumptions can be incorporated—at least in certain cases—in the initial conditions and that they are not incompatible with the "Master Equation"; we emphasise, in addition, that in all these arguments the large number of degrees of freedom of the equation plays an essential role, as in the ergodic theory. The real significance of the Clausius–Boltzmann assumption of molecular chaos is thus explained, but the result is applicable only to a uniform gas; unfortunately, there is nothing at present similar in the general case of a non-uniform state. In conclusion, we see that even if the generalised $\overline{H}$-theorem provides a theoretical framework which is adequate for describing the irreversible evolution of a macroscopic physical system, it has been possible to obtain quantitative results only by introducing *a random element into the evolution of the system itself*, an element which can be reduced in certain cases to a Markovian stochastic process in Γ-space

However, if the introduction of the necessary assumptions for the kinetic method, or of carefully chosen stochastic processes enables us to describe the irreversible behaviour of a non-equilibrium system, it is essential nevertheless that this description remains compatible with the subsequent dynamic evolution expressed by Liouville's equation; this is the sole condition to which the fundamental problem of the statistical mechanics of evolutive systems will be reduced. This is why we review the various researches undertaken on Liouville's equation itself and on its approximate integration, discussing more particularly Brout and Bogolyubov's methods, which appear to open the most interesting prospects. Bogolyubov's method, especially, draws attention to the existence of various time scales in the evolution of a gas; these enable us to take into account, to a certain extent, the mechanism of macroscopic irreversibility and we can associate with them different levels of observation of the state of a system: in this way, the role played in statistical mechanics by certain properties of macroscopic observations, which we have already analysed in the general introduction, can be made more precise. Seeing the importance of the work accomplished in classical theory, we have been obliged to devote the whole of Chapter V to an account of the methods of classical statistical mechanics and we shall not deal with the quantum point of view until Chapter VI.

In addition, the foregoing results are extended without difficulty to the quantum theory where we can repeat point by point the discussions and stages of reasoning of classical theory by calculating the time-proportional transition probabilities with the help of the fundamental assumptions of quantum statistical mechanics, which we have explained in Chapter II. Similarly, we can apply Markovian stochastic processes to the study of irreversible quantum events by setting up a "Master Equation" for the distribution function of the occupation numbers of the individual stationary states of the particles.

Finally, in a last chapter taking the place of a general conclusion, we shall return to an element of irreversibility specifically belonging to wave mechanics, which appears via the Klein lemma: we shall see that its nature enables us to connect it with the irreversibility of the quantum measuring process. After a rapid review of the measuring process, we shall show the fundamental difference existing between macroscopic and quantum irreversibility which appears at a certain stage of the measuring process. We shall see that this difference is reduced to that existing between the square of the probability amplitudes which determine a pure case, and the probabilities in the classical sense which are involved in the definition of statistical ensembles of systems and of mixtures especially.

CHAPTER V

H-Theorems and Kinetic Equations in Classical Statistical Mechanics

I. Mechanical Reversibility and Quasi-Periodicity

1. *Mechanical reversibility*

In this first section we shall study certain properties of mechanical systems which are at the root of serious difficulties for statistical mechanics of non-equilibrium systems, difficulties that we shall try to overcome in subsequent sections. We begin by showing that the fine-grained densities $\varrho(P, t)$ defined in Chapter I are not suitable for describing the irreversible evolution of a system. We recall first of all that the equations of classical mechanics are reversible and this is expressed by the invariance of the Lagrangian equations under the transformation of t to $-t$ and $\dot{q}_i$ to $-\dot{q}_i$. In fact, if the Lagrangian L is conservative and quadratic in the generalised velocities $\dot{q}_i$, the system of Lagrange equations

$$\frac{d}{dt}\frac{\partial L}{\partial \dot{q}_i} - \frac{\partial L}{\partial q_i} = 0$$

can be written also in the form

$$\frac{d}{dt'}\frac{\partial L}{\partial \dot{q}'_i} - \frac{\partial L}{\partial q'_i} = 0,$$

in which we have put $t' = -t$, $\dot{q}'_i(t) = -\dot{q}_i(-t)$, $q'_i(t) = q_i(-t)$; the motions described by $q_i(t)$ and $q'_i(t)$ are thus equally solutions of the Lagrangian equations and they correspond to the same trajectory traversed in the reverse direction with opposite velocities. It follows that to every solution determining a possible motion of the system there always corresponds the reverse motion obtained by replacing t by $-t$ and $\dot{q}_i$ by $-\dot{q}_i$.

This completely general result is the origin of the first principal difficulty encountered in statistical mechanics of non-equilibrium systems which must emphasise a privileged direction of evolution. In fact, if a specified mechanical motion of the system corresponds to an evolution taking place in a certain direction, we see from the above argument that the reverse motion which is theoretically possible involves an evolution in the opposite direction. This statement is at the basis of *Loschmidt's paradox*, which appears to contradict the conclusion of Boltzmann's H-theorem and which makes necessary the introduction of the statistical point of view in the kinetic theory of gases.

This statistical point of view leads us to consider statistical ensembles of systems in Γ-space, defined by an, in general non-stationary, fine-grained density $\varrho(P, t)$, and whose evolution is determined by Liouville's equation

$$\frac{\partial \varrho}{\partial t} = [H, \varrho]. \tag{V.1}$$

This is obtained from the equations of classical mechanics and it retains, as a consequence, their reversible nature; moreover, it expresses only the *conservation of density in phase*, as we can easily ascertain by writing (V.1) in the form:

$$\frac{d\varrho}{dt} = 0, \tag{V.1'}$$

where d/dt is the derivative operator of hydrodynamics "following the motion". We can say also that the evolution of the probability fluid is equivalent to that of an incompressible fluid in Γ-space.

It follows from this property that there is no privileged direction of time for the evolution of such ensembles of systems and that it is not possible to establish an H-theorem for fine-grained densities, without recourse to supplementary assumptions which risk contradicting the principles of classical mechanics.

In addition, we can express this property in another way by introducing a quantity similar to the one we shall use later in the proof of the generalised $\overline{H}$-theorem. Let this quantity be

$$\sigma = \int_{\Gamma} \varrho \ln \varrho \, d\Gamma \tag{V.2}$$

which is equal to the phase average of $\ln \varrho$† and which is made a minimum by microcanonical, canonical and grand-canonical stationary ensembles, as we shall show in the next section. However, if ϱ evolved in such a way as to approach a stationary density, the quantity σ would have to be decreasing and we should have $d\sigma/dt < 0$. However, we can prove without difficulty by using (V.1′) that $d\sigma/dt$ is zero; in fact, we have

$$\frac{d\sigma}{dt} = \int_{\Gamma} (\ln \varrho + 1) \frac{\partial \varrho}{\partial t} \, d\Gamma = 0. \tag{V.3}$$

It follows that the quantity σ remains constant during evolution, which expresses in another way the reversibility of classical mechanics and the impossibility of proving a privileged direction of evolution with fine-grained densities. We shall see that this difficulty is removed by the introduction of coarse-grained statistical densities in the sense of Ehrenfest, which alone allow us to state the H-theorem correctly.

2. *Extremum properties of stationary ensembles*

Before going further, it is interesting to show that the densities defining the stationary ensembles make the expression σ a minimum. In order to prove this result, we must look for expressions of ϱ which make σ a minimum; this minimum is connected with supplementary conditions corresponding to the various macroscopic states considered: we shall therefore use the method of Lagrangian multipliers. To a small variation $\delta\varrho$ of ϱ, according to (V.2) there corresponds a small variation $\delta\sigma$ given by

$$\delta\sigma = \int_{\Gamma} (\ln \varrho + 1) \, \delta\varrho \, d\Gamma. \tag{V.4}$$

(a) In the case where we know that the system is in an energy shell δE, the foregoing integration is carried out only over the region of Γ contained in this shell, and we have the supplementary condition

$$a = \int_{\delta E} \varrho \, d\Gamma = 1, \tag{V.5}$$

† $\eta = \ln \varrho$ is what Gibbs calls the index of probability in phase.

from which we obtain for the variation δa,

$$\delta a = \int_{\delta E} \delta \varrho \, d\Gamma. \tag{V.6}$$

σ will be extremal if we have

$$\delta\sigma + \alpha\, \delta a = 0$$

(where α is a parameter to be determined); thus, by (V.4) and V.6) we must have

$$\int_{\delta E} (\ln \varrho + 1 + \alpha)\, \delta\varrho \, d\Gamma = 0, \tag{V.7}$$

whence

$$\varrho = e^{-1-\alpha} \quad \text{on} \quad \delta E. \tag{V.8}$$

The constant α is determined by putting (V.8) in (V.5); thus it becomes

$$e^{-1-\alpha} = \frac{1}{\int_{\delta E} d\Gamma} \tag{V.9}$$

and in this way we find again the microcanonical ensemble.

(b) If only the average energy of the system is known, the supplementary conditions

$$a' = \int_{\Gamma} \varrho \, d\Gamma = 1, \tag{V.5'}$$

and

$$b = \int_{\Gamma} E\varrho \, d\Gamma = \bar{E}, \tag{V.10}$$

must be added to (V.4), to which correspond the variations

$$\delta a' = \int_{\Gamma} \delta\varrho \, d\Gamma \quad \text{and} \quad \delta b = \int_{\Gamma} E\, \delta\varrho \, d\Gamma. \tag{V.11}$$

σ will be extremal if we have

$$\delta\sigma + \alpha'\, \delta a' + \beta\delta b = 0,$$

or

$$\int_{\Gamma} (\ln \varrho + 1 + \alpha' + \beta E)\, \delta\varrho d\Gamma = 0, \tag{V.12}$$

whence we obtain

$$\varrho = e^{-1-\alpha'-\beta E}. \tag{V.13}$$

As before, we determine α' and β starting from relationship (V.5′) and (V.10); in particular, we have for α':

$$e^{-1-\alpha'} = \frac{1}{\int_\Gamma e^{-\beta E}\, d\Gamma}. \tag{V.14}$$

If we compare it with formula (I.15), we find the canonical ensemble in its usual form by putting

$$-1-\alpha' = \frac{\Psi}{\theta} \quad \text{and} \quad \beta = \frac{1}{\theta}. \tag{V.15}$$

(c) In the case where the number of particles is not fixed and where only its mean value is known, the condition

$$c = \int_\Gamma n\varrho\, d\Gamma = \bar{n} \tag{V.16}$$

must be added to relations (V.5′) and (V.10). The variation

$$\delta c = \int_\Gamma n\, \delta\varrho\, d\Gamma \tag{V.17}$$

corresponds to (V.16) and σ will be extremal if

$$\delta\sigma + \alpha'\, \delta a + \beta\, \delta b + \gamma\, \delta c = 0,$$

or

$$\int_\Gamma (\ln \varrho + 1 + \alpha' + \beta E + \gamma n)\, \delta\varrho\, d\Gamma = 0,$$

whence we derive

$$\varrho = e^{-1-\alpha'-\beta E-\gamma n}. \tag{V.18}$$

We find again the grand-canonical ensemble of formulae (I.17) and (I.18) by putting

$$\beta = \frac{1}{\theta}, \quad \gamma = -\frac{\mu}{\theta}. \tag{V.18′}$$

In this way we have verified that the stationary ensembles (V.8), (V.13) and (V.18) correspond to an extremum of the quantity σ. It is easy now to verify that this extremum is in effect a minimum. For this purpose, let us consider two densities ϱ_1 and ϱ_2, where ϱ_1 is one of the three stationary distributions encountered above and where ϱ_2 is defined by

$$\varrho_2 = \varrho_1 e^{\Delta\eta},$$

$\Delta\eta$ being some phase function. We have always $\delta\varrho = \varrho_2 - \varrho_1 = \varrho_1(e^{\Delta\eta} - 1)$ and, in the microcanonical case where $\varrho_1 =$ constant, we have according to (V.5):

$$\int (e^{\Delta\eta} - 1)\, d\Gamma = 0.$$

If, now, we form the difference

$$\sigma_2 - \sigma_1 = \int (\varrho_2 \ln \varrho_2 - \varrho_1 \ln \varrho_1)\, d\Gamma,$$

the quantity:

$$\int (1 - \ln \varrho_1)(\varrho_1 - \varrho_2)\, d\Gamma$$

can be added, which is zero since ϱ_1 is constant; the difference $\sigma_2 - \sigma_1$ can then be written

$$\sigma_2 - \sigma_1 = \int \varrho_1 [\Delta\eta e^{\Delta\eta} - e^{\Delta\eta} + 1]\, d\Gamma. \qquad \text{(V.19)}$$

This expression is always positive, by virtue of the properties of the function $y = xe^x - e^x + 1$ which is positive for $x \neq 0$ and zero for $x = 0$ (we have, in fact, $y(0) = 0$ and $dy/dx = xe^x$ which always has the sign of x). Thus, we have:

$$\sigma_2 - \sigma_1 \geqq 0,$$

which establishes the property stated for the microcanonical ensemble. Similarly, an expression of the form (V.19) would be obtained for canonical and grand-canonical ensembles by adding to $\sigma_2 - \sigma_1$ the quantities

$$\int [\beta(\Psi - E) + 1](\varrho_1 - \varrho_2)\, d\Gamma$$

$$\text{and} \int [\beta(-\Omega - n\mu + E) - 1](\varrho_2 - \varrho_1)\, d\Gamma,$$

which confirms the minimum properties of these distributions.

3. *Poincaré's theorem*

In addition to the so-called mechanical reversibility, we must point out another important property of classical mechanical systems which is expressed by Poincaré's theorem: *If the hyper-*

surface Σ of constant energy has a finite size, the system will along almost all trajectories return as closely as required to its initial phase after a sufficiently long time (Poincaré, 1890; Carathéodory, 1919).

In order to prove this theorem, we must define more clearly what is understood by the return of the system close to its initial phase P_0 (occupied at the instant t_0). Let τ be an interval of time and $\{P_n\}$ the series of points P_n occupied at the times $t_0 + n\tau$; we can say that the system returns almost to P_0 if any neighbourhood δ_{P_0} of P_0, however small, contains at least one point of the series $\{P_n\}$. The exceptional phases are those for which there exists a neighbourhood of P_0 which does not contain any point of the series $\{P_n\}$: the theorem will be proved if we can prove that these phases form a set of zero measure.

The proof rests on Liouville's theorem and on the finite size of the hypersurface Σ. In fact, we shall cover this with a net whose meshes, demarcating the regions U_n, are so fine that for every point P of Σ and any neighbourhood δ_P of P, there corresponds at least one region U_i which contains the point P and is contained in δ_P: in this way we can have infinitely small meshes and an enumerable infinity of regions U_n. We consider now a region U_i and, in this region, the set of exceptional phases X_i; in the course of time this set is transformed into a series of set $X_i^{(n)}$ (corresponding to the times $t_0 + n\tau$) which cannot have any common point, by hypothesis; therefore, the series $\{X_i^{(n)}\}$ defines an enumerable infinity of disjointed sets, all having the same measure as X_i, according to Liouville's theorem. The measure of their sum is equal to the sum of their measure: it must be finite, since $\mu(\Sigma)$ is finite. Thus, $\mu(X_i) = 0$.

We can repeat this argument for all sets X_i constructed from a region U_i; the set of all the exceptional phases is obviously: $X = \sum_i X_i$ and according to the result for $\mu(X_i)$, we have

$$\mu(X) = \sum_i \mu(X_i) = 0.$$

Thus, Poincaré's theorem shows that the evolution of a mechanical system has almost always a quasi-periodic character: after a sufficiently long time, called the *recurrence time*, the system passes again through a state which is infinitely close to its initial state. We point out in passing, that the proof of this important topological property of the trajectory depends, like the entire ergodic theory, on the theory of the measure of sets (see Chapter I and also the historical review in Appendix I).

If we consider a metric transitive system and a sub-set A of the hypersurface Σ, it is easy to show that the recurrence time T_τ corresponding to a point of A is given by (Kac, 1947a, 1959; Birkhoff, 1931a):

$$T_\tau = \frac{\tau}{m(A)}, \qquad \text{(V.20)}$$

where $m(A)$ is the relative measure of the sub-set $A\,(m(A) = \mu(A)/\mu(\Sigma))$ and where τ is the interval of time which separates two successive observations of the system.†

Since it is almost impossible to be certain that the property of metric transitivity is always satisfied, formula (V.20) is difficult to use and it can in general be replaced, in order to calculate T_τ, with rougher estimates, based on certain stochastic models (see in particular, the work of Smoluchowski (1916a, b) and also Appendix III, formulae (14) and (23)). The recurrence times thus calculated vary within considerable proportions according to the number of degrees of freedom of the system: if the system is macroscopic, they correspond always to enormous durations; they can be observable processes and even very small for systems (or subsystems) of microscopic dimensions. If, for example, we consider the density fluctuations of a system under normal conditions of temperature and pressure, the recurrence time, corresponding to a density excess of 1%, varies from $10^{10^{14}}$ to 10^{-11} sec, according as the linear dimensions of the system are of order 1 cm (macroscopic system) or 10^{-5} cm.

The "recurrence" theorem thus raises a new difficulty for describing the evolution of systems by a mechanical model, the quasi-periodicity of trajectories over the hypersurface seeming to be irreconcilable with the irreversible evolution of a macroscopic system: this is the famous *Zermelo paradox*, which constitutes, with Loschmidt's objection based on mechanical reversibility, one of the fundamental difficulties of the kinetic theory of gases, and of the interpretation of Boltzmann's *H*-theorem. We note, however,

† Naturally, we eliminate the trivial case where $\tau \to 0$; actually, we have also $T_\tau \to 0$, which corresponds to the case of a system for which the representative point has not yet left the region A.

from the few numerical data mentioned above, that the reversibility or the irreversibility of the evolution depends on the dimensions of the observed system: (a) If the system is of microscopic size, perceptible fluctuations about equilibrium are anticipated and may be reproduced at very short intervals of time. (b) On the contrary, if the system is macroscopic the fluctuations, even very small ones, appear only after a period which is so long that it can be considered justifiably as unobservable; the relaxation times of the system are incomparably smaller than this very large period, with the result that the observed evolution is effectively irreversible.

Be that as it may, the foregoing results show that statistical mechanics of non-equilibrium systems must give an answer to the following fundamental question: *How can mechanical reversibility and quasi-periodicity be reconciled with macroscopic irreversibility?* We shall see in what follows, what is the actual state of this problem and we shall explain the proposed solutions (which are far from being definitive) by discriminating the following stages in the reasoning:

1. The definition of coarse-grained distributions enables us to establish that a certain quantity $\bar{H}$ [analogous to σ of (V.2)] is certainly not increasing during evolution: this is the purpose of Section II of this chapter. This generalised $\bar{H}$-theorem, however, can neither prove the tendency towards equilibrium nor describe this evolution effectively; for this, we should be able to integrate Liouville's equation.

2. By depending on the fundamental hypothesis of statistical mechanics (equivalent in certain cases to the "Stosszahlansatz") we can calculate the probabilities for transitions between the cells in phase space and obtain a kinetic equation similar to that of Boltzmann, for the case of a uniform gas. The proof of this kinetic equation, however, supposes in fact the validity *at every instant* of the "Stosszahlansatz" principle, with the result that the evolution of the system is no longer compatible with Liouville's equation, as we shall show in Section III. Thus, Boltzmann's equation is an equation of a statistical nature, but it is based on assumptions whose real significance (in the meaning of the theory of stochastic processes) is far from being clear. It is certain, however, that the probability does not appear only in the definition of an initial statistical ensemble (as is the case in the generalised $\bar{H}$-theorem

and in the ergodic theory), but that it is involved in *the evolution process itself.*

3. This last statement leads us to attempt to describe the irreversible evolution of a system by a Markovian stochastic process in phase space; this method is studied in Section IV. With the transition probabilities provided by the Stosszahlansatz, we set up a linear "Master Equation" from which we can deduce by reduction an equation which is similar to Boltzmann's equation in the uniform case, provided that certain correlations are neglected. The significance of the Boltzmann equation is thus explained in this case, but it seems to be very difficult to extend this method to the case of a non-uniform gas.

4. If the introduction of certain stochastic processes enables us to describe the tendency towards equilibrium of a macroscopic system, it does not follow that its microscopic evolution remains governed by Liouville's deterministic equation; we have thus returned once again to the difficult problem of studying the solutions of this equation. This is the aim of Section V in which, having recalled the method of successive reductions of Liouville's equation leading to the B.B.G.K.Y. equations, we point out the broad lines of Brout's method (in which we are seeking to derive the "Master Equation" from Liouville's equation) and that of Bogolyubov which enables us to obtain formally generalised kinetic equations. In this way, perhaps we can hope to bridge the gap between the macroscopic and microscopic descriptions by depending once again on the large number of degrees of freedom of the system.

II. Coarse-Grained Densities and the Generalised $\bar{H}$-Theorem

1. *Coarse-grained densities*

We have just seen that the statistical description provided by the density ϱ cannot lead to a privileged direction of evolution because of the reversibility of classical mechanics. In order to obtain an *H*-theorem which assures macroscopic irreversibility we are compelled to introduce the coarse-grained density in phase space, by adopting a definition which is appropriate for the macroscopic state of a physical system starting from microscopic mechanical quantities. We discuss here the method of "coarse-graining" of

Γ-space by P. and T. Ehrenfest, since it is the best suited for studying the $\bar{H}$-theorem; we shall encounter other definitions of coarse-grained densities in Section V during the study of the kinetic equations, starting from Liouville's equation [see also the second footnote to § II.4].

Let us suppose that our system consists of N similar particles, each of which has r degrees of freedom described by the r variables $(q_1, q_2, \ldots, q_r)$; we can associate with them an individual phase space, μ-space, defined by $2r$ coordinates $(q_1, \ldots, q_r; p_1, \ldots, p_r)$. The physical state of our system is thus represented by a cluster of N points in μ-space; we note in passing that in the case where $r = 3$ (point particles), μ-space is six-dimensional. Since macroscopic observation is, by its nature, unable to determine these $2r$ coordinates precisely, we shall assume that it enables us to distinguish only whether the molecule has its $2r$ coordinates contained between q_i and $q_i + \delta q_i$, and p_i and $p_i + \delta p_i$ $(i = 1, 2, \ldots, r)$; in this way, we shall be led to dividing μ-space into a set of cells ω_i corresponding to the macroscopic inaccuracies δq_i, δp_i, and each occupying a volume

$$\omega_i = \delta q_1 \ldots \delta q_r \delta p_1 \ldots \delta p_r. \tag{V.21}$$

in μ-space. We shall suppose in what follows that these cells are sufficiently small compared with the measurable macroscopic dimensions but large enough to contain a large number of molecules (P. and T. Ehrenfest, 1911).

If, now, we consider the whole system, its macroscopic state will be described by the set $\{n_i\}$ of the numbers of molecules contained in different cells ω_i: in Γ-space of $2Nr$ dimensions there will correspond to this configuration a cell whose volume is given by

$$(\delta v)_\Gamma = \prod_{i=1}^{N} \omega_i^{n_i}. \tag{V.22}$$

But since the permutations of molecules in the cells of μ-space do not change the macroscopic state of the system, there will be in Γ-space

$$G_n = \frac{N!}{\prod_i n_i!} \tag{V.23}$$

cells $(\delta v)_\Gamma$ corresponding to the same macroscopic situation. Thus, each set of numbers $\{n_i\}$ will define, in Γ-space, a "star" Ω_n whose measure is given by

$$W(\Omega_n) = G_n(\delta v)_\Gamma = \frac{N!}{\prod_i n_i!} \prod_i \omega_i^{n_i}, \quad \text{with} \quad \sum_i n_i = N. \tag{V.24}$$

Thus, to the division of μ-space into cells ω_i there corresponds a division of Γ-space into stars Ω_n which enable us to define the macroscopic state of a system: this state is determined by the star occupied by the representative point of the system in Γ-space and the most precise macroscopic observation can distinguish only in which star the system is located. Let us also point out that to these cells there corresponds a time of observation, which is small but finite, although this is not clearly specified in the definition: in fact, the representative point of the system in Γ-space always needs a finite time for traversing any star Ω_n with the result that transition from one macroscopic state to another can occur generally only after a finite period.

Let us suppose now that we are dealing with a statistical ensemble of systems defined by a fine-grained density ϱ; according to the definition of macroscopic observation adopted above, we shall not observe precise and instantaneous values of this density, but only the respective statistical weights of the stars Ω_n given by the integrals:

$$\int_{\Omega_n} \varrho \, d\Gamma.$$

Under these circumstances the probability of finding a system of the ensemble at a point of Ω_n will be defined by the quantity

$$P_n = \frac{\int_{\Omega_n} \varrho \, d\Gamma}{W(\Omega_n)}. \tag{V.25}$$

This is the coarse-grained density (in Ehrenfest's sense) at the point (q_i, p_i): we see by its definition that it is the mean of the fine-grained density ϱ taken over Ω_n; this density P is therefore constant over each star and it satisfies the relationship

$$\sum_n P_n W(\Omega_n) = 1, \tag{V.26}$$

which can be written also as

$$\int_{\Gamma} P d\Gamma = 1; \qquad \text{(V.26')}$$

we note that the quantity

$$P_n' \equiv P_n W(\Omega_n) \qquad \text{(V.25')}$$

represents, therefore, the probability of finding a system of the ensemble in the star Ω_n.

2. *Definition and properties of* $\bar{H}$

We shall study now the evolution of the coarse-grained density $P(\mathscr{P}, t)$† and we shall try to show that, from some initial value, it tends irreversibly towards a stationary distribution. For this purpose we define a quantity $\bar{H}$, similar to the quantity σ introduced in (V.2), by the relation

$$\bar{H} = \sum_n P_n \ln P_n W(\Omega_n) \qquad \text{(V.27)}$$

or:

$$\bar{H} = \int_{\Gamma} P \ln P \, d\Gamma, \qquad \text{(V.27')}$$

since P_n is constant over the whole volume $W(\Omega_n)$.

We note that the quantity $\bar{H}$, defined by (V.27), depends essentially on the size of the stars Ω_n, which correspond to macroscopic observations. On the other hand, we can replace the definition (V.27') by

$$\bar{H} = \int_{\Gamma} \varrho \ln P \, d\Gamma = \overline{\ln P}, \qquad \text{(V.28)}$$

which is equivalent to (V.27') because of (V.25). Equation (V.28) shows that $\bar{H}$ is the mean value of $\ln P$ taken over the ensemble of systems defined by the statistical density ϱ, which justifies our writing "$\bar{H}$".

We can verify easily that $\bar{H}$ has the same extremum properties as σ. In other words, $\bar{H}$ is made minimum by microcanonical, canonical or grand-canonical distributions according as we are

† In this chapter, we shall denote the representative point of the system by $\mathscr{P}$ in order to avoid any confusion with the coarse-grained density P.

dealing with an isolated system, a system in thermal equilibrium for which $\overline{E}$ is known or with a system for which only $\overline{E}$ and $\bar{n}$ (average number of particles) are specified, respectively. As before, the distributions in question can be written as

$$\left.\begin{array}{ll} P = \text{constant} & \text{over } \delta E, \\ P = 0 & \text{elsewhere} \end{array}\right\} \tag{V.8$'$}$$

for the microcanonical ensemble,

$$P = e^{(\Psi - E)/\theta} \tag{V.13$'$}$$

for the canonical ensemble, and

$$P = \exp\left[\frac{\Omega + \sum_i \mu_i n_i - E}{\theta}\right] \tag{V.18$''$}$$

for the grand-canonical ensemble.

On the other hand, the evolution of $\overline{H}$ with time is different from that of σ because of the introduction of coarse-grained densities which no longer satisfy equation (V.1$'$); if the initial distribution is not a stationary distribution, we have necessarily $dP/dt \neq 0$ (because of the deformation of the volumes in phase space during motion), whence we obtain $d\overline{H}/dt \neq 0$ in contrast to $d\sigma/dt = 0$. We shall show now that evolution probably take splace in a preferential direction by proving the inequality $\overline{H}(t) \leqq \overline{H}(0)$: this is the aim of the $\overline{H}$-theorem for an ensemble of systems, and is called the *generalised $\overline{H}$-theorem.*

3. *The generalised $\overline{H}$-theorem*

Let us suppose now that the initial state of the system (at t_1) is determined by a macroscopic observation of the type defined previously; one can define a statistical ensemble of systems with a fine-grained density ϱ_1 corresponding to this state. The macroscopic observation determines the relative weight of the various stars, so that the density ϱ_1 must be constant over them (in accordance with the fundamental statistical assumptions) and equal to the statistical weights defined by the observation. Thus, we see that, according to these conditions, at the initial time t_1 the fine-grained density ϱ_1 is equal to the coarse-grained density P_1 [using the definition (V.25)]. At t_1, we have for all points of phase space:

$$\left.\begin{aligned} \varrho_1 &= P_1, \\ \bar{H}_1 &= \int_\Gamma \varrho_1 \ln \varrho_1 \, d\Gamma. \end{aligned}\right\} \qquad \text{(V.29)}$$

If ϱ_1 corresponded to a stationary statistical ensemble, for example to a uniform distribution over an energy shell δE, the density $\varrho(t)$ would remain constant and equal to ϱ_1; we should have also $P = \varrho =$ const, with the result that we should find directly an equilibrium state, where $\bar{H}$ would be a minimum according to (V.8′). However, in the general case where the initial observation fixes the system in a state far from equilibrium, the fine-grained density $\varrho(t)$ will vary with time in accordance with (V.1); because of this variation, $\varrho(t)$ and $P(t)$ will be different at any t later than t_1. In fact, even though the representative points of the systems, initially in the same star Ω_n, always occupy a constant volume in Γ-space equal to $W(\Omega_n)$ (according to Liouville's theorem), the shape of the new volume during evolution becomes very different from that of the initial volume, in such a way that it extends over a very large number of stars Ω_n; each of these is therefore occupied at t by numerous "filaments" of representative points corresponding to a large variety of values of the fine-grained density ϱ: the result is, that there is a difference between the fine-grained and coarse-grained densities at t; this difference involves a decrease in $\bar{H}$. Thus, at $t_2 > t_1$, we have

$$\left.\begin{aligned} &\varrho_2 \neq P_2, \\ &\bar{H}_2 = \int_\Gamma P_2 \ln P_2 \, d\Gamma = \int_\Gamma \varrho_2 \ln P_2 \, d\Gamma, \end{aligned}\right\} \qquad \text{(V.30)}$$

according to (V.28). Let us now find the difference $\bar{H}_1 - \bar{H}_2$; according to (V.29) and (V.30) we have

$$\bar{H}_1 - \bar{H}_2 = \int_\Gamma (\varrho_1 \ln \varrho_1 - \varrho_2 \ln P_2) \, d\Gamma. \qquad \text{(V.31)}$$

In order to study this expression, we note first of all that $\varrho_1 \ln \varrho_1$ can be replaced by $\varrho_2 \ln \varrho_2$ under the integration sign, by virtue of

Liouville's theorem; actually, by (V.3) and (V.2) we have:

$$\frac{d}{dt}\int\limits_{\Gamma} \varrho \ln \varrho \, d\Gamma = 0,$$

whence

$$\int\limits_{\Gamma} \varrho_1 \ln \varrho_1 \, d\Gamma = \int\limits_{\Gamma} \varrho_2 \ln \varrho_2 \, d\Gamma.$$

On the other hand, we can add the quantity $- \varrho_2 + P_2$ under the integral sign, which gives a zero contribution by definition; with these modifications (V.31) can be written as

$$\bar{H}_1 - \bar{H}_2 = \int\limits_{\Gamma} (\varrho_2 \ln \varrho_2 - \varrho_2 \ln P_2 - \varrho_2 + P_2) \, d\Gamma. \qquad \text{(V.32)}$$

The $\bar{H}$-theorem will be proved if the integrand I is always positive. Let us verify that this is indeed so by calculating the derivative $\partial I/\partial \varrho_2$ for a given value of P_2. We have

$$\frac{\partial I}{\partial \varrho_2} = \ln \frac{\varrho_2}{P_2}$$

which is positive if $\varrho_2 > P_2$ and negative if $\varrho_2 < P_2$. Since, on the other hand, I and $\partial I/\partial \varrho_2$ are zero for $\varrho_2 = P_2$, we can see that expression I is a minimum for $\varrho_2 = P_2$ and that it is positive if $\varrho_2 \neq P_2$; according to (V.32), we have therefore, at all times

$$\bar{H}_1 - \bar{H}_2 \geqq 0; \qquad \text{(V.33)}$$

this is the generalised $\bar{H}$-theorem.

4. *Discussion of the generalised $\bar{H}$-theorem*

It can be seen from the foregoing developments that this theorem is obtained without any special assumptions, except the fundamental assumption of statistical mechanics, which is necessary for the statistical description of macroscopic phenomena and which cannot contradict the reversibility of the laws of mechanics. The proof rests essentially on the fact that the fine-grained and coarse-grained densities are equal at the moment of the initial observation [equation (V.29)] and that they generally become different during the evolution, whence we derive equation (V.30). This inequality will be valid at all times during the evolution, so that we can assume (without, however, proving it rigorously, which would necessitate integration of the equations of motion) that the quantity $\overline{H}$ continues to decrease until an equilibrium distribution is achieved.

In order to understand the nature of this process we consider the volume of extension in phase corresponding to the representative points located initially in some star Ω_n: as we have mentioned already, this volume is deformed during evolution, while maintaining a constant size (Liouville's theorem) and it is drawn out into long "filaments" which finally extend over very many stars. After a certain time, each Ω_n is thus occupied by points which, initially, happened to be in very many other stars and which correspond to many different values of the fine-grained density ϱ: thus, a mixture of initial conditions is achieved, involving a change of the coarse-grained density and a decrease of $\overline{H}$.

Because of this process, it is probable that the densities ϱ and P differ more and more in the course of time and that $\overline{H}$ will continue to decrease until the system attains a steady state. If the system is isolated, the representative points will remain contained within an energy shell E, $E + \delta E$ and evolution will proceed until the coarse-grained density P becomes uniform over the shell considered. We can deduce from this by considering the evolution of the total system (system + thermostat) and by using an expression for the number of states of the thermostat contained in an energy interval that the density P, corresponding to a system in thermostatic contact with its surrounding, will tend towards a canonical distribution† (Tolman, 1938, pp. 477–501). Once such a steady state is reached, the quantity $\overline{H}$ can no longer decrease, since it has attained its minimum value.

It is important to emphasise that these results are not exact, that they have only a qualitative value and that, in particular, we cannot estimate the time required to reach an equilibrium distribution: we can only assert, by virtue of (V.33), that $\overline{H(t)} \leqq \overline{H(0)}$ and that the process which involves this inequality proceeds continuously during the evolution of the system; we can deduce from this that the coarse-grained density probably approaches a steady state distribution. Moreover, this is completely illustrated by the

† Certain difficulties are encountered, however, in showing the tendency of P towards a canonical ensemble; see for instance Lorentz, 1907.

example (given by Gibbs) of a mixture of water and black ink, initially separated: it is well-known that by stirring it, this liquid will approach a state of mixing with a uniformly grey colour. Since the fine-grained densities of the filaments of water and black ink at a specified point of the liquid do not change during the motion, this stabilisation involves the coarse-grained density which corresponds to our macroscopic observation. We point out, in addition, that the process which we have just described and which enables us to prove the generalised $\bar{H}$-theorem is similar to the process of mixing of initial conditions (corresponding to the various hypersurfaces of constant energy within the same energy shell δE) necessary in ergodic theory for proving Hopf's theorem: in this way we can establish a certain reconciliation between the method of Hopf's ergodic theorem and the less exact method of the generalised $\bar{H}$-theorem. From this point of view, the conditions for ergodicity stated by Hopf's theorem (see Chapter I, § II.4) can be considered as the conditions necessary for the exact proof of an $\bar{H}$-theorem; this shows that the solution of the ergodic problem is essential for a theoretical justification of the methods of statistical mechanics (see also Chapter VI, § III.2).

To summarise, the definition of coarse-grained densities P appears to be the indispensable theoretical basis for describing the irreversible evolution of a macroscopic system;† in particular, it is important never to lose sight of the finite dimensions of the stars in Γ-space, if we wish to avoid the paradoxes of the kinetic theory of gases. In fact, certain systems may behave peculiarly (corresponding to very special initial conditions) and no tendency towards an equilibrium state may occur in such a case; however, because of the finite size of the stars in Γ-space, such systems are very rare compared with all the systems occupying a given star, with the result that their observation can be considered as highly improbable. In conclusion, we emphasise that in the present theory the concept of probability is involved solely in defining the initial

† We note, in addition, that we can resort to other definitions of coarse-grained quantities which are not equivalent to the one by the Ehrenfests; this is the case with the quantities coarse-grained over position which were introduced by Massignon (1957) and which are very suitable for a study of the hydrodynamic properties of a fluid. We shall encounter also, in Kirkwood's method, time-averages of the fine-grained density and, in Bogolyubov's method, a certain synchronisation process (see Chapter V, § V.4).

distribution $P(0)$, but that the evolution takes place according to the deterministic laws of mechanics,† the irreversibility arising from the gross nature of our observations. Thus, it is not possible to have any kind of contradiction between the statistical conclusions of the theory and the reversible behaviour of mechanical systems.

III. Transition Probabilities and Boltzmann's Equation

1. *Definition of* $H(Z, t)$

The foregoing considerations have enabled us to show that the coarse-grained distribution P of a statistical ensemble of systems would probably approach an equilibrium distribution which makes the quantity $\bar{H}$ a minimum. However, the nature of the argument allows us neither to prove exactly the irreversibility of the evolution nor to calculate effectively the speed of approach towards equilibrium; in particular, we cannot estimate the relaxation time of a system which is not in its macroscopic equilibrium position. For this it would be necessary, in fact, to be able to determine at each instant, or at quite small intervals of time τ, the density P which can be calculated only by starting from the expression for the fine-grained density $\varrho(t)$; since this satisfies equation (V.1), we find that we have returned to the problem of the integration of Liouville's equation. The earlier results, based on the generalised $\bar{H}$-theorem, thus provide only a theoretical proof of the principles of statistical mechanics, by highlighting in a qualitative way the tendency of a distribution towards equilibrium, without allowing a quantitative study of it (in particular, we note that the relaxation time depends essentially on the size of the stars Ω_n which is not otherwise specified here: only their finite nature is important).

In order to begin such a study without solving the equations of motion, we shall return to the physical system studied and we shall see how we can describe the evolution in time of its macroscopic state. We know that in the mechanical description a microscopic state of the system is represented by a point $\mathscr{P}$ of Γ-space and that the trajectory of this point in Γ-space defines the (reversible) evolution of the system; but this trajectory, in fact, is not known to us since macroscopic observation does not allow us to define precisely

† This kind of process is called crypto-deterministic by Whittaker (1943).

the exact position in Γ-space of the representative point $\mathscr{P}(t)$ but only to say that $\mathscr{P}(t)$ is located in one of the stars Ω_n. So long as the point $\mathscr{P}(t)$ occupies the same star, the macroscopic state of the system remains unchanged and it is determined completely by the set of numbers $\{n_i\}$. Just as we have seen in the first section, to each of these constellations there corresponds a volume $W(\Omega_n)$ given by (V.24). If we denote by Z the macroscopic state of the system corresponding to the occupation of Ω_n, we can assign a quantity $H(Z)$ to each state Z which is defined, according to Boltzmann's concepts, by the logarithm of the extension in phase (or the statistical weight) corresponding to Z, or

$$H(Z) = -\ln W(Z) = \sum_i n_i \ln n_i + \text{const.} \qquad \text{(V.34)}$$

It is easy to prove that $H(Z)$ is minimum for the Maxwell–Boltzmann distribution: actually, since the system is isolated with an energy which is known within an interval δE and each Ω_n has a weight proportional to its volume, the most probable distribution will be defined by the set $\{n_i\}$ corresponding to the star $\Omega_{\max}$ of δE having the maximum weight; thus it will be determined by the conditions

$$\left.\begin{aligned} \delta \ln P = \delta \ln W(Z) = -\sum_i (\ln n_i + 1)\,\delta n_i = 0, \\ \delta N = \sum_i \delta n_i = 0, \quad \delta E = \sum_i \varepsilon_i \delta n_i = 0, \end{aligned}\right\} \qquad \text{(V.35)}$$

which are satisfied (we use, as before, the method of Lagrangian multipliers) by the Maxwell–Boltzmann formula

$$n_i^{(\max)} = e^{-\alpha-\beta\varepsilon_i}, \qquad \text{(V.36)}$$

where the ε_i denote the energies of molecules located in ω_i (it is assumed in the calculations that the quantities n_i are large enough to enable us to use Stirling's formula).

Moreover, if we consider the volume of a star Ω', close to $\Omega_{\max}$ and corresponding to occupation numbers $n_i' = n_i^{(\max)} + \delta n_i$, we find for this volume by (V.35) and (V.36):

$$W(\Omega') = W(\Omega_{\max}) \exp\left[-\frac{1}{2}\sum_i \frac{(\delta n_i)^2}{n_i^{(\max)}}\right], \qquad \text{(V.36')}$$

in which we have retained only the terms of second order in δn_i. Thus, it can be seen that the maximum $W(\Omega_{\max})$ of the volumes

$W(\Omega_i)$ of the shell δE is a very sharp maximum as soon as the n_i are large. It follows that the volume occupied by non-Maxwellian states is very small in comparison with that occupied by the Maxwellian state. This result, which is essential for statistical mechanics, is associated with the asymptotic properties of N-dimensional space and is expressed in a more precise way by using the central limit theorem following Khinchin (see Appendix I, § 4);† it plays a principal role in the reasonings of the probability ergodic theory (see Chapter I, § III.1 and Chapter IV) and it allows us to assume the return to the Maxwellian state of a non-equilibrium system; we are going to analyse this return in detail in what follows.

Let us return now to the point $\mathscr{P}(t)$: its trajectory traverses in succession various Ω_i (which corresponds to variations in the numbers n_i) and macroscopic observation only enables us to confirm that the system occupies successively a certain sequence of stars; to each of them there corresponds a value of $H(Z)$ which changes suddenly when we note the presence of $\mathscr{P}(t)$ in another star. The set of values of $H(Z)$ in the course of time is thus represented by the function $H(Z, t)$ which evolves stepwise (see Appendix III, Fig. 1a): in order to solve the fundamental problem of statistical mechanics of non-equilibrium systems, we must prove that $H(Z, t)$ will for the majority of the time be in the vicinity of $H_{\min}$ (corresponding to $\Omega_{\max}$) given by (V.36) and that, if at t_0 the quantity H has a value H_0 which is much greater than $H_{\min}$, the function $H(Z,t)$ will decrease rapidly, starting from H_0, to $H_{\min}$. In order to try to show these various points without carrying out a complete integration of the equations of motion, we shall define the transition probabilities $P(\Omega_i, \Omega_j)$ giving the probability that the system initially in Ω_i passes into Ω_j at the end of a small time τ. By deriving (V.34) with respect to time, we have

$$\frac{dH(Z,t)}{dt} = \sum_i (1 + \ln n_i) \frac{dn_i}{dt} = \sum_i (\ln n_i) \frac{dn_i}{dt} \qquad \text{(V.37)}$$

(we have used that $\sum_i n_i = N$) and it can be seen that the study of the behaviour of $H(Z, t)$ thus reduces to that of the numbers dn_i/dt which depend on the probabilities $P(\Omega_i, \Omega_j)$.

† This result must be related also to the phenomenon of the insensitivity of Boltzmann's formula (Appendix III.B).

We shall start by calculating, first of all, these transition probabilities by using the fundamental assumption of statistical mechanics and by assuming that the evolution mechanism is determined exclusively by binary collisions between the atoms (or molecules) of the system. We shall then be able to show the decrease of the most probable value of $H(Z, t)$ at a given instant and we shall see that it is still necessary to make an additional assumption about the evolution of the gas if we wish to obtain the H-theorem and Boltzmann's equation, so that they are valid at any instant. It is with a view to analysing the importance of this latter assumption that we shall try, in Section IV, to describe the successive passage of the representative point $\mathscr{P}(t)$ through the stars Ω_i, Ω_{i+1}, ..., as a stochastic process of the Markov type, with the transition probabilities $P(\Omega_i, \Omega_j)$ given by the "Stosszahlansatz"; such a method, naturally, could find its justification only in the large number of degrees of freedom of the system studied and more particularly in the large value of the numbers n_i. Thus, we are led to study the properties of the "Master Equation" and to justify its application to the problem of the approach to equilibrium of a macroscopic system.

If we apply the methods of statistical mechanics exactly, we can still say that the most precise macroscopic observation only permits us to say that the system exists in an initial star Ω_0 with volume $W(\Omega_0)$: to this star there corresponds a set of points of Γ-space representing an ensemble of systems. To this ensemble there corresponds, in turn, an ensemble of trajectories in Γ-space which describe the mechanical evolution of the different systems of the ensemble considered; at $t_1 = t_0 + \tau$ we have a set of values for $H(Z, t_1)$ and we can associate with them an average value $\mathscr{H}_1$ defined by means of an uniform distribution over the initial ensemble Ω_0 (in accordance with the fundamental assumptions of statistical mechanics); we could define similarly $\mathscr{H}_2$ at $t_2 = t_0 + 2\tau$, ..., and $\mathscr{H}_n$ at $t_n = t_0 + n\tau$. The H-theorem would be established in this case—if we could prove by depending upon the large number of degrees of freedom that:

(a) the dispersion of the values of $H(Z, t_n)$ around $\mathscr{H}_n$ is very small;

(b) the series, $\mathscr{H}_1, \mathscr{H}_2, \ldots, \mathscr{H}_n$ decreases monotonically, starting from H_0, to $H_{\min}$.

The proof of these two theorems encounters the same difficulty as that which we have previously pointed out—we must be able to integrate the equations of motion. We shall show that they are valid, at least for a small interval of time after the initial instant t_0, by taking for the unknown transition probabilites the most probable value of dn_i/dt at t_0 $\left[\text{which we shall denote by } \left(\frac{dn_i}{dt}\right)_P\right]$ calculated by means of the fundamental hypothesis of statistical mechanics; the most probable value $\left(\frac{dn_i}{dt}\right)_P$ is chosen here for reasons of mathematical simplicity and in order to conform to historically established practice. By limiting ourselves to the case in which the variation of the numbers n_i arises solely from binary collisions between molecules, i.e. by accepting the homogeneity of the gas in the absence of external forces, we shall verify that it is always possible to obtain again Boltzmann's equation.

2. *Proof of Boltzmann's equation*

For this purpose we shall assume:

(α) that our physical system is a gas sufficiently dilute that only binary collisions between molecules need be taken into account;

(β) that it is uniformly distributed in a total volume V with a total energy E and that there is no external field;

(γ) that we are dealing with molecules with spherical symmetry which allows us to dispense with taking collision cycles into account; μ-space is then six-dimensional and we can divide it into equal cells ω_i which are a product of the total volume V of the position space and a cell $\bar{\omega}_i$ of momentum space: $\omega_i = V\bar{\omega}_i$.

According to the method of statistical mechanics, we can make correspond to our gas a virtual ensemble of systems, all satisfying the partial specification of the state of the gas and uniformly distributed in accordance with the fundamental assumption of equal phase probability. In what follows, we shall suppose that the state of the gas is specified at t (playing the role of an initial moment) by a macroscopic observation which determines the occupation numbers n_i, n_j of the cells ω_i and ω_j; on the other hand, since the molecules are distributed uniformly in the volume V, the probability of finding the centre of gravity of any molecule in a volume δV is, for the ensemble considered, given by $\delta V/V$. In this way we can calculate the most probable value for the number of collisions which transfer two molecules from the cells (ω_i, ω_j) to the cells (ω_k, ω_l) during an interval of time $\delta t = \tau$ following the initial instant t, by calculating the volume in which the centre of gravity of one of the molecules must be located in order that a single collision would in fact occur with the other molecule during δt. If a_{kl}^{ij} denotes this most probable number per unit time, we have†

$$a_{kl}^{ij}\tau = A_{kl}^{ij} n_i n_j \tau, \qquad \text{(V.38)}$$

where A_{kl}^{ij} is a constant which depends on the parameters defining the collision. We calculate, similarly, the most probable number of inverse collisions $(\omega_k, \omega_l) \to (\omega_i, \omega_j)$:

$$a_{ij}^{kl}\tau = A_{ij}^{kl} n_k n_l \tau, \qquad \text{(V.38')}$$

where A_{ij}^{kl} is equal to A_{kl}^{ij} according to the principle of dynamic reversibility. The coefficients A_{kl}^{ij} are given by

$$A_{kl}^{ij} = \frac{g_{ij}\sigma_d(g_{ij}, \chi)}{V}\, d\Omega, \qquad \text{(V.39)}$$

where g_{ij} is the relative velocity of the molecules i and j, $\sigma_d(g_{ij}, \chi)$ is the differential collision cross section, which is a function of the relative velocity g_{ij} and of the angle of scattering χ, and $d\Omega$ is an element of solid angle in the direction χ. For the rigid sphere model, we have simply

$$\sigma_d(g_{ij}, \chi) = \frac{\delta^2}{4}, \qquad \text{(V.39')}$$

where δ is the molecular diameter.

The fundamental statistical assumption which enables us to calculate a_{kl}^{ij} and a_{ij}^{kl} is here analogous with the assumption of molecular chaos introduced by Boltzmann; formulae (V.38) and (V.38′) are those obtained by starting from the "Stosszahlansatz" and we shall see that Boltzmann's equation can be deduced from them. We define, in fact, a set of functions of the time $f_i(t)$ by the relations

$$n_i = f_i(t)\,\omega_i = V f_i(t)\,\bar{\omega}_i. \qquad \text{(V.40)}$$

† We follow here Tolman (1938, p. 128ff.), to whom we refer for details of the calculations.

The set $\{f_i(t)\}$ defines a particle density in μ-space which depends, according to condition (c), only on the coordinates u, v, and w of momentum space. We shall denote it by $f_i(u, v, w, t)$, where f_i is the value of $f(u, v, w, t)$ for the cell $\bar{\omega}_i$; we have, with this notation:

$$\begin{cases} a_{kl}^{ij} = A_{kl}^{ij} f_i f_j \omega_i \omega_j, \\ a_{ij}^{kl} = A_{ij}^{kl} f_k f_l \omega_k \omega_l. \end{cases} \tag{V.41}$$

Now, according to Liouville's theorem, we have

$$A_{kl}^{ij} \omega_i \omega_j = A_{ij}^{kl} \omega_k \omega_l = B\begin{pmatrix} ij \\ kl \end{pmatrix} \tag{V.42}$$

whence

$$a_{kl}^{ij} = B\begin{pmatrix} ij \\ kl \end{pmatrix} f_i f_j, \tag{V.43}$$

$$a_{ij}^{kl} = B\begin{pmatrix} ij \\ kl \end{pmatrix} f_k f_l. \tag{V.44}$$

According to the definitions of a_{kl}^{ij} and of $\left(\frac{dn_i}{dt}\right)_P$ we can write also:

$$\begin{cases} a_{kl}^{ij} = -\left(\frac{dn_i}{dt}\right)_P = -\omega_i \frac{df_i}{dt} = \left(\frac{dn_k}{dt}\right)_P = \omega_k \frac{df_k}{dt}, \\ \quad\;\; = -\left(\frac{dn_j}{dt}\right)_P = -\omega_j \frac{df_j}{dt} = \left(\frac{dn_l}{dt}\right)_P = \omega_l \frac{df_l}{dt}. \end{cases} \tag{V.45}$$

We are now able to calculate the most probable value of $\frac{dH(Z, t)}{dt}$ by means of formulae (V.43), (V.44), (V.45) and (V.37). If we are interested only in the most probable variations of $H(Z, t)$ produced by collisions of the type considered, $(\omega_i, \omega_j) \to (\omega_k, \omega_l)$, and of the opposite type, $(\omega_k, \omega_l) \to (\omega_i, \omega_j)$, we have by denoting this variation by $\left(\frac{dH\begin{pmatrix} ij \\ kl \end{pmatrix}}{dt}\right)_P$:

$$\left(\frac{dH\begin{pmatrix} ij \\ kl \end{pmatrix}}{dt}\right)_P = \left(\frac{dH}{dt}\right)_{\substack{ij \\ kl}} + \left(\frac{dH}{dt}\right)_{\substack{kl \\ ij}}, \tag{V.46}$$

with

$$\left(\frac{dH}{dt}\right)_{\substack{ij\\kl}} = Bf_i f_j(-\ln f_i - \ln f_j + \ln f_k + \ln f_l) = Bf_i f_j \ln \frac{f_k f_l}{f_i f_j}$$

(V.47)

and

$$\left(\frac{dH}{dt}\right)_{\substack{kl\\ij}} = Bf_k f_l \ln \frac{f_i f_j}{f_k f_l}, \qquad \text{(V.48)}$$

whence we obtain easily

$$\left(\frac{dH\begin{pmatrix}ij\\kl\end{pmatrix}}{dt}\right)_P = B(f_i f_j - f_k f_l) \ln \frac{f_k f_l}{f_i f_j}. \qquad \text{(V.49)}$$

This latter expression is of the form $B(x - y) \ln (y/x)$: it is thus always negative according to the properties of f_i, and this proves the *H*-theorem in this case; this result is extended without difficulty to other collisions in such a way that it remains valid for the total variation of $H(Z, t)$.

Formula (V.49), by identifying the mean values of $\mathscr{H}_i$ defined previously with the most probable values of $H(Z, t)$, allows us to check points (a) and (b) of section 1 for a short interval of time τ after the instant of the initial observation; thus, the programme stated above would be fulfilled if the results of (V.49) could be extended to every instant t: but this would necessitate an additional assumption which does not enter into the framework of the fundamental postulate of statistical mechanics. We shall return presently to this point; we note first that this additional assumption is at the basis of Boltzmann's equation.

In fact, we can easily deduce Boltzmann's equation giving the variation of $f(u, v, w, t)$ due to collisions by using formulae (V.43), (V.44) and (V.45) for all possible groups of cells (ω_i, ω_j) and (ω_k, ω_l); finally, we can write

$$\frac{\partial f}{\partial t} = -\iiiint\!\!\int (ff_1 - f'f_1')\, \psi(g, \chi)\, du_1\, dv_1\, dw_1\, d\Omega, \qquad \text{(V.50)}$$

where $\psi(g, \chi) = g\sigma_d(g, \chi)$ depends only on the relative velocity g of the two molecules before the collision and on the angle χ

between this relative velocity and the direction of the solid angle $d\Omega$. As usual, we have put

$$f = f(u, v, w, t) \qquad f_1 = f(u_1, v_1, w_1, t) \quad \text{before the collision;}$$

$$f' = f(u', v', w', t), \quad f_1' = f(u_1', v_1', w_1', t) \quad \text{after the collision.}$$

Strictly speaking, equation (V.50) is valid only for a short interval of time τ after the macroscopic observation, for it is by virtue of the result of this observation that we have constructed the statistical ensemble enabling us to calculate expressions (V.38–V.38′). If we accept that Boltzmann's equation is valid at every instant, we can deal quantitatively with the problem of the irreversible evolution of a gas which satisfies conditions (α), (β), and (γ), but at the price of an additional assumption which we shall study presently. Starting from (V.50) we obtain also:

$$\frac{dH}{dt} = -\frac{1}{4}\iiiint\iiiint(\ln ff_1 - \ln f'f_1')(ff_1 - f'f_1') \times \qquad \text{(V.51)}$$

$$\times\ \psi(g, \chi)\, du\, dv\, dw\, du_1\, dv_1\, dw_1\, d\Omega,$$

whence we obtain, for the same reasons as before

$$\frac{dH}{dt} \leqq 0; \qquad \text{(V.52)}$$

there can be no equality sign unless we have identically

$$ff_1 - f'f_1' = 0. \qquad \text{(V.53)}$$

This is the principle of detailed balancing in classical statistical mechanics (it assumes this form only for molecules with spherical symmetry); we know that (V.53) has the Maxwell–Boltzmann distribution as its only solution:

$$f(u, v, w) = ae^{-b[(u-u_0)^2+(v-v_0)^2+(w-w_0)^2]}. \qquad \text{(V.53′)}$$

We note that condition (β) allows us here to simplify the calculation of $\left(\frac{dn_i}{dt}\right)_P$, since the change in the numbers n_i then comes exclusively from binary collisions. We know that, in the general case, we introduce into the first term of (V.50) terms which express the effect of a density gradient or of external forces on the gas.

Thus, we end with the complete Boltzmann equation:

$$\frac{\partial f}{\partial t} + u\frac{\partial f}{\partial x} + v\frac{\partial f}{\partial y} + w\frac{\partial f}{\partial z} + X\frac{\partial f}{\partial u} + Y\frac{\partial f}{\partial v} + Z\frac{\partial f}{\partial w}$$

$$= \iiiint\!\!\int (f'f_1' - ff_1)\,\psi(g,\chi)\,du_1\,dv_1\,dw_1\,d\Omega, \qquad \text{(V.50')}$$

where now $f \equiv f(u, v, w, x, y, z, t)$ and where X, Y and Z are the components of the specific external forces, assumed to be velocity-independent; in order to obtain the collision term, we must assume further the principle of molecular disorder (or the principle of molecular chaos in Boltzmann's sense), which postulates the absence of correlations between the positions and velocities of two particles prior to collision: the number of collisions of a given kind is then proportional to the product ff_1. Thus, it is not possible to arrive at equation (V.50') by simply following the method which we have just used to obtain (V.50), since this would involve an additional assumption expressing that the effects of the density gradient and the external forces, on the one hand, and those of binary collisions, on the other hand, can be considered separately. There exists, in fact, "interference" between the first and second term of (V.50') as we shall see in Section V in connection with Bogolyubov's method: the interpretation of Boltzmann's equation (V.50') is also made more critical by it.

Nevertheless, it is important to emphasise that in its complete form Boltzmann's equation makes possible a quantitative description of the various transport phenomena and of the tendency towards equilibrium in agreement with experimental results, thanks to various approximation methods due to Maxwell (1867, 1879a), Lorentz (1905). Hilbert (1912), Enskog (1917), Chapman (see Chapman and Cowling, 1953), and Grad (1949, 1952, 1958) (see also Jancel and Kahan, 1964). In addition, we mention that we can prove similarly an H-theorem starting from (V.50'), by means of certain limit conditions and for external forces which can be derived from a potential.

3. *Statistical interpretation of Boltzmann's equation*

An analysis of the foregoing results gives rise to many important points:

(a) Knowledge of Boltzmann's function $f(u, v, w, t)$ is equivalent to a knowledge of the occupation numbers n_i, with the result that if

$f(u, v, w, t_0)$ is given we know in which star Ω_0 of Γ-space the system exists; expressions (V.50) and (V.51) for $\partial f/\partial t$ and dH/dt refer to the most probable evolution of $H(Z, t)$ and of $f(u, v, w, t)$ beginning with the appropriate initial expressions. The relation between $f(u, v, w, t)$ and the coarse-grained density in Γ-space is thus made more precise; this relationship makes certain that the evolution described by means of $f(u, v, w, t)$ is irreversible.

(b) In fact, Boltzmann's equation is deduced from considerations which depend on the division of μ-space into *finite* cells ω_i; when we use the differential form (V.50) or (V.51) we assume that these cells are sufficiently small on a macroscopic scale to be replaced by the volume element $du dv dw$: but we must never lose sight of the fact that each of these volume elements must be large enough to contain a finite and large number n_i of molecules and large enough that f be a continuous function of (u, v, w).† These points emphasise the essentially statistical nature of Boltzmann's equation, since knowledge of f does not constitute a precise definition of the mechanical state of the system.

(c) By depending on the foregoing points, we can avoid the usual paradoxes of the kinetic theory of gases: it is well known that these originate from either the reversibility principle of classical mechanics [Loschmidt's paradox (Loschmidt, 1876a, b, 1877): to every evolution of a system such that H is decreasing, there corresponds the inverse mechanical evolution in which H increases with time], or Poincaré's recurrence theorem [Zermelo's paradox (Zermelo, 1896a, b): after a sufficiently long time, the system passes through states close to the states which it has occupied previously with the result that H must increase after a certain period].

We must note first of all that these paradoxes both rely on theorems of mechanics which deal with an exact dynamic state of the system: thus, they cannot contradict the results of the H-theorem, which depend on the most probable behaviour of an ensemble of such mechanical systems. This, according to the nature of the proof itself, is essentially a theorem of statistical mechanics; thus, it is applicable only to a system whose macroscopic state is known. To this macroscopic state there corresponds a very large number of microscopic states capable of containing states which give rise

† For a definition of the kinetic method, see also Section V of this chapter.

to an increase of H. Theorem (V.49) then shows us that the number of these latter is certainly very small relative to the states corresponding to a decrease of H, if the system is sufficiently remote from equilibrium; the decrease of H can thus be considered as very probable, at least for a small duration following the initial observation (obviously, this must be related to the properties of the sharp maximum of $W(\Omega_{max})$). The reversibility principle involves, on the contrary, only a single microscopic state of the system and the effective realisation of an inverse evolution would be possible only on condition that the speeds of all the molecules were accurately known and that they could all be reversed at a given instant. Thus, the two theories are applicable to very different physical situations: by introducing the concept of the preparation of a system in a given macroscopic state, the H-theorem simply expresses the fact that it is much easier, when the macroscopic state considered is sufficiently far from equilibrium, to prepare the system in a state which gives rise to a decrease of H than to subject it to initial conditions such that H increases.

With regards to Zermelo's paradox, this concerns the evolution of an isolated system over a very long duration; according to Poincaré's theorem, every decrease of H must be compensated at the end of an interval of time which may be very long, by a corresponding increase, with the result that increases and decreases of H are produced with equal frequency for an isolated system. This conclusion cannot be avoided since, if we consider the microcanonical ensemble representing an isolated system, there is corresponding to every state in this ensemble always an inverse state for which the direction of evolution is found to be reversed; moreover, this is not a contradiction of the fact that the most probable evolution of an isolated system at a given instant is a decrease of H, since the statistical nature of theorem (V.49) does not exclude the possibility of an increase of H for certain states. Statistical and mechanical considerations, on the contrary, are complementary to one another and enable us [as shown by P. and T. Ehrenfest (1911); see also Appendix III, § 5] to describe the evolution of $H(Z, t)$ as a series of fluctuations above the value H_{min}; moreover, these fluctuations are the more unusual the more they deviate from the equilibrium position; they are, in fact, unobservable events because of the enormous duration of the time of recurrence (see § I.3 of this chapter).

(d) The considerations of the previous section enable us to describe the evolution of a macroscopic system, either for a very long or very short duration after the instant of the initial observation. It remains now for us to study the irreversible tendency of the system towards its equilibrium state, which involves intermediate durations of the order of relaxation times. If, for this purpose, we use Boltzmann's equation (V.51), we have already seen that it introduces surreptitiously a supplementary assumption which we shall now analyse. We note first of all that formulae (V.49) and (V.51) rely essentially on the fundamental postulate of statistical mechanics which, in the case we are considering, is equivalent to the molecular chaos hypothesis (Stosszahlansatz); but this theory is applicable only to the instant at which the initial macroscopic observation is completed and this enables us to define the statistical ensembles used. It follows that the most probable numbers of collisions given by (V.38) and (V.38′) are valid only during a short interval of time τ which follows the instant of observation; the same is, therefore, true for the value of $\left(\frac{dH}{dt}\right)_P$. We conclude from this that if we wish to describe the irreversible evolution of a system and its tendency towards equilibrium (relaxation time) by equations (V.50) and (V.51), we must make not only the assumption of molecular chaos (which provides us with the numbers a_{kl}^{ij}) but we must accept also that *this assumption is valid at every instant*: let us show that this latter condition is capable of contradicting the laws of mechanics, which was not the case for the assumption of molecular chaos (at a given instant t_0).

We suppose that, in fact, an initial observation at t_0 had shown us that the system occupied a star Ω_0 to which, for the function H, there corresponds a certain value $H(t_0)$ which is appreciably different from $H_{\min}$. Since nothing was known about the system prior to the instant of observation, we associate with this condition the virtual ensemble of systems represented by the points of Ω_0 and in this way we can calculate the most probable value of dH/dt at the instant t_0, or $\left(\left(\frac{dH}{dt}\right)_{t_0}\right)_P$. We deduce from it that the most probable value of H at an instant t_1 (close to t_0) is given by

$$H(t_1) = H(t_0) + \left(\left(\frac{dH}{dt}\right)_{t_0}\right)_P (t_1 - t_0) \qquad \text{(V.54)}$$

and it is assured that $H(t_1) < H(t_0)$; the most probable star at t_1 is a star Ω_1 which corresponds to $H(t_1)$. If, however, we wish to proceed further we must assume that a new observation on the system is made at t_1, which will show us that it occupies a new star Ω_1', which is generally different from but close to Ω_1 and such that the value of $H'(t_1)$ which corresponds to it is very probably less than $H(t_0)$. If we wish, starting from this new situation, to calculate the value of $\left(\left(\frac{dH}{dt}\right)_{t_1}\right)_P$ in order to deduce the ultimate evolution of the system, we must construct a new statistical ensemble compatible with the observation at t_1. By proceeding as before, it would be shown that $\left(\left(\frac{dH}{dt}\right)_{t_1}\right)_P$ is always negative and that the most probable value of H at an instant t_2 satisfies the relationship $H(t_2) < H(t_1)$; the decrease of H during an evolution of long duration would thus be proved, on condition that the series of successive observations of the physical system† is taken into account. (We note that it is not necessary for the observation to be actually made, since we are dealing here with a macroscopic observation to which corresponds an objective state of the system.) It must be emphasised, however, that it is no longer valid to consider solely, in order to determine the new ensemble at t_1, the results of the observation at this instant: in fact, we now know something more about the system—we know that it occupied the star Ω_0 at t_0, with the result that we would have to assume—as elements of the ensemble in question—only systems of Ω_1' which had occupied Ω_0 initially; it is clear that this selection can be made only after integration of the equations of motion. We conclude from this, that if the results of (V.38)–(V.38′) (which are valid only at t_0) are applied to the instant t_1, we must assume that the "macroscopic system has no memory" and that everything proceeds as if we were able to ignore its previous development.

Nevertheless, it is obvious that such an assumption will naturally lead to a contradiction of the laws of mechanics and of the consequence of Liouville's equation; indeed, it introduces a random element into the evolution itself and no longer only into the initial conditions. The question of knowing how we can reconcile a

† The same point of view has also been adopted by ter Haar (see in particular ter Haar, 1955).

deterministic mechanical evolution of a system with such a model will be dealt with in Section V, but we can mention at this point that it is reasonable to assume that in the case where the observations do not provide a precise specification of the state of the system and where the number of degrees of freedom is very large, the states to be left out of the ensemble are quite small in number and distributed in such a way that the sign of dH/dt is not modified by their elimination.

Because of these arguments, we can see finally that the use of Boltzmann's equation in studying the irreversible evolution of dilute gases is valid only provided that we accept that the effectively observed state is well represented by the most probable state [determined by (V.54)] and that we may apply to each stage of evolution suitable statistical assumptions without taking account of the previous states occupied by the gas.

We shall end this section by comparing Boltzmann's H-theorem, which we established in (V.49), with the generalised $\bar{H}$-theorem of the preceding section. If we return to the definition of $\bar{H}$ according to (V.27), it can be seen that we can write

$$\bar{H} = \sum_n G_n \frac{P'_n}{G_n(\delta v)_\Gamma} \ln \frac{P'_n}{G_n(\delta v)_\Gamma} (\delta v)_\Gamma = \sum_n P'_n \ln \frac{P'_n}{G_n(\delta v)_\Gamma} \qquad \text{(V.27'')}$$

since we have, from (V.25') and (V.24), $P'_n = P_n G_n(\delta v)_\Gamma$. On the other hand, according to (V.34), we can put

$$\ln \frac{1}{G_n(\delta v)_\Gamma} = H_n,$$

where H_n is Boltzmann's H, associated with a system in the macroscopic state n. We can then write

$$\bar{H} = \sum_n P'_n H_n + \sum_n P'_n \ln P'_n = \bar{H}_{\text{syst}} + \sum_n P'_n \ln P'_n \qquad \text{(V.54')}$$

and we can see that $\bar{H}$ is the sum of two terms: $\bar{H}_{\text{syst}}$, which is the average, for all members of the ensemble, of the values of Boltzmann's H corresponding to each state n, and $\sum_n P'_n \ln P'_n$, which is related to the distribution of the systems of the ensemble over the different stars Ω_n. In the special case where $P'_n = 1$, which represents a system in the state n, we have simply $\bar{H} = H_n$.

In order to compare the two H-theorems, we now differentiate

(V.54′) with respect to t; this gives us

$$\frac{d\bar{H}}{dt} = \frac{d\bar{H}_{\text{syst}}}{dt} + \sum_n \frac{dP'_n}{dt} \ln P'_n ,$$

whence it results that the decrease of $\bar{H}$ originates from two terms:

(a) the term in $d\bar{H}_{\text{syst}}/dt$, which decreases by virtue of Boltzmann's H-theorem; actually, since H_n is decreasing for each system of the ensemble, the same is the case for the average value of H_n in the ensemble;

(b) the term in $\sum_n \frac{dP'_n}{dt} \ln P'_n$ which decreases similarly because of the qualitative reasoning of Section II and this decrease expresses a progressive uniformisation of the ensemble over the different possible conditions Ω_n. We have thus established the relation between Boltzmann's H-theorem and the generalised $\bar{H}$-theorem corresponding to Gibbs' ideas.

IV. Stochastic Processes and *H*-Theorems

1. *Stochastic evolution of a distribution* $P(n_1, ..., n_i, ...; t)$ *in* Γ*-space. "Master Equation"*

We have just seen that if we wish to obtain quantitative results relative to the irreversible evolution of a system, we have so far available only Boltzmann's equation (valid for dilute gases). However, we have seen that this involves assumptions which can contradict the laws of mechanics and which depend essentially on the application at each instant of the principle of molecular chaos; thus, we introduce ar andom element into the evolution itself of a macroscopic system; we must analyse this now. We shall show in this section that the evolution of the system is comparable, at least in the case of spatial homogeneity, to a Markovian stochastic process: the tendency towards an equilibrium distribution can then be established, thanks to recent advances in the theory of stochastic processes (Chandrasekhar, 1943; Moyal, 1949; Kac, 1959; Siegert, 1949; Wang and Uhlenbeck, 1945)† which we must first of all discuss briefly.

Since our aim is to reconcile the dynamic reversibility and the irreversible evolution of a system, it is important to note that the characteristic aspect of Markovian steady-state processes is

† For mathematical papers dealing with these problems, we refer to the references at the end of the book.

precisely that of combining the quasi-periodic motion of the individual elements of an ensemble with a monotonic tendency of the initial probability function towards an equilibrium distribution: it is because of this property that we try now to compare the evolution of a macroscopic system with those processes for which we shall derive the fundamental equation.

Suppose, then, that a stochastic variable X takes a discrete set of values X_i at discrete instants of observation $t_s = s\tau$. If $P(X_j|X_i, s)$ is the conditional probability that we have $X = X_i$ at the instant $s\tau$, if $X = X_j$ at $t = 0$, this probability will satisfy Smoluchowski's equation

$$P(X_j|X_i, s) = \sum_l P(X_j|X_l, s-1)\, P(X_l|X_i, 1) \qquad \text{(V.55)}$$

in the case of a Markovian process.

The conditional probabilities $P(X_l|X_i, 1)$ are called transition probabilities and, if we denote them by $Q(X_l, X_i)$, we have

$$\sum_i Q(X_l, X_i) = 1.$$

We can then rewrite (V.55) in the form

$$P(X_i, s) - P(X_i, s-1)$$
$$= \sum_l{}' \, [P(X_l, s-1)\, Q(X_l, X_i) - P(X_i, s-1)\, Q(X_i, X_l)], \quad \text{(V. 55}_1\text{)}$$

where we have omitted for simplicity the initial state X_j and where $\sum_l{}'$ signifies that the sum is taken over all states except the one corresponding to $l = i$. This latter equation is easily interpreted: in fact, the time variation in the number of states X_i is equal to the gains due to transitions $X_l \to X_i$, less the losses due to transitions $X_i \to X_l$. If all the variables are continuous, Smoluchowski's equation can be written (with the same notation) as

$$\frac{\partial P(X, t)}{\partial t} = \int dY [P(Y, t)\, Q(Y, X) - P(X, t)\, Q(X, Y)]. \quad \text{(V.55}_2\text{)}$$

It can then be shown in the discrete case, under very general conditions, that the conditional probability $P(X_i, s)$ tends, as $s \to \infty$, in a monotonic way towards an equilibrium distribution, although each series of observations $X_1, X_2, \ldots,$ shows a quasi-periodic behaviour without a privileged direction of evolution; this result, expressing the ergodic properties of Markovian chains, probably extends to the continuous case.

In order to define more precisely the relations existing between the evolution of a gas and such a stochastic process, it is convenient to consider the total Γ-space of the system. Let us suppose that the system be sufficiently dilute that binary collisions between molecules are relatively rare. In the absence of collisions, the representative point of the system $\mathscr{P}(t)$ describes a straight line over a constant-energy hypersurface in Γ-space; at each collision, it makes a sudden jump on the hypersurface, then describes a new straight line, and so on. This motion resembles a "random walk" over the hypersurface, which seems to justify the description of the irreversible evolution of the system by a Markovian stochastic process in Γ-space, since the random walk problem is a well-known special case of Markovian processes.

If we return to the macroscopic description of the system, Γ-space appears to be divided into stars Ω_i (corresponding to the set $\{n_i\}$) and the system passes successively and in a random way (since we do not know its precise mechanical state) during its evolution from one star to another. Thus, we shall try to describe such an evolution by a Poisson type Markovian process in Γ-space; in order that such a problem is well defined, it is sufficient to know the probabilities for a transition per unit time which correspond to the probabilities $P(\Omega_i, \Omega_j)$ about which we spoke in the previous section. In the case where the gas is sufficiently dilute and spatially uniform, the obvious choice is to take as the expression for these probabilities the most probable number of transitions, calculated on the assumption of molecular chaos: the irreversible evolution of a gas will be described therefore by a Markovian process with transition probabilities given by (V.38) and (V.38′). There will be a distribution function corresponding to the random motion of the point $\mathscr{P}(t)$ in Γ-space, and it will depend on the quantities n_i (defining the stars Ω_i). If $P(n_1, n_2, \ldots, n_i, \ldots; t)$ denotes this distribution function it will obey the following equation, which is analogous to (V.55$_1$):

$$\frac{\partial P(n_1, \ldots, n_i, \ldots; t)}{\partial t} = \frac{1}{2} \sum_{(ij),(kl)} [A_{ij}^{kl}(n_k + 1)(n_l + 1) P(n_1, \ldots, n_{k+1}, \ldots, n_{l+1}, \ldots, n_{i-1}, \ldots, n_{j-1}, \ldots; t) - A_{kl}^{ij} n_i n_j P(n_1, \ldots, n_i, \ldots; t)]. \tag{V.56}$$

This is the "Master Equation": the first term on the right-hand side corresponds to the molecules which enter per unit time into the star defined by the numbers n_i and the second term corresponds to molecules which leave this same star; we see that (V.56) is a linear equation, which constitutes an essential difference with Boltzmann's equation. The distribution $P(n_1, ..., n_i, ...; t)$ is equivalent to the coarse-grained density in Ehrenfest's sense but its evolution with time is different from the evolution of the coarse-grained density: in fact, $P(n_1, ..., n_i, ...; t)$ is the distribution corresponding to a Markovian stochastic process [obeying (V.56)] whilst the evolution of the coarse-grained density is determined by Liouville's equation (which expresses the laws of mechanics) as well as by the size of the stars in Γ-space.

If we denote the sum over all the n_i by $\sum_n$, so that $\sum_i n_i = N$, we must have

$$\sum_n P(n_1, ..., n_i, ...; t) = 1. \qquad \text{(V.57)}$$

Actually, by taking $\sum_n$ of (V.56) we can prove that this normalisation is conserved in the course of time; we then have

$$\frac{\partial}{\partial t} \sum_n P(n_1, ..., n_i, ...; t) = 0, \qquad \text{(V.58)}$$

in which we have used the microreversibility $A_{kl}^{ij} = A_{ij}^{kl}$ and the symmetry of (V.56) in (ij) and (kl). If, now, we define the average values of the n_q by

$$\bar{n}_q = \sum_n n_q P(n_1, ..., n_i, ...; t), \qquad \text{(V.59)}$$

we obtain, according to (V.56):

$$\frac{d\bar{n}_q}{dt} = \frac{1}{2} \sum_{(ij),(kl)} A_{ij}^{kl} \times$$
$$\sum_n [n_q(n_k + 1)(n_l + 1) P(n_1, \ldots, n_{k+1}, \ldots, n_{l+1} \ldots n_{i-1}, \ldots, n_{j-1}, \ldots; t)$$
$$- n_q n_i n_j P(n_1, ..., n_i, ...; t)]. \qquad \text{(V.60)}$$

We can verify easily that the cubic terms in the n_i in (V.60) always cancel out, so that we can write, according to definitions

(V.59),

$$\frac{d\bar{n}_q}{dt} = \frac{1}{2} \sum_{j(kl)} A_{jq}^{lk} (\overline{n_k n_l} - \overline{n_q n_j}). \tag{V.61}$$

By comparison with equations (V.44)–(V.45), it can be seen that (V.61) assumes a form which is equivalent to Boltzmann's equation, provided we put

$$\begin{cases} \overline{n_k n_l} = \overline{n_k} \cdot \overline{n_l}, \\ \overline{n_q n_j} = \overline{n_q} \cdot \overline{n_j}, \end{cases} \tag{V.62}$$

which amounts to neglecting correlations between the values of n_i: this can be regarded as not too restrictive if the values of n_i are very large. In the same way, we would obtain a hierarchic chain of coupled linear equations for the time-derivatives of higher-order moments; Boltzmann's non-linear equation appears therefore as an approximation to the first equation of this chain of linear equations.†

On the other hand, Siegert (1949) has shown that the Master Equation (V.56) leads, as $t \to +\infty$, to the microcanonical equilibrium distribution, which justifies the application of (V.56) to the study of the irreversible evolution of a macroscopic system. Moreover, the foregoing argument shows us that we can obtain Boltzmann's equation starting from the distribution $P(n_1, ..., n_i, ...; t)$ provided that:

(a) the average occupation number value $\bar{n}_q$, calculated by means of $P(n_1, ..., n_i, ...; t)$, is identified with the most probable value defined in the previous section;

(b) the absence of correlations between the n_i is assumed.

Having obtained these first results, the final development of the theory can be undertaken by following several different routes.

1. The foregoing scheme can be used with a view to examining thoroughly the real significance of the molecular chaos hypothesis, which is expressed here by the condition (V.62): the first problem, then, is to study the compatibility of this condition with the Master Equation. Since this equation is linear and only involves first-order time-derivatives, its solution is completely defined in principle

† This is associated, naturally, with the derivation of Boltzmann's equation, starting from the B.B.G.K.Y. chain of equations.

if we are given the initial distribution $P(n_1, \ldots, n_i, \ldots; 0)$ which we can always choose in such a way that (V.62) is satisfied at the initial instant. Having determined the general solution, $P(n_1, \ldots, n_i, \ldots; t)$ of (V.56) corresponding to this initial distribution it would be neccessary then to prove that condition (V.62) is conserved, that it is satisfied completely at every instant, and that it is thus a consequence of the Master Equation, provided that it is introduced by virtue of the initial condition on $P(n_1, \ldots, n_i, \ldots, 0)$. Since the proof of this property is very tricky in the general case, we shall only consider this problem within the framework of Kac's simplified model, which we shall deal with in the second part of this section.

2. On the other hand, we must try to reconcile the Master Equation and Liouville's equation. Actually, we know that the Master Equation results from the introduction of a random element into the evolution of the gas: thus, it is essential to define precisely to what degree of approximation this description is compatible with Liouville's deterministic equation, which remains the exact equation governing the mechanical evolution of a system. We shall see, by explaining in the next section the interesting attempt of Brout, that it is not impossible to achieve such a programme, provided that we deal with spatially uniform systems; this condition, unfortunately, is very restrictive since it involves the impossibility of deriving the complete Boltzmann equation (with the flow terms) from the Master Equation.

3. Finally, we can envisage numerous applications of this general theory to various special models, such as Ehrenfest's so-called "dog-flea" and "wind-wood" models (Ehrenfest, 1907; Smoluchowski, 1912; ter Haar and Green, 1953, 1955; Green and ter Haar, 1955). The interest in these models is that they deal with extremely simple cases, in which the foundations of the theory of irreversible processes can be discussed more easily, due especially to a thorough comparison between the coarse-grained density $P(q_i, p_i, t)$ and the distribution $P(n_i, \ldots, n_i, \ldots; t)$; in particular, it would be possible thus to define precisely the role of the size of the stars Ω_i which, in any case, will correspond always to large values of n_i, if we wish to justify the application of stochastic processes to the description of the irreversible evolution of macroscopic systems.

2. *Kac's model* (Kac, 1954, 1959)

We shall conclude this section by studying briefly a model proposed by Kac; this enables us to demonstrate, in a very special case, the conservation during evolution of the property of molecular chaos. We consider a perfect gas, spatially homogeneous, in which the only evolution factors are the collisions between molecules; if $\boldsymbol{v}_1, \ldots, \boldsymbol{v}_n$ are the velocities of the n molecules, the representative point of the system moves in the $3n$-dimensional Γ-space over a hypersphere with $3n - 3$ dimensions, defined by the relations

$$\begin{cases} \sum_i \boldsymbol{v}^2 = nw^2, \\ \sum_i \boldsymbol{v}_i = \boldsymbol{c}, \end{cases} \tag{V.63}$$

which express the conservation of energy and momentum during collisions. (Conservation of momentum is not realised if we take into account wall effects; this restriction actually arises from the fact that the reduced Boltzmann equation is not strictly valid if we take into account the external force corresponding to the presence of a container.)

If $\boldsymbol{R}$ is a point of this hypersphere, we shall denote by $\Phi(\boldsymbol{R}, t)\, d^{3n-3}\boldsymbol{R}$ the probability that the representative point of the system is in the element $d^{3n-3}\boldsymbol{R}$ around the point $\boldsymbol{R}$ at the instant t, so that $\Phi(\boldsymbol{R}, t)$ represents the probability density as function of $\boldsymbol{R}$ over the hypersurface (V.63). By assuming now that the evolution of the system can be described by a random walk over the hypersphere (V.63), with the transition probabilities determined by the hypothesis of molecular chaos, $\Phi(\boldsymbol{R}, t)$ will satisfy the Master Equation

$$\frac{\partial \Phi(\boldsymbol{R}, t)}{\partial t} = \frac{1}{2V} \sum_{i,j} \int \{\Phi(\Theta_{ij}\boldsymbol{R}, t) - \Phi(\boldsymbol{R}, t)\}\, \psi_{ij}\, d^2\Omega \equiv L\Phi. \tag{V.64}$$

In (V.64), $\boldsymbol{R}' = \Theta_{ij}\boldsymbol{R}$ represents the rotation experienced by $\boldsymbol{R}$ during the impact of the two molecules i and j, whose line of centres forms the axis of the solid angle element $d^2\Omega$; Θ_{ij} is thus an operator which depends on the collision parameters (g_{ij} and χ) and the probability for such a collision is $\frac{\psi_{ij}}{V}\, d^2\Omega$ by virtue of the hypothesis of molecular chaos (V.38) (V is the volume occupied by the gas and we have: $\psi_{ij} = g_{ij}\, \sigma_d(g_{ij}, \chi)$).

As before, we note that we are dealing with a linear equation (L is a linear operator). In order to establish the relation with Boltzmann's equation it is necessary to introduce "successive contractions" (or "marginal distributions") of the function $\Phi(\boldsymbol{R}, t)$. We have, by definition, for the distribution function with k particles, $F_k^{(n)}$:

$$F_k^{(n)} = \underbrace{\int \dots \int}_{n-k} d^3v_{k+1} \dots d^3v_n \Phi(\boldsymbol{R}, t), \qquad \text{(V.65)}$$

and, according to (V.64)–(V.65), the integro-differential equation satisfied by $F_1^{(n)}$ can be written as

$$\frac{\partial F_1^{(n)}}{\partial t} = \frac{n-1}{V} \int d^3v_2 \int \{F_2^{(n)}(v_1', v_2', t) - F_2^{(n)}(v_1, v_2, t)\} \, \psi(g, \chi) \, d^2\Omega. \qquad \text{(V.66)}$$

We see that (V.66) is equivalent to Boltzmann's equation, if we identify the function f of the kinetic theory of gases with $\frac{n}{V} F_1^{(n)}$ and if we put

$$F_2^{(n)}(v_1, v_2, t) = F_1^{(n)}(v_1, t) \, F_1^{(n)}(v_2, t). \qquad \text{(V.67)}$$

We restate in (V.67) the assumption about the absence of correlations which we have encountered in (V.62), but described here with the marginal distributions in place of the average values. We shall study now the compatibility between relation (V.67) and the Master Equation which determines the time evolution of $\Phi(\boldsymbol{R}, t)$. We note immediately that this relation amounts to imposing a certain condition on $\Phi(\boldsymbol{R}, t)$; however, because of the linearity of (V.64), $\Phi(\boldsymbol{R}, t)$ is unambiguously determined once we know $\Phi(\boldsymbol{R}, 0)$. It follows that the property postulated in (V.67) must be incorporated in the initial conditions $\Phi(\boldsymbol{R}, 0)$ and that it must be conserved with time. The non-linear nature of Boltzmann's equation in this case would originate from the initial form of the distribution function $\Phi(\boldsymbol{R}, 0)$; thus, in the case of spatial homogeneity, the real significance of Boltzmann's equation would be stated explicitly: it would correspond to a Markovian stochastic process with transition probabilities given by the "Stosszahlansatz" and an initial distribution satisfying (V.67).

It is difficult to prove the conservation in time of the property (V.67) in the general case of equation (V.64). Kac has been able to prove this conservation for a considerably simplified model: we first of all abandon the conservation of momentum, which amounts to the removal of the second equation of (V.63) (motion now takes place over the hypersphere S_{3n-1} with $(3n-1)$ dimensions and with radius $w\sqrt{n}$); we next restrict ourselves to the case in which the collision probability ψ_{ij}/V is constant and equal to $\nu/2\pi n$; finally, we simplify the rotations Θ_{ij} in six dimensions by replacing them with rotations in two dimensions $A_{ij}(\theta)$ in the (ij) plane. Equation (V.64) can then be written in the following form ($\boldsymbol{R}$ is now a point on the sphere S_{3n-1}):

$$\frac{\partial \Phi(\boldsymbol{R}, t)}{\partial t} = L_r \Phi, \tag{V.68}$$

where L_r is the reduced linear operator defined by

$$L_r = \frac{\nu}{4\pi n} \sum_{i,j} \int_0^{2\pi} d\theta (e^{i\theta L_{ij}} - 1), \tag{V.69}$$

and where L_{ij} is the operator of infinitesimal rotations in the (ij) plane. We derive from (V.69) the analogue of Boltzmann's equation by "contraction" and by adding an assumption of the type (V.67). Because of the linearity of (V.68), it is necessary to introduce this assumption into the initial distribution $\Phi(\boldsymbol{R}, 0)$ and to establish its conservation with time. By virtue of the properties of the operator L_r, we can show that if a series of distributions $\Phi_n(\boldsymbol{R}, 0)$ which are quadratically summable on S_{3n-1} possesses "Boltzmann's property",

$$\lim_{n\to\infty} F_k^{(n)}(x_1, \ldots, x_k; 0) = \prod_{j=1}^{k} \lim_{n\to\infty} F_1^{(n)}(x_j, 0), \tag{V.70}$$

the functions $\Phi_n(\boldsymbol{R}, t)$ which are solutions of (V.68) also satisfy "Boltzmann's property", that is,

$$\lim_{n\to\infty} F_k^{(n)}(x_1, \ldots, x_k; t) = \prod_{j=1}^{k} \lim_{n\to\infty} F_1^{(n)}(x_j, t). \tag{V.70′}$$

This expresses the "propagation" in time of Boltzmann's property; the result is that the non-linear nature of Boltzmann's equation originates from the initial conditions satisfied by the function $\Phi(\boldsymbol{R}, 0)$ which elucidates in this case the contents of the hypothesis of molecular chaos.† Unfortunately, this proof is valid only for

† We note once again the essential role played by the large number of degrees of freedom of the system (transition to the limit $n \to \infty$).

the very special model which we have considered and can be extended only with difficulty to cases which are of physical interest (we mention, for example, that it is inapplicable to the case of the elastic hard spheres); nevertheless, it indicates an interesting field of research.

We can show also without difficulty, by starting from (V.68) an *H*-theorem which expresses the ergodic properties of the Markovian process considered. Indeed, we have

$$\frac{d}{dt}\int \Phi(\boldsymbol{R}, t) \ln \Phi(\boldsymbol{R}, t)\, d\Sigma \leqq 0, \qquad \text{(V.71)}$$

the equality holding for the microcanonical distribution

$$\Phi(\boldsymbol{R}, t) = \frac{1}{\mu(S_{3n-1})}, \qquad \text{(V.72)}$$

where $\mu(S_{3n-1})$ is the measure of the $(3n - 1)$-dimensional hypersurface and where the convergence $\Phi(\boldsymbol{R}, t) \to \dfrac{1}{\mu(S_{3n-1})}$ is a weak convergence in Hilbert space associated with the functions $\Phi(\boldsymbol{R}, t)$. We can deduce from (V.71) an *H*-theorem which is analogous to that of Boltzmann and we can show similarly that for the contracted distribution $f_1^{(n)}$ we have

$$\lim_{t\to\infty} f_1^{(n)}(x, t) \sim \frac{1}{\sqrt{2\pi}} e^{-x^2/2}. \qquad \text{(V.73)}$$

In addition, we emphasise that we can prove, without additional difficulties, the validity of these ergodic properties for the Master Equation (V.64) itself, contrary to what occurs for the conservation in time of Boltzmann's property.

V. Integration of the Liouville Equation

Throughout the foregoing sections we have been halted frequently by the problem of the integration of Liouville's equation; this has been the case first in studying the time evolution of coarse-grained densities, then for discussing the validity of the theory of molecular chaos in Boltzmann's equation, and finally in justifying the Master Equation. For this reason, we shall devote the present section to a more detailed study of Liouville's equation, by pointing out the principal methods used for deriving kinetic equations of the

Boltzmann type or for justifying the description of the evolution of a system by a Markovian stochastic process; in doing this we shall mainly emphasise the various hypotheses which introduce an element of irreversibility into the evolution. Having discussed the derivation of the reduced chain of equations (the so-called B.B.G.K.Y. chain of equations), we shall analyse in succession the essential principles of the methods of Born and Green, Yvon, Kirkwood, and Bogolyubov, and we shall conclude by discussing Brout's method, which establishes a bridge between the Master Equation and Liouville's equation.

1. *The B.B.G.K.Y. equations* (Bogolyubov, Born and Green, Kirkwood, Yvon)

We derive these from Liouville's equation by successive integrations (process of marginalisation) and we are guided by the following remarks: in general, observed macroscopic quantities are independent of the distribution function for n particles $\varrho_n(\boldsymbol{x}_1, \ldots, \boldsymbol{x}_n, t)$,† but depend only on the distribution function for s particles, where s is nearly always small. Frequently, even, it is sufficient to know the distribution function of a single particle ($s = 1$), i.e. the distribution function in μ-space. Since these functions are obtained by integration of the distribution $\varrho_n(\boldsymbol{x}_1, \ldots, \boldsymbol{x}_n, t)$ in Γ-space over the $n - s$ remaining variables (marginalisation), they satisfy the differential equations derived from Liouville's equation by integrating over these $n - s$ variables. We are thus led to defining the specific distribution functions for s particles, $\varrho_s(\boldsymbol{x}_1, \ldots, \boldsymbol{x}_s, t)$, and the corresponding generic function $f_s(\boldsymbol{x}_1, \ldots, \boldsymbol{x}_s, t)$ by

$$\left.\begin{aligned} \varrho_s(\boldsymbol{x}_1, \ldots, \boldsymbol{x}_s, t) &= \frac{1}{V_s} F_s(\boldsymbol{x}_1, \ldots, \boldsymbol{x}_s, t) \\ &= \underbrace{\int \cdots \int}_{n-s} \varrho_n(\boldsymbol{x}_1, \ldots, \boldsymbol{x}_n, t)\, d^6\boldsymbol{x}_{s+1} \ldots d^6\boldsymbol{x}_n, \\ f_s(\boldsymbol{x}_1, \ldots, \boldsymbol{x}_s, t) &= \frac{n!}{(n-s)!} \varrho_s(\boldsymbol{x}_1, \ldots, \boldsymbol{x}_s, t); \end{aligned}\right\} \qquad (V.74)$$

we have introduced the functions F_s in order to emphasise the role

† In what follows $\boldsymbol{x}_i$ stands for $\boldsymbol{r}_i, \boldsymbol{p}_i$, that is, it is a point in the 6-dimensional μ-space (monatomic gas).

of the volume V containing the system, using Bogolyubov's notation. According to these definitions, $\varrho_s(\boldsymbol{x}_1, \ldots, \boldsymbol{x}_s, t)\, d^6\boldsymbol{x}_1 \ldots d^6\boldsymbol{x}_s$ represents the probability of finding the s particles considered in the element of volume $d^6\boldsymbol{x}_1 \ldots d^6\boldsymbol{x}_s$ at t, whatever the dynamic state of the other $n - s$ particles.

The differential equations satisfied by $\varrho_s(\boldsymbol{x}_1, \ldots, \boldsymbol{x}_s, t)$ are then obtained by integrating Liouville's equation (V. 1) over the $n - s$ variables $\boldsymbol{x}_{s+1}, \ldots, \boldsymbol{x}_n$. We shall assume in what follows that the Hamiltonian of the system is of the form

$$H = \sum_{1 \leqq i \leqq n} H(\boldsymbol{x}_i) + \sum_{1 \leqq i < j \leqq n} \Phi_{ij}, \tag{V.75}$$

where $\Phi_{ij} \equiv \Phi(|\boldsymbol{r}_i - \boldsymbol{r}_j|)$ represents the interaction potential between the pair of particles (i, j) and $H(\boldsymbol{x}_1)$, which is given by

$$H(\boldsymbol{x}_i) = \frac{\boldsymbol{p}_i^2}{2m_i} + U(\boldsymbol{r}_i), \tag{V.76}$$

is the sum of the kinetic energy of the particle i and the potential energy $U(\boldsymbol{r}_i)$ due to the external forces and to wall effects. With this notation, Liouville's equation (V.1) can be written as

$$\frac{\partial \varrho_n}{\partial t} = [H, \varrho_n] = \sum_{1 \leqq i \leqq n} [H(\boldsymbol{x}_i), \varrho_n] + \sum_{1 \leqq i < j \leqq n} [\Phi_{ij}, \varrho_n], \tag{V.77}$$

or according to (V.76),

$$\frac{\partial \varrho_n}{\partial t} = - \sum_{1 \leqq i \leqq n} \left(\frac{\boldsymbol{p}_i}{m_i} \cdot \nabla_{r_i} \varrho_n \right) + \sum_{1 \leqq i \leqq n} (\nabla_{r_i} U_i \cdot \nabla_{p_i} \varrho_n) + \sum_{1 \leqq i < j \leqq n} (\nabla_{r_i} \Phi_{ij} \cdot \nabla_{p_i} \varrho_n). \tag{V.77'}$$

By integrating (V.77) over the $n - s$ variables $\boldsymbol{x}_{s+1}, \ldots, \boldsymbol{x}_n$ and by taking account of the symmetry of ϱ_n in the variables $\boldsymbol{x}_1, \ldots \boldsymbol{x}_n$ as well as of the relations

$$\int [H(\boldsymbol{x}_i), \varrho_n]\, d^6\boldsymbol{x}_i = 0, \quad \iint [\Phi_{ij}, \varrho_n]\, d^6\boldsymbol{x}_i\, d^6\boldsymbol{x}_j = 0, \tag{V.78}$$

we obtain finally

$$\frac{\partial \varrho_s}{\partial t} = [H_s, \varrho_s] + (n - s) \int \sum_{1 \leqq i \leqq s} [\Phi_{i,s+1}, \varrho_{s+1}]\, d^6\boldsymbol{x}_{s+1}, \tag{V.79}$$

in which we have put

$$H_s = \sum_{1 \leq i \leq s} H(\boldsymbol{x}_i) + \sum_{1 \leq i < j \leq s} \Phi_{ij}. \qquad \text{(V.80)}$$

The wall effects occur in (V.79) as a result of the potentials $U(\boldsymbol{r}_i)$; it can be shown, as we shall see in discussing Bogolyubov's method, that these effects are negligible if the volume V of the container becomes very large; thus, we shall assume starting from now that the potentials $U(\boldsymbol{r}_i)$ involve only external forces but no wall effects.

The cases $s = 1$ and $s = 2$ are particularly important for what follows; putting $X_1 = -\nabla_{r_1} U(\boldsymbol{r}_1)$ for the external force acting on the particle 1, we have the equations

$$\frac{\partial \varrho_2}{\partial t} = [H(\boldsymbol{x}_1) + H(\boldsymbol{x}_2) + \Phi_{12}, \varrho_2] + (n-2) \int \sum_{1 \leq i \leq 2} [\Phi_{i3}, \varrho_3] \, d^6\boldsymbol{x}_3 \qquad (s = 2), \qquad \text{(V.81)}$$

$$\frac{\partial \varrho_1}{\partial t} = -\left(\frac{\boldsymbol{p}_1}{m_1} \cdot \nabla_{r_1} \varrho_1\right) - \left(\boldsymbol{X}_1 \cdot \nabla_{p_1} \varrho_1\right) + (n-1) \int (\nabla_{r_1} \Phi_{12} \cdot \nabla_{p_1} \varrho_2) d^6\boldsymbol{x}_2 \qquad (s = 1). \qquad \text{(V.82)}$$

We can see from (V.79) that the equation of s-th order is related to the $(s+1)$-st-order equation through the distribution function ϱ_{s+1} occurring in the integral term; the system (V.79) represents therefore a chain of coupled equations equivalent to Liouville's equation; these are the B.B.G.K.Y. equations. In particular, the integral term of equation (V.82), satisfied by ϱ_1, expresses the binary interactions of particles ("collisions") and, consequently, it depends on the distribution function $\varrho_2(\boldsymbol{x}_1, \boldsymbol{x}_2, t)$. It is by carrying out certain approximations on this term that we can try to derive Boltzmann's equation from (V.82), provided we let the generic function $f_1 = n\varrho_1$ correspond to the distribution function $f(x, t)$ from the kinetic theory of gases.

Before pointing out the various methods proposed for this derivation, we must make a preliminary comment concerning the two ways of introducing statistical considerations in mechanics, i.e. concerning the concept of distribution functions. The first method, which is that of statistical mechanics, consists in considering an ensemble of systems identical with the physical system being studied; this ensemble is completely determined by a probability density in Γ-space represented by the function $\varrho_n(\boldsymbol{x}_1, \ldots, \boldsymbol{x}_n, t)$; thus, one has implicitly in mind the possibility of experiments which are repeated on each system of the ensemble. The quantities observed macroscopically are then the average values of the mechanical observables taken over this ensemble of systems with the distribution function ϱ_n. This is the point of view that we have adopted generally throughout this book; in this case, the distribution function $\varrho_1(\boldsymbol{x}_1, t)$ is simply a marginal probability density which is deduced from ϱ_n by "successive contractions".†

We have encountered already the second method during the introduction of the $f_i(t)$ functions in Section III of this chapter: this is the method of the kinetic theory of gases which consists in defining in μ-space a distribution function $f(\boldsymbol{x}, t)$ representing the density of particles located in a volume element $d^6\boldsymbol{x}$ of μ-space, which is "neither too large nor too small". This latter definition is essential to the theory, since the element $d^6\boldsymbol{x}$ considered must be large enough to contain a very large number of particles and small enough for variations of macroscopic quantities to be negligible on this scale. The existence of a distribution function in this way is, in some way, the fundamental postulate of the kinetic theory of gases, to which a precise expression can be given only in the limiting case as $n \to \infty$.

This function f must be distinguished from the marginal distribution $n\varrho_1$, since it actually represents the numerical density related to a single physical system and not a probability density associated with a statistical ensemble of systems; it describes a different physical situation corresponding to the observation of a single system which involves a large number of particles (and even an infinitely large number). According to what we have seen in Section III, it is reasonable to assume that the transition from the probability density $n\varrho_1(\boldsymbol{x}_1, t)$ to the distribution function $f(\boldsymbol{x}, t)$ of the kinetic theory of gases can be carried out only by a certain process of averaging, similar to that which leads to the definition of coarse-grained densities in Ehrenfest's sense. This remark is the basis of the different methods of deriving Boltzmann's equation, which we shall study now.

† We point out, that from this point of view equation (V.82) determines the distribution function of a single particle in interaction with its surroundings; this interaction, represented by the integral term of (V.82) is of a disordered nature and lends itself to a description of a stochastic character.

2. *The methods of Yvon* (1935, 1937a, b) *and of Born and Green* (1946, 1947)

These two methods, which are essentially similar, consist in introducing truncated distribution functions defined as functions of a parameter δ, whose order of magnitude is equal to the range r_0 of the intermolecular forces. Following Grad (1958), we define the truncated distribution function $\varrho_1^\delta(\boldsymbol{x}_1, t)$ by

$$\varrho_1^\delta(\boldsymbol{x}_1, t) \equiv \int_D \varrho_n(\boldsymbol{x}_1, \ldots, \boldsymbol{x}_n, t)\, d^6\boldsymbol{x}_2 \ldots d^6\boldsymbol{x}_n, \tag{V.83}$$

where D is a $6(n-1)$-dimensional region defined in the following way: if $\boldsymbol{r}_1$ is the position of the particle 1, a region D_i of μ_i-space is assigned to each particle i such that

$$|\boldsymbol{r}_i - \boldsymbol{r}_1| \geqq \delta \tag{V.84}$$

and we put $D = D_2 \times D_3 \times \ldots \times D_n$ (the $\times$ sign denotes here the direct product of the different regions); thus, D contains all states of the particles 2, ..., n for which none of these particles is at a distance less than δ from particle 1. Similarly, we define

$$\varrho_2^\delta(\boldsymbol{x}_1, \boldsymbol{x}_2, t) \equiv \int_{D'} \varrho_n(\boldsymbol{x}_1, \boldsymbol{x}_2, \ldots, \boldsymbol{x}_n, t)\, d^6\boldsymbol{x}_3 \ldots d^6\boldsymbol{x}_n, \tag{V.85}$$

with $D' = D_3 \times D_4 \times \ldots \times D_n$; ϱ_2^δ is not symmetrical in $\boldsymbol{x}_1$ and $\boldsymbol{x}_2$ since the integration region contains only $\boldsymbol{x}_1$ and not $\boldsymbol{x}_2$. Having done this, we obtain the differential equation satisfied by ϱ_1^δ by integrating Liouville's equation (V.77′) over the region D; using the identities

$$\left.\begin{aligned}
&\int_{|\boldsymbol{r}_2-\boldsymbol{r}_1|>\delta} (\nabla_{\boldsymbol{r}_2} \cdot \boldsymbol{A}(\boldsymbol{r}_1, \boldsymbol{r}_2))\, d^3\boldsymbol{r}_2 = -\oint_{|\boldsymbol{r}_2-\boldsymbol{r}_1|=\delta} (\boldsymbol{A} \cdot d^2\boldsymbol{S}),\\
&\left(\nabla_{\boldsymbol{r}_1} \cdot \left[\int_{|\boldsymbol{r}_2-\boldsymbol{r}_1|>\delta} \boldsymbol{A}(\boldsymbol{r}_1, \boldsymbol{r}_2)\, d^3\boldsymbol{r}_2\right]\right)\\
&= \left(\int_{|\boldsymbol{r}_2-\boldsymbol{r}_1|>\delta} \nabla_{\boldsymbol{r}_1} \cdot \boldsymbol{A}(\boldsymbol{r}_1, \boldsymbol{r}_2)\, d^3\boldsymbol{r}_2\right) - \oint_{|\boldsymbol{r}_2-\boldsymbol{r}_1|=\delta} (\boldsymbol{A} \cdot d^2\boldsymbol{S}),
\end{aligned}\right\} \tag{V.86}$$

we get finally (with $p_1 = m_1 v_1$)

$$\frac{\partial \varrho_1^\delta}{\partial t} + (v_1 \cdot \nabla_{r_1}) \varrho_1^\delta + \left(\frac{X_1}{m_1} \cdot \nabla_{v_1}\right) \varrho_1^\delta$$

$$= (n-1) \oint_{S_2} \varrho_2^\delta ((v_2 - v_1) \cdot d^2 S) d^3 v_2$$

$$+ \frac{n-1}{m_1} \left(\nabla_{v_1} \cdot \left(\int_{D_2} \nabla_{r_1} \Phi_{12} \right) \varrho_2^\delta \, d^3 r_2 \right), \qquad \text{(V.87)}$$

where S_2 represents the sphere $|r_2 - r_1| = \delta$. In the case where the intermolecular force becomes negligible for $|r_2 - r_1| > \delta$, (V.87) is reduced to

$$\frac{\partial \varrho_1^\delta}{\partial t} + (v_1 \cdot \nabla_{v_1}) \varrho_1^\delta + \left(\frac{X_1}{m_1} \cdot \nabla_{v_1}\right) \varrho_1^\delta$$

$$= (n-1) \oint_{S_2} \varrho_2^\delta ((v_2 - v_1) \cdot d^2 S) \, d^3 v_2; \qquad \text{(V.87')}$$

we obtain a similar equation for ϱ_2^δ by integrating over D'. This equation shows us that the number of "free" particles (separated from other particles by a distance at least equal to the range of the intermolecular forces) only changes due to the loss of particle pairs which penetrate their sphere of mutual influence or to the gain of particle pairs which move away sufficiently from one another. Thus, equation (V.87′) has, in common with Boltzmann's equation, the property that the variation of ϱ_1^δ is determined by "complete" collisions. Concerning the term of equation (V.87) which we neglected, this represents the correction to be applied to (V.87′) for taking "grazing" collisions into account in the case where the range of the intermolecular forces is relatively large.

We shall now be able to derive Boltzmann's equation formally from equation (V.87′), provided that we introduce a certain number of supplementary assumptions. First of all, we rewrite the collision term of (V.87′) using the normal notation; we introduce the polar coordinates (b, ε) in the plane through the centre of S_2 perpendicular to the relative velocity $g_{21} = v_2 - v_1$, with $0 \leqq b \leqq \delta$ and $0 \leqq \varepsilon \leqq 2\pi$. In order to evaluate the scalar product $(g_{21} \cdot d^2 S)$, we must discriminate on S_2 the two hemispheres S_2^+ for which $(g_{21} \cdot d^2 S) > 0$ (particles which are moving away from one another) and S_2^- for which $(g_{21} \cdot d^2S) < 0$ (particles which are approaching each other). For these conditions we have

$$(g_{21} \cdot d^2S) = gb\, db\, d\varepsilon \text{ on } S_2^+, \quad (g_{21} \cdot d^2S) = -\, gb\, db\, d\varepsilon \text{ on } S_2^-$$
$$(g = |g_{21}|); \tag{V.88}$$

on the other hand, to every point (b, ε) of the plane there corresponds a point of S_2^+ that we shall call $r_2^+(b, \varepsilon)$ and a point of S_2^- that we shall denote by $r_2^-(b, \varepsilon)$. Equation (V.87′) then becomes

$$\frac{\partial \varrho_1^\delta}{\partial t} + (v_1 \cdot \nabla_{r_1})\, \varrho_1^\delta + \left(\frac{X_1}{m_1} \cdot \nabla_{v_1}\right) \varrho_1^\delta$$
$$= (n-1) \int [\varrho_2^\delta(x_1, x_2^+) - \varrho_2^\delta(x_1, x_2^-)]\, gb\, db\, d\varepsilon\, d^3v_2, \tag{V.89}$$

where we have written x_2^+ for (p_2, r_2^+) and x_2^- for (p_2, r_2^-).

Having done this, we can proceed from equation (V.89) to Boltzmann's equation by making the following four assumptions:

(a) We assume that the function $n\varrho_1^\delta(x_1, t)$ can be identified with the function $f(x, t)$ from the kinetic theory of gases; this amounts to considering the number of particles colliding at t as negligible compared with n.

(b) It is supposed that the gas density is low enough to take into account only binary collisions, the number of ternary and higher order collisions being negligible compared with that of the binary collisions. If then we consider equation (V.81), satisfied by $\varrho_2(x_1, x_2, t)$ in the case where $|r_2 - r_1| < \delta$, it can be seen that the integral term which contains the ternary collisions is negligible in comparison with the Poisson bracket expressing the binary interaction between particles 1 and 2; it follows, therefore, that equation (V.81) reduces, for the duration of a binary collision, to Liouville's equation in the 12-dimensional space of particles 1 and 2. The solution of this is ϱ_2 = constant over a trajectory in this 12-dimensional phase space corresponding to the two-body problem; thus, $\varrho_2(x_1, x_2^+, t)$ can be replaced by $\varrho_2(x_1', x_2', t')$, where x_1' and x_2' are the points occupied by particles 1 and 2 at t', the moment at which the collision commences, in such a way that these particles are located at x_1 and x_2^+ at time t at the end of the binary collision considered.

(c) We apply the principle of molecular chaos, assuming that we can replace $\varrho_2(x_1, x_2^-)$ by $\varrho_1(x_1)\, \varrho_1(x_2^-)$ and $\varrho_2(x_1', x_2')$ by

$\varrho_1(\boldsymbol{x}_1') \varrho_1(\boldsymbol{x}_2')$. This expresses how two particles on the point of colliding are statistically independent, whereas they can be correlated after collision; if this distinction were omitted, and if we had put simply $\varrho_2(\boldsymbol{x}_1, \boldsymbol{x}_2) = \varrho_1(\boldsymbol{x}_1) \varrho_1(\boldsymbol{x}_2)$ we should see the collision term of (V.89) disappear.

(*d*) We assume, finally, that the variations of ϱ_1 are sufficiently slow to be able to replace $\boldsymbol{r}_2^-$, $\boldsymbol{r}_1'$, $\boldsymbol{r}_2'$ by $\boldsymbol{r}_1$ and t' by t; this substitution is justified easily if we note that the various positions $\boldsymbol{r}$ and the time t' differ from $\boldsymbol{r}_1$ and t only by amounts of order δ.

By carrying out these various transformations in (V.89) and by replacing $n - 1$ by n, we have finally

$$\frac{\partial f}{\partial t} + (\boldsymbol{v}_1 \cdot \nabla_{r_1}) f + \left(\frac{\boldsymbol{X}_1}{m} \cdot \nabla_{v_1}\right) f$$
$$= \iiint [f(\boldsymbol{v}_2', \boldsymbol{r}_1, t) f(\boldsymbol{v}_1', \boldsymbol{r}_1, t) - f(\boldsymbol{v}_2, \boldsymbol{r}_1, t) f(\boldsymbol{v}_1, \boldsymbol{r}_1, t)] g b \, db \, d\varepsilon \, d^3\boldsymbol{v}_2; \tag{V.90}$$

which is Boltzmann's equation.† The method ew have just described is similar to those proposed by Yvon, on the one hand, and by Born and Green, on the other hand; it has the advantage of showing clearly that Boltzmann's equation in its usual form (V.90) ceases to be valid if the density or the range of the intermolecular forces become too large. However, the foregoing proof involves certain approximations whose precise justification is difficult; this is the case particularly for the assumption of molecular chaos and the question remains open of knowing whether Boltzmann's irreversible equation can be deduced from Liouville's reversible equation without external interference of a stochastic nature.

Actually, it is this property of irreversibility which distinguishes Boltzmann's equation from Liouville's equation and hence from the chain of equations (V.79) to which Liouville's equation is equivalent. This difference can be illustrated by comparing the invariance during motion of the quantity σ, defined in (V.2), with the monotonic decrease of Boltzmann's H-function. In order to elucidate this point, we define the quantities σ_Γ and σ_μ by

$$\sigma_\Gamma = \int \varrho_n \ln \varrho_n \, d\Gamma, \quad \sigma_\mu = \int \varrho_1 \ln \varrho_1 \, d^6\boldsymbol{x}_1, \tag{V.91}$$

† The role of the primed and unprimed variables is reversed here in relation to equation (V.50′), but nothing is changed obviously because of the microscopic reversibility.

where ϱ_n and ϱ_1 are the probability densities in Γ-space and μ-space, respectively. [Of course, σ_Γ is identical with σ of (V.2).] According to (V.74) we have necessarily the relation

$$\sigma_\Gamma \geqq n\sigma_\mu; \tag{V.92}$$

the equality only occurs if the particles are independent, which involves

$$\varrho_n(\boldsymbol{x}_1, \ldots, \boldsymbol{x}_n) = \varrho_1(\boldsymbol{x}_1)\,\varrho_1(\boldsymbol{x}_2) \ldots \varrho_1(\boldsymbol{x}_n). \tag{V.93}$$

Even though σ_Γ remains constant during the evolution described by Liouville's equation (V.77), the same is by no means true for σ_μ, whose time-variation cannot be studied in a simple manner. We note, however, that if we choose the initial distributions $\varrho_1(\boldsymbol{x}_1, 0)$ and $\varrho_n(\boldsymbol{x}_1, \ldots, \boldsymbol{x}_n, 0)$ so that (V. 93) is satisfied, we shall have initially $\sigma_\Gamma(0) = n\sigma_\mu(0)$; since σ_Γ remains constant during motion we can see, by (V.92), that $\sigma_\mu(t)$ must always remain smaller than $\sigma_\mu(0)$. However, in order to proceed further and obtain Boltzmann's *H*-theorem, we must replace equation (V.82) by (V.90), i.e. we must justify the kinetic method by proceeding from ϱ_1 functions to ϱ_1^δ functions and by accepting the set of assumptions (a)–(d); of these, it is obviously assumption (c) of molecular chaos that is essential for ensuring the irreversibility and monotonic decrease of *H*. In order to find out whether this assumption is compatible with Liouville's equation, it would be necessary to show that the property of molecular chaos is conserved in time, just as we proved at the conclusion of Section IV for a special case of the "Master Equation".

We shall again find in the account of Kirkwood's method the assumptions which we have just stated; more precisely, we shall see once again that the function f of the kinetic theory of gases must be defined by a certain averaging process which relates it to coarse-grained densities and is related to the necessarily finite nature of the volume elements in μ-space.

3. *Kirkwood's method* (1946, 1947)

This method depends essentially on replacing the concept of a coarse-grained density introduced in Section II by that of the time-average of probability densities over an interval τ. Let us consider, in fact, the time-average over the interval τ of an observable

$G(\boldsymbol{r}^{(n)}, \boldsymbol{p}^{(n)})$† which does not depend on the time explicitly; by introducing the evolution operator $T_t^{(n)}$‡ of Chapter I, this time-average can be written as:

$$\tilde{G} = \frac{1}{\tau}\int_0^\tau G(T_\theta^{(n)}\boldsymbol{r}^{(n)}, T_\theta^{(n)}\boldsymbol{p}^{(n)})\, d\theta, \tag{V.94}$$

where the operator $T_\theta^{(n)}$ moves the representative point $(\boldsymbol{r}^{(n)}, \boldsymbol{p}^{(n)})$ at t to the representative point at $t+\theta$. Let us assume now—and this is the essential assumption—that a physical quantity observed in a system in evolution must be identified with the phase average of the time-average $\tilde{G}(\boldsymbol{r}^{(n)}, \boldsymbol{p}^{(n)})$, or

$$\bar{\tilde{G}} = \frac{1}{\tau}\iint d^{3n}\boldsymbol{r}^{(n)}\, d^{3n}\boldsymbol{p}^{(n)} \int_0^\tau G(T_\theta^{(n)}\boldsymbol{r}^{(n)}, T_\theta^{(n)}\boldsymbol{p}^{(n)})\, \varrho_n(\boldsymbol{r}^{(n)}, \boldsymbol{p}^{(n)}, t)\, d\theta, \tag{V.95}$$

which can be written, by virtue of Liouville's theorem, as

$$\bar{\tilde{G}} = \iint G(\boldsymbol{r}^{(n)}, \boldsymbol{p}^{(n)})\, \tilde{\varrho}_n(\boldsymbol{r}^{(n)}, \boldsymbol{p}^{(n)}, t)\, d^{3n}\boldsymbol{r}^{(n)}\, d^{3n}\boldsymbol{p}^{(n)}, \tag{V.96}$$

in which we have put

$$\tilde{\varrho}_n(\boldsymbol{r}^{(n)}, \boldsymbol{p}^{(n)}, t) = \frac{1}{\tau}\int_0^\tau \varrho_n(\boldsymbol{r}^{(n)}, \boldsymbol{p}^{(n)}, t+\theta)\, d\theta. \tag{V.97}$$

The quantity observed macroscopically is thus identified with the average of $G(\boldsymbol{r}^{(n)}, \boldsymbol{p}^{(n)})$ calculated with the density $\tilde{\varrho}_n$, which is, itself, the time-average over the interval τ of the probability density $\varrho_n(\boldsymbol{r}^{(n)}, \boldsymbol{p}^{(n)}, t)$. We note that the mean density $\tilde{\varrho}_n$ similarly obeys Liouville's equation†† and that, in consequence, the introduction of the time-average is insufficient in itself alone for making irreversibility appear; this can be obtained only by means of supple-

† We introduce here the notation $\boldsymbol{r}^{(n)} = (\boldsymbol{r}_1, \boldsymbol{r}_2, \ldots, \boldsymbol{r}_n)$, $\boldsymbol{p}^{(n)} = (\boldsymbol{p}_1, \boldsymbol{p}_2, \ldots, \boldsymbol{p}_n)$; $\boldsymbol{r}^{(n)}$ and $\boldsymbol{p}^{(n)}$ are vectors in configuration space and momentum space, respectively. This is in particular the notation used by de Boer (1949) and by Hirschfelder, Curtiss and Bird (1954).

‡ We write here $T_t^{(n)}$ in order to emphasise that we are dealing with the operator corresponding to the n-body problem.

†† Actually, Liouville's equation is linear and the same is true for the transformation which defines $\tilde{\varrho}_n$ starting from ϱ_n.

mentary assumptions which we shall state later. It must be noted also that if we consider the averages taken over all the coordinates and momenta as in (V.96), we can then prove the interchangeability of the two operations of time-average and phase-average, which can be written as $\bar{\tilde{G}} = \tilde{\bar{G}}$. Kirkwood has shown, on the one hand, that these two averages are not interchangeable in the case where the integrations are not carried out over all variables and that, on the other hand, this non-interchangeability is connected with dissipative processes originating during return to equilibrium.

Having stated this, we can identify the function $f(\boldsymbol{r}, \boldsymbol{p}, t)$ of the kinetic theory of gases with the time average $\tilde{f}_1(\boldsymbol{r}_1, \boldsymbol{p}_1, t)$ of the generic marginal distribution, defined by

$$\tilde{f}_1(\boldsymbol{r}_1, \boldsymbol{p}_1, t) = n \iint \tilde{\varrho}_n(\boldsymbol{r}^{(n)}, \boldsymbol{p}^{(n)}, t)\, d^{3(n-1)}\boldsymbol{r}^{(n-1)}\, d^{3(n-1)}\boldsymbol{p}^{(n-1)}. \quad \text{(V.98)}$$

We obtain the equation satisfied by $\tilde{f}_1$ by taking the time-average of equation (V.82) over the interval τ; it becomes

$$\frac{\partial \tilde{f}_1}{\partial t} + (\boldsymbol{v}_1 \cdot \nabla_{\boldsymbol{r}_1})\tilde{f}_1 + \left(\frac{X_1}{m_1} \cdot \nabla_{\boldsymbol{v}_1}\right)\tilde{f}_1$$

$$= \frac{1}{\tau}\int_0^\tau\!\!\iint (\nabla_{\boldsymbol{r}_1}\Phi_{12} \cdot \nabla_{\boldsymbol{p}_1} f_2(T_\theta^{(n)}\boldsymbol{r}_1, T_\theta^{(n)}\boldsymbol{r}_2, T_\theta^{(n)}\boldsymbol{p}_1, T_\theta^{(n)}\boldsymbol{p}_2))\, d^3\boldsymbol{r}_2\, d^3\boldsymbol{p}_2\, d\theta, \quad \text{(V.99)}$$

and this equation is identical with that of Boltzmann, providing that we make the following three assumptions:

(a) The density is sufficiently low that only binary collisions are considered; Liouville's equation applied to the collisions of two particles can be written in this case as

$$f_2(\boldsymbol{r}_1, \boldsymbol{r}_2, \boldsymbol{p}_1, \boldsymbol{p}_2, t) = f_2(T_\theta^{(2)}\boldsymbol{r}_1, T_\theta^{(2)}\boldsymbol{r}_2, T_\theta^{(2)}\boldsymbol{p}_1, T_\theta^{(2)}\boldsymbol{p}_2, t+\theta).$$

(b) The interval τ must be large compared with the duration of a collision but small on the scale of the recurrence time in order to avoid the difficulties arising from Poincaré's recurrence theorem; this must enable us to make the integral on the right-hand side of (V.99) proportional to τ, so that τ disappears from the final equation. Very probably it is this assumption which introduces the irreversibility in (V.99); in addition, this must be compared with the definition of functions truncated at a distance δ, since it leads similarly to an equation in which only "complete" collisions are involved and thus enables us to assume molecular chaos.

(c) Finally, we suppose that the deviations from equilibrium are sufficiently small to be able to replace the time-average of the product $f_1(\boldsymbol{p}_1)f_1(\boldsymbol{p}_2)$ by the product of the time-averages.

By means of these three assumptions and after quite long calculations, Kirkwood succeeded in obtaining Boltzmann's equation for the function f_1, or

$$\frac{\partial \tilde{f}_1}{\partial t} + (\boldsymbol{v}_1 \cdot \nabla_{r_1})\tilde{f}_1 + \left(\frac{\boldsymbol{X}_1}{m_1} \cdot \nabla_{v_1}\right)\tilde{f}_1$$
$$= \iiint [\tilde{f}_1(\boldsymbol{p}_1')\tilde{f}_1(\boldsymbol{p}_2') - \tilde{f}_1(\boldsymbol{p}_1)\tilde{f}_1(\boldsymbol{p}_2)]\, gb\, db\, d\varepsilon\, d^3\boldsymbol{p}_2 . \tag{V.100}$$

Summarising, we can say that the methods which have just been analysed emphasise the various problems which are raised by deriving Boltzmann's equation, starting from Liouville's equation, yet without contributing definitive solutions; in particular, it seems to be natural to identify the distribution function $f(\boldsymbol{x}, t)$ of the kinetic theory of gases with a certain average of the marginal probability densities of statistical mechanics by taking either a coarse-grained density or a time-average over an interval τ of the distribution functions. We shall proceed now to the study of an entirely different method proposed by Bogolyubov.

4. *Bogolyubov's method* (1946a, b)

The essential features of Bogolyubov's method are the definition of three time-scales distinguishing several stages in the evolution of the gas and the expansion of distribution functions in a power series in the concentration.

We start again with the system of equations (V.79) in which we make the transition to the limit $n \to \infty$; from the various ways of proceeding, we choose the one in which we have also $V \to \infty$ in such a way that the concentration $n/V = c$ remains constant. By using the F_s functions of (V.74), the system (V.79) can then be written as

$$\frac{\partial F_s}{\partial t} = [H_s, F_s] + c \int \sum_{1 \leq i \leq s} [\Phi_{i,s+1}, F_{s+1}]\, d^6\boldsymbol{x}_{s+1}, \tag{V.101}$$

where the integration is carried out over the whole of μ_{s+1}-space and where the contribution arising from wall effects disappears as a consequence of transition to the limit (actually, these terms are of the form $[U_V(\boldsymbol{r}_i), F_s]$ and are non-vanishing only if the $\boldsymbol{r}_i$ are in the vicinity of the boundary surface which is made to move to infinity). The F_s functions which satisfy (V.101) are thus asymptotic expressions for the distribution functions of a system containing a very large number of particles enclosed in a macroscopic volume V. By dealing with the case of a low concentration and with short range interaction forces, we can try to solve the system (V.101) by finding for F_s an expansion in series of powers of the concentration (similar to the well-known virial expansion); we can then put

$$F_s = F_s^0 + cF_s^1 + c^2F_s^2 + \dots. \tag{V.102}$$

Elementary considerations show, moreover, that the expansion (V.102) is in effect one in powers of the "dimensionless density" r_0^3c, where r_0 is the range of the intermolecular forces. Putting (V.102) into (V.101) and comparing coefficients of the same powers of c, we obtain

$$\frac{\partial F_s^0}{\partial t} = [H_s, F_s^0], \tag{V.103}$$

$$\frac{\partial F_s^1}{\partial t} = [H_s, F_s^1] + \int \Big[\sum_{1 \leq i \leq s} \Phi_{i,s+1}, F_{s+1}^0 \Big] d^6 \boldsymbol{x}_{s+1}. \tag{V.104}$$

In order to simplify, we shall assume in what follows that the external forces determined by the potentials $U(\boldsymbol{r}_i)$ vanish; the $H(\boldsymbol{x}_i)$ thus contain, according to (V.76), only the kinetic energy terms. In order formally to solve this system, we introduce the operator $T_t^{(s)}$ corresponding to the s-body problem, i.e. to the motion of s point particles of mass m with a Hamiltonian H_s. If $(\boldsymbol{x}_1, \dots, \boldsymbol{x}_s)$ represents the dynamic state at the initial time, the operator $T_t^{(s)}$ has the effect of replacing these variables by

$$(\boldsymbol{X}_1(\boldsymbol{x}_1, \dots, \boldsymbol{x}_s, t), \dots, \boldsymbol{X}_s(\boldsymbol{x}_1, \dots, \boldsymbol{x}_s, t))$$

which represent the state of the system at t; we can write symbolically the following two relations for a function $\psi(\boldsymbol{x}_1, \dots, \boldsymbol{x}_s)$:

$$\left.\begin{aligned} T_t^{(s)}\psi(\boldsymbol{x}_1, \dots, \boldsymbol{x}_s) &= \psi(\boldsymbol{X}_1, \dots, \boldsymbol{X}_s); \\ \frac{\partial}{\partial t} T_{-t}^{(s)}\psi(\boldsymbol{x}_1, \dots, \boldsymbol{x}_s) &= [H_s, T_{-t}^{(s)}\psi(\boldsymbol{x}_1, \dots, \boldsymbol{x}_s)]. \end{aligned}\right\} \tag{V.105}$$

With this notation, the solutions of (V.103) and (V.104) can formally be written as

$$F_s^0(x_1, \ldots, x_s, t) = T_{-t}^{(s)} F_s^0(x_1, \ldots, x_s, 0), \qquad \text{(V.106)}$$

$$F_s^1(x_1, \ldots, x_s, t) = T_{-t}^{(s)} F_s^1(x_1, \ldots, x_s, 0) + \int_0^t \Big\{ T_{-(t-\theta)}^{(s)} \times$$

$$\times \int \Big[\sum_{1 \leq i \leq s} \Phi_{i+1}, F_{s+1}^0(x_1, \ldots, x_{s+1}, \theta) \Big] d^6 x_{s+1} \Big\} d\theta, \qquad \text{(V.107)}$$

whence, for the distribution function $F_s(x_1, \ldots, x_s, t)$,

$$F_s(x_1, \ldots, x_s, t)$$

$$= T_{-t}^{(s)} F_s(x_1, \ldots, x_s, 0) + c \int_0^t \Big\{ T_{-(t-\theta)}^{(s)} \int \Big[\sum_{1 \leq i \leq s} \Phi_{i,s+1},$$

$$T_{-\theta}^{(s+1)} F_{s+1}(x_1, \ldots, x_{s+1}, 0) \Big] d^6 x_{s+1} \Big\} d\theta + c^2 \ldots . \qquad \text{(V.108)}$$

The physical interpretation of this expansion is simple; to a first approximation we have in fact $F_s(t) = T_{-t}^{(s)} F_s(0)$, with the result that the variation in time of F_s takes place as if the s particles were moving independently of other particles. The correction term in c takes account of the interaction with the $(s + 1)$-st particle; this is an interaction which is added on starting from s particles of the complex considered.

In studying the limit of applicability of such an expansion we shall see two characteristic times of evolution of the gas appear. We choose, in fact, as the unit of length the range r_0 of the intermolecular forces and as the unit of time the average duration of a collision $\tau_c = r_0/\overline{V}$ (where $\overline{V}$ is the mean speed of the particles). With these units, the functions appearing in our equations are of order unity and the term in c in (V.108) has an order of magnitude equal to that of the principal terms multiplied by $(r_0^3 ct/\tau_c)$. The condition for the applicability of the expansion (V.108) is then

$$t \ll \frac{\tau_c}{r_0^3 c} \equiv t_0 \qquad \text{(V.109)}$$

where t_0 is a time equivalent to the time of free flight (time interval between two collisions) $\lambda/\overline{V}$, where λ is the mean free path. τ_c and t_0 are the two characteristic times that we have mentioned: for gases of low density we have always $t_0 \gg \tau_c$, the ratio t_0/τ_c being of order

10^2 to 10^3; according to the foregoing, we can see that (V.108) is also an expansion in powers of τ_c/t_0.

The special consequence of this is that (V.108) can be used only for an interval $t \ll t_0$, during which the momentum distribution $F_1(\boldsymbol{p}, t)$ of a molecule does not deviate appreciably from the initial distribution, since this distribution does not depend directly on molecular interactions. It follows that the foregoing method cannot be applied to the establishment of a kinetic equation, since we must follow in this case the variation of the distribution function F_1 over a time which is sufficiently long to reveal the irreversible tendency towards an equilibrium distribution (for example, the Maxwell distribution).

However, the expansion (V.108) will enable us now to study the correlation functions F_s with $s \geqq 2$ over a time which is large compared with τ_c (in fact, we have generally $1/r_0^3 c \gg 1$) and to show that they satisfy certain limiting conditions expressing a synchronisation process between the F_s and F_1 terms. For this purpose, we give several definitions concerning the evolution of a complex of s particles in the case where the repulsive interaction potential is decreasing monotonically and becomes negligible for $r > r_0$. Since the particles are all separated by a distance greater than r_0, their motion tends towards a uniform and rectilinear motion, so that we have

$$\left\{\begin{array}{l} |T_t^{(s)}\boldsymbol{r}_i - T_t^{(s)}\boldsymbol{r}_j| \to \infty, \\ T_{-t}^{(s)}\boldsymbol{p}_i \to \boldsymbol{P}_i^{(s)}(\boldsymbol{x}_1, \ldots, \boldsymbol{x}_s), \quad \text{as} \quad t \to +\infty; \end{array}\right. \tag{V.110}$$

the speed of the tendency towards uniform motion assures the validity of the relation

$$\int_0^t \{\boldsymbol{P}_i^{(s)} - T_{-\theta}^{(s)}\boldsymbol{p}_i\}\, d\theta \to \int_0^\infty \{\boldsymbol{P}_i^{(s)} - T_{-\theta}^{(s)}\boldsymbol{p}_i\}\, d\theta$$

$$\text{with} \quad t\{\boldsymbol{P}_i^{(s)} - T_{-t}^{(s)}\boldsymbol{p}_i\} \to 0 \quad \text{as} \quad t \to +\infty;$$

We derive from this

$$T_{-t}^{(s)}\boldsymbol{r}_i + \frac{T_{-t}^{(s)}\boldsymbol{p}_i}{m}\, t \to \boldsymbol{R}_i^{(s)}, \tag{V.111}$$

with

$$\boldsymbol{R}_i^{(s)} = \boldsymbol{r}_i + \frac{1}{m}\int_0^\infty \{\boldsymbol{P}_i^{(s)} - T_{-\theta}^{(s)}\boldsymbol{p}_i\}\, d\theta. \tag{V.112}$$

In this notation, $\boldsymbol{P}_i^{(s)}$ is the initial momentum of the ith particle in the "s-tuple" collision which leads to the state $(\boldsymbol{x}_1, \ldots, \boldsymbol{x}_s)$ at t and $\boldsymbol{R}_i^{(s)}$ is the position that this particle would occupy at t if it had continued its motion with the initial momentum $\boldsymbol{P}_i^{(s)}$. Since the real motion is distinguished from the uniform motion only during the time in which the interparticle distances are of order r_0, it follows that for $|\boldsymbol{r}_i - \boldsymbol{r}_j| \sim r_0$ the duration of the approach process (V.110) and (V.111) is of order $\tau_c = r_0/\overline{V}$.

Now let us apply these results to the study of the correlation functions F_s $(s \geqq 2)$. We know that they can be chosen arbitrarily at the initial time, provided that we take into account that the correlations decrease as the inter-particle distances increase; thus, if the particles are further away than r_0, the functions F_s can be expressed practically by products of the F_1. With the previous notation, the F_s functions would then satisfy the asymptotic condition

$$T_{-\theta}^{(s)}\{F_s(\boldsymbol{x}_1, \ldots, \boldsymbol{x}_s, 0) - \prod_{1 \leqq i \leqq s} F_1(\boldsymbol{x}_i, 0)\} \to 0 \quad \text{as} \quad \theta \to +\infty \; (s \geqq 2), \tag{V.113}$$

which can be written also as

$$F_s(\boldsymbol{x}_1, \ldots, \boldsymbol{x}_s, t) - \prod_{1 \leqq i \leqq s} F_1(\boldsymbol{R}_i^{(s)}, \boldsymbol{P}_i^{(s)}, t) \to 0, \quad \text{as} \quad t \to +\infty. \tag{V.114}$$

The result of this is that in the principal terms of the correlation functions F_s a process of synchronisation with F_1 is produced, with the result that the F_s functions can be approximated by the products of $F_1(\boldsymbol{R}_i^{(s)}, \boldsymbol{P}_i^{(s)}, t)$ which are determined completely by a knowledge of the distribution function F_1 at t. The duration of this process is of order τ_c and it can be seen that two processes can be distinguished in the evolution of the gas: a slow process, associated with the variation of the momentum distribution of a single particle, with a characteristic duration of order t_0 (the mean free flight time); a rapid process, corresponding to the synchronisation of the correlation distribution functions $(s \geqq 2)$, with a characteristic duration of order τ_c (the duration of a "collision"). The fast process can be described by the expansion (V.108) which is applicable for $t \gg \tau_c$; on the contrary, the slow process can be described by this method only at its start.

Summarising, if we consider the evolution of the gas described by the distribution ϱ_n, we can distinguish several stages in this evolution:

(1) The initial stage, in which all the correlation functions $F_s(\boldsymbol{x}_1, ..., \boldsymbol{x}_s, t)$ ($s \geqq 2$), determined by starting from the arbitrary function $\varrho_n(\boldsymbol{x}_1, ..., \boldsymbol{x}_n, 0)$ vary rapidly over a duration of order τ_c; only the function $F_1(\boldsymbol{x}, t)$ varies slowly, as we have seen already.

(2) The kinetic stage; after a time of order τ_c, the final evolution of the gas is determined by the time-variation of F_1; since the correlations F_s depend on F_1 at equilibrium, we can assume that because of (V.114) they are of the form

$$F_s(\boldsymbol{x}_1, ..., \boldsymbol{x}_s, t) = F_s(\boldsymbol{x}_1, ..., \boldsymbol{x}_s, F_1), \tag{V.115}$$

i.e. that they depend on the time only through F_1, whatever the initial distribution. On the other hand, the kinetic equation satisfied by F_1 must take the form

$$\frac{\partial F_1}{\partial t} = A(\boldsymbol{x}; F_1); \tag{V.116}$$

it is this which describes the time-evolution of the gas in the kinetic stage. The solutions which satisfy (V.115) are obviously not the most general ones, but the procedure for using them is similar to that leading to the normal solutions of Boltzmann's equation (Chapman–Enskog method). We can say that at the beginning of the time τ_c there is a kind of "contraction" in the description of the gas since it depends only on a much smaller number of variables (namely, F_1 in place of all F_s). This is similar to Kirkwood's average process: in the two cases a large part of the initial information contained in the function $\varrho_n(\boldsymbol{x}_1, ..., \boldsymbol{x}_n, 0)$ is lost after an interval of time of order τ_c.

(3) The hydrodynamic stage; this involves a third characteristic time $L/\overline{V}$, where L is a macroscopic length over which we can establish a significant variation of macroscopic quantities such as the density, the macroscopic velocity of the gas or the temperature. For gases of low density and with macroscopic quantities which are slowly variable, we have $L/\overline{V} \gg t_0 \gg \tau_c$, which justifies the use of the Chapman–Enskog approximation method for solving Boltzmann's equation (in this way we obtain the so-called "normal" solutions of this equation).

Having reached now the kinetic stage, we shall derive the kinetic equation by relying on relations (V.115) and (V.116). For this purpose, we shall again turn to a series expansion of the F_s functions by putting

$$F_s(\boldsymbol{x}_1, \ldots, \boldsymbol{x}_s, t) = F_s^0(\boldsymbol{x}_1, \ldots, \boldsymbol{x}_s, F_1) + cF_s^1(\boldsymbol{x}_1, \ldots, \boldsymbol{x}_s, F_1) + \ldots; \tag{V.117}$$

similarly, the kinetic equation (V.116) assumes the form

$$\frac{\partial F_1}{\partial t} = A_0(\boldsymbol{x}_1; F_1) + cA_1(\boldsymbol{x}_1; F_1) + c^2 A_2(\boldsymbol{x}_1; F_1) + \ldots \tag{V.118}$$

and the coefficients $A_0, A_1, \ldots$, must be such that equation (V.101) is satisfied. If we write this equation for $s = 1$, we have immediately

$$\begin{cases} A_0(\boldsymbol{x}_1; F_1) = [H_1, F_1] = -\left(\dfrac{\boldsymbol{p}_1}{m_1} \cdot \nabla_{r_1}\right) F_1, \\ A_1(\boldsymbol{x}_1; F_1) = \int [\Phi_{12}, F_2^0 (\boldsymbol{x}_1, \boldsymbol{x}_2; F_1)]\, d^6\boldsymbol{x}_2, \\ \cdots\cdots\cdots\cdots\cdots\cdots\cdots\cdots\cdots\cdots \\ A_r(\boldsymbol{x}_1; F_1) = \int [\Phi_{12}, F_2^{(r-1)}(\boldsymbol{x}_1, \boldsymbol{x}_2; F_1)]\, d^6\boldsymbol{x}_2. \end{cases} \tag{V.119}$$

We can see that the term A_0 corresponds to the flow term of Boltzmann's equation (independent of "collisions") and that the coefficients A_r can be calculated if we know the various approximations of the function for two particles F_2; thus, we must calculate these approximations starting from equations (V.101) for $s \geqq 2$. If we note that the time-derivative of F_s, by virtue of (V.115), (V.116) and (V.118), can be written as

$$\frac{\partial F_s}{\partial t} = \frac{\delta F_s}{\delta F_1}\frac{\partial F_1}{\partial t} = \frac{\delta F_s}{\delta F_1}(A_0 + cA_1 + c^2 A_2 + \ldots)$$
$$\equiv D_0 F_s + cD_1 F_s + c^2 D_2 F_s + \ldots,$$

where $\delta F_s/\delta F_1$ is the functional derivative of F_s with respect to F_1 and where we have put

$$D_r F_s \equiv \frac{\delta F_s}{\delta F_1} A_r(\boldsymbol{x}_1; F_1), \tag{V.120}$$

equations (V.101) then give [by taking (V.117) and by equating coefficients of the same powers of c]

$$[H_s, F_s^0] - D_0 F_s^0 = 0, \tag{V.121}$$

$$\left\{\begin{array}{l} [H_s, F_s^1] - D_0 F_s^1 = D_1 F_s^0 - \int \Big[\sum_{1 \leqq i \leqq s} \Phi_{i,s+1}, F_{s+1}^0\Big] d^6 \boldsymbol{x}_{s+1} \\ \quad \equiv -\psi_s(\boldsymbol{x}_1, \ldots, \boldsymbol{x}_s; F_1), \\ \ldots\ldots\ldots\ldots\ldots\ldots\ldots\ldots\ldots\ldots, \end{array}\right. \tag{V.122}$$

with an obvious definition of $\psi_s(\boldsymbol{x}_1, \ldots, \boldsymbol{x}_s; F_1)$ which depends only on F_s^0. In order to solve these differential equations in F_s^0, F_s^1, ..., we must state the corresponding limiting conditions. These are deduced from (V.113) by taking (V.115); in this way we obtain

$$T_{-\theta}^{(s)}\{F_s(\boldsymbol{x}_1, \ldots, \boldsymbol{x}_s; T_\theta^{(1)} F_1) - \prod_{1 \leqq i \leqq s} (T_\theta^{(1)} F_1)\} \to 0, \quad \text{as} \quad \theta \to +\infty, \tag{V.123}$$

which, taking account of the expansion (V.117), resolves into

$$\left\{\begin{array}{l} T_{-\theta}^{(s)}\{F_s^0(\boldsymbol{x}_1, \ldots, \boldsymbol{x}_s; T_\theta^{(1)} F_1) - \prod_{1 \leqq i \leqq s} (T_\theta^{(1)} F_1)\} \to 0, \quad (\theta \to +\infty). \\ T_{-\theta}^{(s)} F_s^1(\boldsymbol{x}_1, \ldots, \boldsymbol{x}_s; T_\theta^{(1)} F_1) \to 0, \end{array}\right. \tag{V.124}$$

Given these limiting conditions (V.124), we can find the solutions of equations (V.121)–(V.122) and we can determine in this way the expression for F_s^r, and consequently for A_r, as functionals of the function $F_1(\boldsymbol{x})$. By following a reasoning similar to that carried out for formulae (V.106) and (V.107) and by using the notation of (V.110) and (V.111), we obtain at once

$$F_s^0(\boldsymbol{x}_1, \ldots, \boldsymbol{x}_s; F_1) = \prod_{1 \leqq i \leqq s} F_1(\boldsymbol{R}_i^{(s)}, \boldsymbol{P}_i^{(s)}), \tag{V.125}$$

whence, from (V.119),

$$A_1(\boldsymbol{x}_1; F_1) = \int \Big[\Phi_{12}, \prod_{1 \leqq i \leqq 2} F_1(\boldsymbol{R}_i^{(2)}, \boldsymbol{P}_i^{(2)})\Big] d^6 \boldsymbol{x}_2. \tag{V.126}$$

Similarly, by taking the second condition of (V.124) into account and by proceeding to the limit $\theta = +\infty$, we have next:

$$F_s^1(\boldsymbol{x}_1, \ldots, \boldsymbol{x}_s; F_1) = \int_0^\infty T_{-\theta}^{(s)} \psi_s(\boldsymbol{x}_1, \ldots, \boldsymbol{x}_s; T_\theta^{(1)} F_1)\, d\theta, \tag{V.127}$$

whence, for the coefficient A_2,

$$A_2(\boldsymbol{x}_1; F_1) = \int \left[\Phi_{12}, \int_0^\infty T_{-\theta}^{(2)} \psi_2(\boldsymbol{x}_1, \boldsymbol{x}_2; T_\theta^{(1)} F_1)\, d\theta \right] d^6\boldsymbol{x}_2. \quad \text{(V.128)}$$

By proceeding in the same way for the other approximations, we can write for F_s the following formal expression:

$$F_s = \prod_{1 \leq i \leq s} F_1(\boldsymbol{R}_i^{(s)}, \boldsymbol{P}_i^{(s)}, t) + c \int_0^\infty T_{-\theta}^{(s)} \psi_s(\boldsymbol{x}_1, \ldots, \boldsymbol{x}_s; T_\theta^{(1)} F_1)\, d\theta + \ldots, \quad \text{(V.129)}$$

and for the kinetic equation:

$$\frac{\partial F_1(\boldsymbol{x}_1, t)}{\partial t} = [H_1, F_1] + c \int [\Phi_{12}, F_1(\boldsymbol{R}_1^{(2)}, \boldsymbol{P}_1^{(2)}, t)\, F_1(\boldsymbol{R}_2^{(2)}, \boldsymbol{P}_2^{(2)}\, t)] d^6\boldsymbol{x}_2 + c^2 \int \left[\Phi_{12}, \int_0^\infty T_{-\theta}^{(2)} \psi_2(\boldsymbol{x}_1, \boldsymbol{x}_2; T_\theta^{(1)} F_1)\, d\theta \right] d^6\boldsymbol{x}_2 + \ldots. \quad \text{(V.130)}$$

If we neglect terms starting with c^2 (containing ternary and higher order interactions), we obtain the kinetic equation in first approximation

$$\frac{\partial F_1(\boldsymbol{r}_1, \boldsymbol{p}_1, t)}{\partial t} = -\left(\frac{\boldsymbol{p}_1}{m_1} \cdot \nabla_{\boldsymbol{r}_1} \right) F_1 + c \int [\Phi_{12}, F_1(\boldsymbol{R}_1^{(2)}, \boldsymbol{P}_1^{(2)}, t) F_1(\boldsymbol{R}_2^{(2)}, \boldsymbol{P}_2^{(2)}, t)] d^6\boldsymbol{x}_2. \quad \text{(V.131)}$$

By analysing the integral term of this equation, which expresses the effect of binary "collisions", we can compare (V.131) with Boltzmann's equation, the function f of the kinetic theory of gases being equated here with $cF_1(\boldsymbol{x}_1, t)$. In the special case of spatial homogeneity, the collision term is calculated by noting that, because of the definition of the $\boldsymbol{P}_i^{(s)}$, we have

$$[\Phi_{12}, F_1(\boldsymbol{P}_1^{(2)}, t)\, F_1(\boldsymbol{P}_2^{(2)}, t)] = -\left[\frac{\boldsymbol{p}_1^2}{2m_1} + \frac{\boldsymbol{p}_2^2}{2m_2}, F_1(\boldsymbol{P}_1^{(2)}, t)\, F_1(\boldsymbol{P}_2^{(2)}, t) \right], \quad \text{(V.132)}$$

so that the distribution function $F_1(\boldsymbol{p}_1, t)$ satisfies the equation

$$\frac{\partial F_1(\boldsymbol{p}_1, t)}{\partial t} = c \iiint [F_1(\boldsymbol{p}'_1, t)\, F_1(\boldsymbol{p}'_2, t) - F_1(\boldsymbol{p}_1, t)\, F_1(\boldsymbol{p}_2, t)]\, g_{21} b\, db\, d\varepsilon\, d^3\boldsymbol{p}_2, \tag{V.133}$$

where $\boldsymbol{p}'_1$ and $\boldsymbol{p}'_2$ are the momenta "before the collision", expressed as a function of the momenta $\boldsymbol{p}_1, \boldsymbol{p}_2$ "after the collision"; (V.133) is obviously identical with Boltzmann's equation in the case of spatial homogeneity and in the absence of external forces.

In the general case of a spatially inhomogeneous distribution, we calculate the collision term by a similar procedure, starting for example, from the equation $[H_2, F_2^{(0)}] - D_0 F_2^{(0)} = 0$. In this way we obtain a much more complicated expression than previously and one which is not reduced simply to the Boltzmann collision term; it depends in particular on the value of the distribution function $F(\boldsymbol{r}_1, \boldsymbol{p}_1, t)$ taken at different points $\boldsymbol{R}_1^{(2)}$ and $\boldsymbol{R}_2^{(2)}$ of physical space, as well as on its derivatives with respect to the space coordinates. As a result, there is a coupling in the Boltzmann equation between the collision term and the flow term connected with the density gradient; this relation expresses the existence of a transfer of molecular properties by means of collisions and not only by the free motion of the particles. The importance of this coupling depends, naturally, on the range and nature of the inter-particle forces; in particular, the collision term is simplified considerably for the model of elastic hard spheres of diameter δ—in the latter case we have

$$\frac{\partial F_1(\boldsymbol{r}_1, \boldsymbol{p}_1, t)}{\partial t} + \left(\frac{\boldsymbol{p}_1}{m_1} \cdot \nabla_{\boldsymbol{r}_1}\right) F_1 = c \iiint [F_1(\boldsymbol{r}_1, \boldsymbol{p}'_1, t)\, F_1(\boldsymbol{r}_1 + \delta\boldsymbol{k}, \boldsymbol{p}'_2, t) - F_1(\boldsymbol{r}_1, \boldsymbol{p}_1, t)\, F_1(\boldsymbol{r}_1 - \delta\boldsymbol{k}, \boldsymbol{p}_2, t)]\, g_{21} b\, db\, d\varepsilon\, d^3\boldsymbol{p}_2, \tag{V.134}$$

where $\boldsymbol{k}$ is a unit vector in the direction of $\boldsymbol{g}_{21}$. It can be seen that the collision term depends on the value of the distribution function at the points $\boldsymbol{r}_1, \boldsymbol{r}_1 - \delta\boldsymbol{k}$ and $\boldsymbol{r}_1 + \delta\boldsymbol{k}$ and that the relation considered obviously depends on the particle diameters. The function $F_1(\boldsymbol{r}_1, \boldsymbol{p}_1, t)$ thus satisfies, to a first approximation, an equation of the Boltzmann type, provided that the model of perfectly elastic hard spheres is adopted and that the variation of F_1 over a distance of order of the molecular diameter is neglected.† Finally, we point out that the coupling between the two terms of the Boltzmann equation would assume a still more complicated form in the presence of external forces, since these would be involved in the definition of $\boldsymbol{P}_i^{(s)}$ and $\boldsymbol{R}_i^{(s)}$, as well as in H_s and D_0; thus, we can give a precise meaning to the assumption formulated at the end of Section III, according to which we can consider separately the effects of free flow and the effects of collision.

Summarising, we can see that Bogolyubov's method allows us to obtain a kinetic equation which generalises Boltzmann's equation and which reduces to it—to a first approximation—in a certain number of cases. It depends on two essential points: on the one hand, the synchronisation process between the correlation functions F_s and the function F_1, which corresponds to a kind of "coarse-graining" and which, applied to (V.113), leads to the limiting condition (V.124) playing the role of the principle of molecular chaos (this process, moreover, introduces a characteristic time of order of the duration of a collision τ_c; it would be interesting to compare it with the time τ introduced by Kirkwood in order to define the time-average of the distribution functions); on the other hand, the expansion in powers of the concentration—similar in the case of equilibrium to the virial expansion and to the cluster functions method of Ursell–Mayer (Ursell, 1927; Mayer, 1937)—which enables us to find the order of magnitude of the contribution from collisions of various orders and, by confining ourselves to a first approximation, to consider only the effect of binary collisions.

As regards the irreversibility of the evolution, it appears to be connected first of all with the process of synchronisation between F_s and F_1 (coarse-graining) and to the limiting conditions (V.113) and (V.124)‡ corresponding to the assumption of molecular chaos; in addition, we note that this latter assumption is essential in order to be able to extract from the B.B.G.K.Y. chain of equations, which are totally reversible, an irreversible equation involving only F_1. The interest in Bogolyubov's method, therefore, is to enable us to make more precise the meaning and the limit of validity of the

† Such an approximation would no longer be valid for dense gases; concerning this, we can compare equation (V.134) with that proposed by Enskog for studying dense gases (see Chapman and Cowling, 1953, chapter 16).

‡ The role of these limiting conditions has been studied in detail by Cohen and Berlin (1960; see also Cohen, 1962).

necessary assumptions for establishing Boltzmann's equations, namely, the introduction of coarse-grained densities or of the time-averages of the probability densities, the limitation to binary collisions and the principle of molecular chaos. This method relates them to the general theory and to the fundamental physical parameters such as the "collision duration" τ_c, the range r_0 of the intermolecular forces, the concentration c and the mean free flight time t_0.

5. *Liouville's equation and the Master Equation. Brout's method*

We shall conclude this section by showing briefly the essential characteristics of the method proposed by Brout (1956), for the purpose of establishing a bridge between Liouville's deterministic equation and the "Master Equation". We shall all the time consider a system composed of a very large number of particles n (n can therefore be considered as infinitely large) and we shall suppose for simplicity that there are no external forces. Liouville's equation, (V.77′), can in this case† be written as

$$\frac{\partial \varrho_n}{\partial t} + \sum_{1 \leqq i \leqq n} (\boldsymbol{v}_i \cdot \nabla_{\boldsymbol{r}_i}) \varrho_n + \sum_{1 \leqq i < j \leqq n} \left(\frac{\boldsymbol{F}_{ij}}{m_i} \cdot \nabla_{\boldsymbol{v}_i} \right) \varrho_n = 0, \quad \text{(V.135)}$$

in which we have put $\boldsymbol{F}_{ij} = -\nabla_{\boldsymbol{r}_i} \Phi_{ij}$. If $\varrho_n(\boldsymbol{v}_1, \dots, \boldsymbol{v}_n, \boldsymbol{r}_1, \dots, \boldsymbol{r}_n, 0)$ denotes an initial distribution, we have for the distribution at time t:

$$\varrho_n(\boldsymbol{v}_1, \dots, \boldsymbol{v}_n, \boldsymbol{r}_1, \dots, \boldsymbol{r}_n, t) = \varrho_n(\boldsymbol{v}_1^{-t}, \dots, \boldsymbol{v}_n^{-t}, \boldsymbol{r}_1^{-t}, \dots, \boldsymbol{r}_n^{-t}, 0), \quad \text{(V.136)}$$

where $\boldsymbol{v}_1^{-t}, \dots, \boldsymbol{v}_n^{-t}, \boldsymbol{r}_1^{-t}, \dots, \boldsymbol{r}_n^{-t}$ are the velocities and positions of the particles at $-t$, corresponding to $\boldsymbol{v}_1, \dots, \boldsymbol{v}_n, \boldsymbol{r}_1, \dots, \boldsymbol{r}_n$ at time 0; the parameters $\boldsymbol{v}^{-t}$ and $\boldsymbol{r}^{-t}$ are determined, obviously, by the movement of the n particles under the effect of the interaction forces $\boldsymbol{F}_{ij}$. It will be useful for us to introduce also the velocities $[\boldsymbol{v}_i^{-t}]$ and the positions $[\boldsymbol{r}_i^{-t}]$ of the particles, in the case where they are not subjected to these interaction forces; in this case we have a rectilinear and uniform motion, whence

$$[\boldsymbol{v}_i^{-t}] = \boldsymbol{v}_i, \quad [\boldsymbol{r}_i^{-t}] = \boldsymbol{r}_i - \boldsymbol{v}_i t. \quad \text{(V.137)}$$

† We neglect the effect of the walls, although they are involved in the integration limits for the $\boldsymbol{r}$.

Having done this, we define the functions

$$\varrho(1, \ldots, i_1', \ldots, i_2', \ldots, i_\nu', \ldots, n, t) \qquad \text{(V.138)}$$

$$= \varrho(\boldsymbol{v}_1, \ldots, \boldsymbol{v}_{i_1}^{-t}, \ldots, \boldsymbol{v}_{i_2}^{-t}, \ldots, \boldsymbol{v}_{i_\nu}^{-t}, \ldots, \boldsymbol{v}_n$$

$$[\boldsymbol{r}_1^{-t}], \ldots, \boldsymbol{r}_{i_1}^{-t}, \ldots, \boldsymbol{r}_{i_2}^{-t}, \ldots, \boldsymbol{r}_{i_\alpha}^{-t}, \ldots, [\boldsymbol{r}_n^{-t}], 0),$$

where $\boldsymbol{v}_1, \ldots, \boldsymbol{v}_{i_1}^{-t}, \ldots, \boldsymbol{v}_{i_\nu}^{-t}, \ldots, \boldsymbol{v}_n, [\boldsymbol{r}_1^{-t}], \ldots, \boldsymbol{r}_{i_1}^{-t} \ldots, \boldsymbol{r}_{i_\nu}^{-t}, \ldots, [\boldsymbol{r}_n^{-t}]$ are the velocities and positions of the n particles of the system at $-t$, given that only the forces of interaction $F_{i_\varkappa i_\lambda}$, with $1 \leqq \varkappa, \lambda \leqq \nu$, are non-vanishing, while all other interactions are zero; the primed indices $i_1', \ldots, i_\nu'$ denote the particles which are effectively interacting. Since a "collision" involves always two particles at least, we complete the foregoing definition by putting

$$\varrho(1, \ldots, i_1', \ldots, n) = \varrho(1, \ldots, i_1, \ldots, n) = \varrho(\boldsymbol{v}_1, \ldots, \boldsymbol{v}_n, [\boldsymbol{r}_1^{-t}], \ldots, [\boldsymbol{r}_n^{-t}], 0). \qquad \text{(V.139)}$$

We introduce now the following cluster functions:

$$C^0 = \varrho(1, 2 \; \ldots \; n)$$

$$C_i^1 = \varrho(1, 2, \ldots, i', \ldots, n) - \varrho(1, 2, \ldots, i, \ldots, n) = 0,$$

$$\begin{aligned} C_{ij}^2 = {} & \varrho(1, 2, \ldots, i', \ldots, j', \ldots, n) \\ & - \varrho(1, 2, \ldots, i', \ldots, n) \\ & - \varrho(1, 2, \ldots, i, \ldots, j', \ldots, n) + \varrho(1, 2, \ldots, n), \end{aligned}$$

$$\cdots\cdots\cdots\cdots\cdots\cdots \qquad \text{(V.140)}$$

$$\begin{aligned} C_{i_1, \ldots, i_\nu}^\nu = {} & \varrho(1, \ldots, i_1', \ldots, i_\nu', \ldots, n) \\ & - \sum_{\varkappa=1}^{\nu} \varrho(1, \ldots, i_1', \ldots, i_{\varkappa-1}', i_\varkappa, i_{\varkappa+1}', \ldots, i_\nu', \ldots, n) \\ & + \sum_{1 \leqq \varkappa < \lambda \leqq \nu} \varrho(1, \ldots, i_1', \ldots, i_{\varkappa-1}', i_\varkappa, i_{\varkappa+1}', \ldots, \\ & \qquad i_{\lambda-1}', i_\lambda, i_{\lambda+1}', \ldots, i_\nu', \ldots, n) \\ & + \ldots \\ & + (-1)^n \varrho(1, 2, \ldots, n), \end{aligned}$$

and we have the fundamental identity

$$\begin{aligned}
&\varrho(\boldsymbol{v}_1, \ldots, \boldsymbol{v}_n, \boldsymbol{r}_1, \ldots, \boldsymbol{r}_n, t) \\
&= \varrho(\boldsymbol{v}_1^{-t}, \ldots, \boldsymbol{v}_n^{-t}, \boldsymbol{r}_1^{-t}, \ldots, \boldsymbol{r}_n^{-t}, 0) \qquad \text{(V.141)} \\
&= C^0 + \sum_{i=1}^{n} C_i^1 + \sum_{1 \leq i < j \leq n} C_{ij}^2 + \ldots = \sum_{\nu=0}^{n} \sum_{i_1 < i_2 < \ldots < i_\nu} C_{i_1, i_2, \ldots, i_\nu}^{\nu}.
\end{aligned}$$

The different cluster terms have the following meaning: if no particle interacts in the interval $(0, t)$, then all the C^ν are zero and we have $\varrho(t) = \varrho(1, 2, \ldots, n)$. If only the two particles 1 and 2 interact in the same interval, the clusters C^0 and C_{12}^2 are then non-vanishing and we have $\varrho = C^0 + C_1^1 + C_2^1 + C_{12}^2 = \varrho(1', 2', \ldots, n)$. Similarly, if the three particles 1, 2 and 3 interact, we have

$$\varrho = C^0 + C_1^1 + C_2^1 + C_3^1 + C_{12}^2 + C_{13}^2 + C_{23}^2 + C_{123}^3,$$

and so on. The more particles there are interacting, the more terms are contained in ϱ, the highest order of the cluster being equal to the total number of particles which interact. The expansion (V.141) is thus of the same nature as that of Ursell–Mayer for the equilibrium case; moreover, we can contract equation (V.141) by integrating over a group of $n - s$ variables and then we obtain equations which are identical with those of Green (1956b).

In order to find the Master Equation, we shall assume now that the interaction is represented completely by the hard spheres model; we shall suppose in addition that, since the system is spatially homogeneous, the distribution function ϱ_n is independent, at the initial time, of the distance r_{ij} of the two particles i and j, when we have $r_{ij} > \delta$ (δ is the molecular diameter). This assumption can be put in the form

$$\varrho(\boldsymbol{v}_1, \ldots, \boldsymbol{v}_n, \boldsymbol{r}_1, \ldots, \boldsymbol{r}_n, 0) = \Phi(\boldsymbol{v}_1, \ldots, \boldsymbol{v}_n, 0)\, \Psi(\boldsymbol{r}_1, \ldots, \boldsymbol{r}_n, 0), \qquad \text{(V.142)}$$

where, for dilute gases, we have in the zeroth approximation of the virial expansion

$$\Psi(\boldsymbol{r}_1, \ldots, \boldsymbol{r}_n, 0) \simeq \frac{1}{V^n}, \quad \text{for all } r_{ij} \geqq \delta \quad (V = \text{total volume}). \qquad \text{(V.143)}$$

Since we have assumed that the system was spatially homogeneous at the initial time, we can put in addition:

$$\left.\begin{aligned}&\Psi(\boldsymbol{r}_1, \ldots, \boldsymbol{r}_n, 0) = \\ &\frac{\prod\limits_{1 \leqq i<j \leqq n} \{1 - \delta(|\boldsymbol{r}_i - \boldsymbol{r}_j|)\}}{\underbrace{\int\limits_V \cdots \int\limits_V}_{n} \prod\limits_{1 \leqq i<j \leqq n} \{1 - \delta(|\boldsymbol{r}_i - \boldsymbol{r}_j|)\}\, d^3\boldsymbol{r}_1 \ldots d^3\boldsymbol{r}_n}, \\ &\qquad \text{where } \delta(x) = \begin{cases} 1 \text{ if } 0 \leqq x < \delta, \\ 0 \text{ if } x \geqq \delta. \end{cases}\end{aligned}\right\} \tag{V.144}$$

Now we shall attempt to calculate:

$$\Phi(\boldsymbol{v}_1, \ldots, \boldsymbol{v}_n, t) = \underbrace{\int\limits_V \cdots \int\limits_V}_{n} \varrho_n(\boldsymbol{v}_1, \ldots, \boldsymbol{v}_n, \boldsymbol{r}_1, \ldots, \boldsymbol{r}_n, t)\, d^3\boldsymbol{r}_1 \ldots d^3\boldsymbol{r}_n, \tag{V.145}$$

by using the expansion (V.141). First of all, we have

$$\underbrace{\int\limits_V \cdots \int\limits_V}_{n} C^0\, d^3\boldsymbol{r}_1 \ldots d^3\boldsymbol{r}_n = \Phi(\mathbf{v}_1, \ldots, \mathbf{v}_n, 0), \tag{V.146}$$

and, since $C_i^0 = 0$, we must then calculate terms of the form $\overbrace{\int \ldots \int}^{n} C_{ij}^2\, d^3\boldsymbol{r}_1 \ldots d^3\boldsymbol{r}_n$. If we suppose, for simplicity, that $i = 1$ and $j = 2$, we have according to (V.137):

$$\begin{aligned}
&\underbrace{\int\limits_V \cdots \int\limits_V}_{n} C_{12}^2 d^3\boldsymbol{r}_1 \ldots d^3\boldsymbol{r}_n \\
&= \underbrace{\int\limits_V \cdots \int\limits_V}_{n} [\varrho(1', 2', \ldots, n) - \varrho(1, 2, \ldots, n)]\, d^3\boldsymbol{r}_1 \ldots d^3\boldsymbol{r}_n \\
&= \underbrace{\int\limits_V \cdots \int\limits_V}_{n} \{\Phi(\boldsymbol{v}_1^{-t}, \boldsymbol{v}_2^{-t}, \boldsymbol{v}_3, \ldots, \boldsymbol{v}_n)\, \Psi(\boldsymbol{r}_1^{-t}, \boldsymbol{r}_2^{-t}, [\boldsymbol{r}_3^{-t}], \ldots, [\boldsymbol{r}_n^{-t}]) \\
&\qquad - \Phi(\boldsymbol{v}_1, \ldots, \boldsymbol{v}_n)\, \Psi([\boldsymbol{r}_1^{-t}], \ldots, [\boldsymbol{r}_n^{-t}])\}\, d^3\boldsymbol{r}_1 \ldots d^3\boldsymbol{r}_n \\
&= \underbrace{\int\limits_V \cdots \int\limits_V}_{n} [\Phi(\boldsymbol{v}_1^{-t}, \boldsymbol{v}_2^{-t}, \boldsymbol{v}_3, \ldots, \boldsymbol{v}_n)\, \Psi(\boldsymbol{r}_1^{-t}, \boldsymbol{r}_2^{-t}, \boldsymbol{r}_3, \ldots, \boldsymbol{r}_n) \\
&\qquad - \Phi(\boldsymbol{v}_1, \ldots, \boldsymbol{v}_n)\, \Psi(\boldsymbol{r}_1, \ldots, \boldsymbol{r}_n)]\, d^3\boldsymbol{r}_1 \ldots d^3\boldsymbol{r}_n.
\end{aligned} \tag{V.147}$$

Since we have

$$\underbrace{\int_V \cdots \int_V}_{n-2} \Psi(\boldsymbol{r}_1, \boldsymbol{r}_2, \ldots, \boldsymbol{r}_n, 0)\, d^3\boldsymbol{r}_3 \ldots d^3\boldsymbol{r}_n = \varrho(\boldsymbol{r}_1, \boldsymbol{r}_2, 0) \simeq \frac{1}{V^2},$$

we can write

$$\underbrace{\int_V \cdots \int_V}_{n} C_{12}^2\, d^3\boldsymbol{r}_1 \ldots d^3\boldsymbol{r}_n$$

$$= \frac{1}{V^2} \int\limits_{S(\boldsymbol{v}_1, \boldsymbol{v}_2)} [\Phi(\boldsymbol{v}_1^{-t}, \boldsymbol{v}_2^{-t}, \boldsymbol{v}_3, \ldots, \boldsymbol{v}_n) - \Phi(\boldsymbol{v}_1, \boldsymbol{v}_2, \ldots, \boldsymbol{v}_n)]\, d^3\boldsymbol{r}_1\, d^3\boldsymbol{r}_2, \tag{V.148}$$

where $S(\boldsymbol{v}_1, \boldsymbol{v}_2)$ is a region of (6-dimensional) μ-space for which $C_{12}^2 \neq 0$. It is clear that $\boldsymbol{r}_1$ (or $\boldsymbol{r}_2$) can be taken arbitrarily in V (since we neglect wall effects); on the other hand, the condition $C_{12}^2 \neq 0$ implies that $\boldsymbol{r}_2$ must be chosen in such a way that the particles located at $(\boldsymbol{r}_1, \boldsymbol{v}_2)$ and $(\boldsymbol{r}_2, \boldsymbol{v}_2)$ at $t = 0$, make exactly one collision during the interval $(-t, 0)$ (we take an interval of time much larger than the mean duration of a collision, which enables us to consider all collisions as complete). In the special case of the hard spheres model, we obtain easily

$$\underbrace{\int_V \cdots \int_V}_{n} C_{12}^2 d^3\boldsymbol{r}_1 \ldots d^3\boldsymbol{r}_n$$

$$= \frac{1}{2} \frac{t}{V} \int [\Phi(\Theta_{12}\boldsymbol{R}, t) - \Phi(\boldsymbol{R}, t)]\, \psi_{12}\, d^2\Omega = tL_{12}\Phi(\boldsymbol{R}, 0), \tag{V.149}$$

in which we have used the notation of Section IV. Taking account of all binary collisions, we can then write

$$\Phi(\boldsymbol{R}, t) = \Phi(\boldsymbol{R}, 0) + t \sum_{1 \leq i < j \leq n} L_{ij}\Phi(\boldsymbol{R}, 0) + \ldots . \tag{V.150}$$

We note that the operator $L = \sum_{1 \leq i < j \leq n} L_{ij} \equiv \frac{1}{2} \sum_{i,j} L_{ij}$ is precisely the linear operator of the Master Equation and that the solution of (V.64) can be written formally as

$$\Phi(\boldsymbol{R}, t) = e^{Lt}\Phi(\boldsymbol{R}, 0) = \left(1 + tL + \frac{t^2}{2!}L^2 + \ldots\right)\Phi(\boldsymbol{R}, 0). \tag{V.151}$$

Thus, we see that if we neglect powers of L greater than or equal to 2 in (V.151), we fall back on the expansion (V.150), limited to terms in C^2.

In order to treat this problem in its entirety, we should have to consider all the C^ν terms (with $\nu \geqq 2$) of the fundamental expansion (V.141); the transition from Liouville's equation to the Master Equation would then be assured if we could establish, by means of certain conditions, that the contribution of terms in C^ν are expressed by successive powers of L, $\frac{t^2}{2!} L^2$, $\frac{t^3}{3!} L^3$ .... Brout has been able to prove effectively such a result by considering only binary collisions and by means of certain approximations (almost instantaneous collisions, ...).

In the general case, the study of the various C^ν with $\nu \geqq 2$ is particularly complicated; by relying on the analysis of the different types of ternary collisions (Green, 1956b) it has been possible to show that the terms in C^3, combined with certain terms of C^4, make a contribution equal to $\frac{t^2}{2!} L^2\Phi(\boldsymbol{R}, 0)$, provided that the gas is assumed to be sufficiently dilute and that collisions of a "cyclic" nature are neglected as well as certain wall effects. Subsequently, we can extend these arguments to terms going from C^n to C^{2n} and we can attempt to show that they provide a contribution of the form $\frac{t^n}{n!} L^n\Phi(\boldsymbol{R}, 0)$.

It is obvious, however, that the study of the general case comes up against almost insuperable difficulties: in the first place, we must analyse different collision configurations, which leads us to introduce diagrams similar to those used by Feynman in quantum field theory; then we must take certain limits ($n \to \infty$, $V \to \infty$, $c = \frac{n}{V} \to 0$, $t \to \infty$ with ct fixed), which again increases the difficulty of the problem.† Be that as it may, it does not appear to be impossible to derive the Master Equation from Liouville's equation

† In this connection, we draw attention to recent papers by Prigogine and Balescu (1959a, b; 1960). In addition, we note that this diagrammatic method appears to be applicable with success to numerous problems of statistical mechanics.

in certain limiting cases, so that the method which has just been outlined opens up an interesting field of research. We must emphasise, however, that the Master Equation has the inconvenience of being very difficult to generalise in the case of spatial inhomogeneity; it follows that it is not always known to what extent Boltzmann's complete equation can be deduced from a certain stochastic process.

VI. Prigogine's Theory of Irreversible Processes

1. *Introduction*

Given the importance of recent work inspired by the theory of Prigogine and his colleagues, we shall now briefly discuss this method, emphasising the characteristic features. As in previous models, this method is applicable to a system whose Hamiltonian can be put in the form: $H = H_0 + \lambda V$, where λ is a parameter which characterises the order of magnitude of the perturbation V. It involves the following essential concepts, which we shall analyse in the following:

(a) The definition of a "resolvant operator" associated with Liouville's equation and the use of its expansion in powers of λ.

(b) Replacement of the distribution function f_s and of the phase density $\varrho(P, t)$ by their Fourier transforms with respect to space coordinates; these transforms are related to various correlations as well as to the deviation of the state of the system from spatial uniformity.

(c) Derivation of the chain of equations satisfied by the Fourier components of the phase density $\varrho(P, t)$; in this way we obtain a "dynamics of correlations" in which the interactions described by V have the effect of inducing transitions between the various Fourier coefficients. By using the expansion of the resolvant operator in powers of λ, these transitions can be represented by "graphs" in more and more complicated diagrams, corresponding to the various ways of interacting in the system. The analysis of these different types of graphs enables us to study the time-evolution of the system and we shall see that, in certain limited cases which are of practical interest, we shall arrive at generalised kinetic equations.

2. *Liouville's operator in classical statistical mechanics and the formal solutions of Liouville's equation*

We begin by introducing Liouville's classical operator $\hat{L}$ which can be associated with Liouville's equation (V.1). This operator, which was first introduced by Koopman and then studied by von Neumann, is a Hermitian operator $\hat{L}$ whose action on a phase function $f(P)$ is described by the definition:

$$\hat{L}f = i[H, f]. \tag{V.152}$$

Under these conditions, equation (V.1) becomes

$$\frac{\partial \varrho}{\partial t} = -i\hat{L}\varrho, \tag{V. 153}$$

and for the formal solution we have

$$\varrho(P, t) = e^{-i\hat{L}t}\varrho(P, 0), \tag{V.154}$$

where $e^{-i\hat{L}t}$ is a unitary operator (because of the hermiticity of $\hat{L}$) defined by its series expansion.

Similarly, the time-evolution of any phase function $f(P)$ can be expressed by:

$$\frac{df}{dt} = -[H, f] = i\hat{L}f, \tag{V.155}$$

the formal solution of which can be written as:

$$f(P, t) = f(P_t) = e^{i\hat{L}t}f(P, 0); \tag{V.156}$$

we note, in passing, that the operator $e^{i\hat{L}t}$ is none other than the operator denoted by $\hat{U}_t$ in Chapter I.

Equations (V.154) and (V.156) therefore introduce into classical mechanics a kind of Heisenberg representation of the time-evolution of a system. As we have seen in Chapter I, the average value $\overline{f(P_t)}$ of a physical quantity $f(P)$ is given by the scalar product $(f, \varrho(t))$ and, by virtue of the unitarity of the operator of evolution, we have

$$\begin{aligned} \overline{f(P_t)} &= (f, \varrho(t)) = (f, e^{-i\hat{L}t}\varrho(0)) \\ &= (e^{i\hat{L}t}f, \varrho(0)) = (f(P_t), \varrho(0)); \end{aligned} \tag{V.157}$$

thus, it can be proved that the calculation of average values at time t can be carried out in two ways, but that it suffices to make a single statistical assumption concerning the distribution of the initial phases of the system.

We shall find similar solutions in quantum theory with the definition of a Liouville quantum operator (Chapter VI, § V.2); we note also that the operator $e^{-i\hat{L}t}$ is frequently called the propagator or Green function associated with Liouville's equation. We shall, especially in what follows, make use of the Laplace transform with respect to time of this propagator, which is defined by:

$$\int_0^\infty dt\, e^{-st} e^{-i\hat{L}t} = \frac{1}{s + i\hat{L}}; \tag{V.158}$$

$(s + i\hat{L})^{-1}$ is the resolvant operator, also called the propagator or Green function, associated with Liouville's equation. If we define the Laplace transform $g(s)$ of $\varrho(t)$ by

$$g(s) = \int_0^\infty dt\, e^{-st} \varrho(t), \tag{V.159}$$

the Laplace transform of Liouville's equation (V.153) can then be written as:

$$sg(s) - \varrho(0) = -i\hat{L}g(s),$$

the formal solution of which is

$$g(s) = (s + i\hat{L})^{-1}\, \varrho(0); \tag{V.160}$$

thus, we prove the important role played by the operator $(s + i\hat{L})^{-1}$. It will be convenient for us to introduce a complex variable:

$$z = is \tag{V.161}$$

and to reserve the name resolvant operator for the operator $\hat{R}(z)$, defined by

$$\hat{R}(z) = \frac{1}{\hat{L} - z}. \tag{V.162}$$

Under these conditions, equation (V.160) will become

$$g(z) = -i\hat{R}(z)\, \varrho(0). \tag{V.163}$$

If we consider cases in which the Hamiltonian H can be put in the form

$$H = H_0 + \lambda V, \tag{V.164}$$

the operator $\hat{L}$ will be equal to the sum of the two terms

$$\hat{L} = \hat{L}_0 + \lambda \hat{L}', \qquad \text{(V.165)}$$

where $\hat{L}_0$ corresponds to the free motion of the particles and $\lambda\hat{L}'$ to the interaction term λV. By defining a resolvant operator $\hat{R}_0(z)$ associated with Liouville's unperturbed operator $\hat{L}_0$ and by making use of the operator identity

$$\hat{A}^{-1} - \hat{B}^{-1} = \hat{A}^{-1}(\hat{B} - \hat{A})\,\hat{B}^{-1},$$

we can write

$$\hat{R}(z) - \hat{R}_0(z) = -\lambda \hat{R}_0(z)\,\hat{L}'\hat{R}(z), \qquad \text{(V.166)}$$

whence, by iteration, we obtain:

$$\hat{R}(z) = \sum_{n=0}^{\infty} (-\lambda)^n\, \hat{R}_0(z)\,[\hat{L}', \hat{R}_0(z)]^n. \qquad \text{(V.167)}$$

This relation can serve as the basis for a perturbation expansion and the formal solution of (V.163) can now be written as:

$$g(z) = -i \sum_{n=0}^{\infty} (-\lambda)^n\, \hat{R}_0(z)\,[\hat{L}', \hat{R}_0(z)]^n. \qquad \text{(V.168)}$$

In order to obtain a solution to Liouville's equation (V.153), we must return ultimately to the initial function $\varrho(t)$ by inverting the Laplace transform (V.159). According to well-known methods (see for example, Doetsch, 1943), this becomes

$$\varrho(t) = \frac{1}{2\pi i} \int_{\gamma - i\infty}^{\gamma + i\infty} ds e^{ts} (s + i\hat{L})^{-1}\, \varrho(0), \qquad \text{(V.169)}$$

where the integration contour is a straight line parallel to the imaginary axis, having to its left all the singularities of the operator $(s + i\hat{L})^{-1}$; in this case $\varrho(t)$ vanishes for $t < 0$, which corresponds to the "causal" solution.

Thus, the crucial role played by the singularities of the operator $(s + i\hat{L})^{-1}$ can be seen. Since $\hat{L}$ is a Hermitian operator with real eigenvalues, $(s + i\hat{L})^{-1}$ has all its singularities on the imaginary axis. For a finite system, these singularities are isolated, whilst for an infinite system (with $N, V \to \infty$) there is a finite discontinuity on the imaginary axis associated with the continuous spectrum of $\hat{L}$; in this latter case, the causal solution is given by the function $g^+(s)$, which is analytic in the right-hand half-plane and is continued analytically in the left-hand half-plane.

If we use the variable z, we have

$$\varrho(t) = -\frac{1}{2\pi i}\int_{-\infty+i\gamma}^{+\infty+i\gamma} dz\, e^{-izt}\hat{R}(z)\,\varrho(0), \tag{V.170}$$

where the integration contour is now a straight line parallel to the real axis, above all the singularities of $R(z)$, located on the real axis. Because of the presence of the exponential e^{-izt}, we can replace this straight line by a closed integration contour C, comprising a straight line located just above the real axis and running from right to left, and a sufficiently large semicircle in the lower half-plane. Taking equation (V.167) into consideration we have finally:

$$\begin{aligned}\varrho(t) &= -\frac{1}{2\pi i}\oint_C dze^{-itz}\hat{R}(z)\,\varrho(0)\\ &= -\frac{1}{2\pi i}\oint_C dze^{-izt}\sum_{n=0}^{\infty}(-\lambda)^n\,\hat{R}_0(z)\,[\hat{L}',\hat{R}_0(z)]^n\varrho(0),\end{aligned} \tag{V.171}$$

which we shall use as the basic formula in the perturbation calculations; we note also that we must use the function $g^+(z) = -i\hat{R}^+(z)\varrho(0)$, which is analytic in the upper half-plane and is continued analytically in the lower half-plane.

3. *Fourier series expansion of the distribution functions*

A. We shall proceed now to a study of the Fourier components of the distribution functions f_s and of the phase density $\varrho(P, t)$. We note first of all that Liouville's equation, written in the form of equation (V.153) represents a formal analogy with Schrödinger's equation of quantum mechanics. In particular, we can look for the eigenvalues and eigenfunctions of the linear Hermitian operator $\hat{L}$, with suitable boundary conditions; as for the operators of quantum theory, we shall encounter either discrete spectra (case of finite systems) or continuous spectra (case of infinite systems), or the two combined. If $\varphi_k(P)$ is the orthonormal base, assumed to be complete, of the eigenfunctions of $\hat{L}$, the formal solution of equation (V. 153) can be written as

$$\varrho(P, t) = \sum_k a_k(t)\, \varphi_k(P), \tag{V.172}$$

where the $a_k(t)$ are of the form:

$$a_k(t) = c_k e^{-i\lambda_k t}, \tag{V.173}$$

and the λ_k are the eigenvalues corresponding to the functions φ_k.

As in quantum theory, we can change the "representation" and it will be convenient to introduce the representation corresponding to the free motion of the particles. In this case we have:†

$$H_0 = \sum_{i=1}^{N} \frac{\boldsymbol{P}_i^2}{2m}, \quad \hat{L}_0 = -i \sum_{j=1}^{N} \left(\frac{\boldsymbol{P}_j}{m} \cdot \nabla_{\boldsymbol{r}_j} \right) \tag{V.174}$$

and the eigenfunctions and eigenvalues of $\hat{L}_0$ are given by:‡

$$\varphi_{\{k\}} = V^{-N/2} \exp i \sum_j (\boldsymbol{k}_j \cdot \boldsymbol{r}_j) \tag{V.175}$$

$$\lambda_{\{k\}} = \sum_j \left(\boldsymbol{k}_j \cdot \frac{\boldsymbol{p}_i}{m} \right); \tag{V.176}$$

It can be seen that the eigenfunctions associated with free motion of the particles are plane waves, whose "wave vectors" $\boldsymbol{k}_j$ are real vectors determined by the conditions at the boundaries. Since we are interested only in "large" systems for which we must let V (at the same time as N) approach infinity, these conditions at the boundaries generally play a negligible role in the theory and we can choose the simplest of them. We shall take, therefore, periodic boundary conditions, such that

$$\varphi_{\{k\}}(\boldsymbol{r}_1 \dots \boldsymbol{r}_N) = \varphi_{\{k\}}(\boldsymbol{r}_1 + V^{-\frac{1}{3}}, \dots, \boldsymbol{r}_N + V^{-\frac{1}{3}}),$$

which, for the $\boldsymbol{k}_j$, involve the relations:

$$\boldsymbol{k}_j = \frac{2\pi \boldsymbol{n}_j}{V^{-\frac{1}{3}}}, \tag{V.177}$$

where the $\boldsymbol{n}_j$ are vectors whose components are integers. The set of eigenfunctions $\varphi_{\{k\}}$ constitutes a complete orthonormal base in

† In future, N will denote the total number of particles of the system.
‡ The symbol $\{k\}$ denotes the set of wave vectors $\boldsymbol{k}_1, \boldsymbol{k}_2, \dots, \boldsymbol{k}_N$.

configuration space $(\boldsymbol{r}_1, \ldots, \boldsymbol{r}_N)$, but not in the total phase space of the system. It follows that in equations (V.172) and (V.173), the coefficients $a_k(t)$ or c_k are functions of the moments $(\boldsymbol{p}_1, \ldots, \boldsymbol{p}_N)$: the $c_k(\boldsymbol{p}_1, \ldots, \boldsymbol{p}_N)$ are none other than the multiple Fourier expansion coefficients with respect to the space variables of the density $\varrho(P)$. In the case of free motion of the particles we shall have, therefore:

$$\varrho(\boldsymbol{p}_1 \ldots \boldsymbol{p}_N, \boldsymbol{r}_1 \ldots \boldsymbol{r}_N; t) = \sum_{\{k\}} c_{\{k\}}(\boldsymbol{p}_1 \ldots \boldsymbol{p}_N) \exp\left\{i \sum_j \left(\boldsymbol{k}_j \cdot \left[\boldsymbol{r}_j - \frac{\boldsymbol{p}_j}{m} t\right]\right)\right\}, \tag{V.178}$$

an equation which expresses simply that $\varrho(P, t)$ is an arbitrary function of the $\boldsymbol{p}_j$ and of the $\boldsymbol{r}_j - \boldsymbol{p}_j t/m$.

If, now, we return to the general case in which the Hamiltonian is of the form $H = H_0 + \lambda V$, we can continue in the representation associated with the free Hamiltonian, which means again considering the Fourier expansion of the phase density $\varrho(P, t)$. In equation (V.178), the coefficients $c_{\{k\}}$ thus become functions of time because of the existence of the perturbation λV; we can write now:

$$\varrho(P, t) = \sum_{\{k\}} c_{\{k\}}(\boldsymbol{p}_1 \ldots \boldsymbol{p}_N; t) \exp\left\{i \sum_j \left(\boldsymbol{k}_j \cdot \left[\boldsymbol{r}_j - \frac{\boldsymbol{p}_j}{m} t\right]\right)\right\}. \tag{V.179}$$

In this expression, the effects due to H_0 (free motion) are separate from those resulting from the interaction represented by the perturbation λV; thus, we have the equivalent of the interaction representation used in quantum theory. We obtain easily, starting from Liouville's equation (V.153) and definitions (V.172) and (V.175), the equations satisfied by the coefficients $c_{\{k\}}(t)$; we have:†

$$\frac{\partial c_{\{k\}}(t)}{\partial t} = -i\lambda \sum_{\{k\}} \langle \{\boldsymbol{k}\} | \hat{L}'(t) | \{\boldsymbol{k}'\} \rangle \, c_{\{k'\}}(t), \tag{V.180}$$

with

$$\hat{L}'(t) = e^{i\hat{L}_0 t} \hat{L}' e^{-i\hat{L}_0 t}; \tag{V.181}$$

† We shall omit showing the dependence of the Fourier coefficients on the moments $\boldsymbol{p}_1, \ldots, \boldsymbol{p}_N$, when no ambiguity is possible. Thus in (V.180) we write $c_{\{k\}}(t)$ instead of $c_{\{k\}}(\boldsymbol{p}_1, \boldsymbol{p}_2, \ldots, \boldsymbol{p}_N; t)$.

the $\langle\{\boldsymbol{k}\}|\hat{L}'(t)|\{\boldsymbol{k}\}\rangle$ denote the matrix elements of the operator $\hat{L}'(t)$ with respect to the eigenfunctions $\varphi_{\{k\}}$ and $\varphi_{\{k'\}}$. Since the operator $\hat{L}_0$ is diagonal in the representation chosen, equation (V. 180) can be rewritten as

$$\frac{\partial c_{\{k\}}(t)}{\partial t} = -i\lambda \sum_{\{k'\}} \exp\left\{i \sum_j \left(\boldsymbol{k}_j \cdot \frac{\boldsymbol{p}_j}{m} t\right)\right\} \times$$
$$\times \langle\{\boldsymbol{k}\}|\hat{L}'|\{\boldsymbol{k}'\}\rangle \exp\left\{-i \sum_j \left(\boldsymbol{k}'_j \cdot \frac{\boldsymbol{p}_j}{m} t\right)\right\} c_{\{k'\}}(t), \quad \text{(V.182)}$$

where the matrix elements $\langle\{\boldsymbol{k}\}|\hat{L}'|\{\boldsymbol{k}'\}\rangle$ are given by:

$$\langle\{\boldsymbol{k}\}|\hat{L}'|\{\boldsymbol{k}'\}\rangle = V^{-N} \int_V \exp\left\{-i \sum_j (\boldsymbol{k}_j \cdot \boldsymbol{r}_j)\right\} L' \times$$
$$\times \exp\left\{i \sum_j (\boldsymbol{k}'_j \cdot \boldsymbol{r}_j)\right\} d^3\boldsymbol{r}_1\, d^3\boldsymbol{r}_2 \ldots d^3\boldsymbol{r}_N. \quad \text{(V.183)}$$

It will happen frequently that we shall introduce the whole time-dependence in the Fourier components of the density $\varrho(P, t)$. Thus, we obtain new coefficients $\varrho_{\{k\}}(t)$, which are related to $c_{\{k\}}(t)$ by:

$$c_{\{k\}}(t) = \exp\left\{i \sum_j \left(\boldsymbol{k}_j \cdot \frac{\boldsymbol{p}_j}{m} t\right)\right\} \varrho_{\{k\}}(t) \quad \text{(V.184)}$$

and which are defined by the usual formulae:

$$\varrho_{\{k\}}(t) = V^{-\frac{N}{2}} \int_V \varrho(p, t) \exp\left\{-i \sum_j (\boldsymbol{k}_j \cdot \boldsymbol{r}_j)\right\} d^3\boldsymbol{r}_1 \ldots d^3\boldsymbol{r}_N. \quad \text{(V.185)}$$
$$= \int_V \varrho \varphi^*_{\{k\}} d^3\boldsymbol{r}_1 \ldots d^3\boldsymbol{r}_N.$$

The expansion of equation (V.179) can then be simply written as:

$$\varrho(P, t) = \sum_{\{k\}} \varrho_{\{k\}}(t)\, \varphi_{\{k\}} = V^{-N/2} \sum_{\{k\}} \varrho_{\{k\}}(t) \exp\{i \sum_j (\boldsymbol{k}_j \cdot \boldsymbol{r}_j)\} \quad \text{(V.186)}$$

and the time-evolution of the Fourier coefficients $\varrho_{\{k\}}(t)$ is determined by the set of equations

$$\frac{\partial \varrho_{\{k\}}(t)}{\partial t} + i \sum_j \left(\boldsymbol{k}_j \cdot \frac{\boldsymbol{p}_j}{m}\right) \varrho_{\{k\}}(t) = -i\lambda \sum_{\{k'\}} \langle\{\boldsymbol{k}\}|\hat{L}'|\{\boldsymbol{k}'\}\rangle\, \varrho_{\{k\}}(t). \quad \text{(V.187)}$$

B. Having thus defined the Fourier coefficients of the phase density $\varrho(P, t)$, it now remains to study the physical significance of the new quantities that we have just introduced. We begin first of all by establishing certain fundamental properties of these coefficients.

If we consider systems in which the interactions are pair-interactions which depend only on the distance between the particles, we can write:

$$V = \sum_{i<j} V_{ij}(|\boldsymbol{r}_i - \boldsymbol{r}_j|) \tag{V.188}$$

and

$$\hat{L}' = \sum_{i<j} \hat{L}'_{ij} = i \sum_{i<j} (\nabla_{r_i} V_{ij} \cdot (\nabla_{p_i} - \nabla_{p_j})). \tag{V.189}$$

Thus, for the matrix elements of equation (V. 183), we have:

$$\langle\{\boldsymbol{k}\}|\hat{L}'|\{\boldsymbol{k}'\}\rangle = \sum_{i<j} \langle\{\boldsymbol{k}\}\,|\hat{L}'_{ij}|\,\{\boldsymbol{k}'\}\rangle, \tag{V.190}$$

with, taking into account the orthogonality relationships satisfied by the $\varphi_{\{k\}}$,

$$\langle\{\boldsymbol{k}\}|\hat{L}'_{ij}|\{\boldsymbol{k}'\}\rangle =$$

$$-V^{-\frac{1}{2}}((\boldsymbol{k}_i - \boldsymbol{k}'_i) \cdot V_{|\boldsymbol{k}_i - \boldsymbol{k}'_i|}\,(\nabla_{p_i} - \nabla_{p_j}))\,\delta^{\mathrm{Kr}}(\boldsymbol{k}_i - \boldsymbol{k}_j - \boldsymbol{k}'_i - \boldsymbol{k}'_j) \times$$

$$\times \prod_{q \neq ij} \delta^{\mathrm{Kr}}(\boldsymbol{k}_q - \boldsymbol{k}'_q). \tag{V.191}$$

In equation (V.191), $\delta^{\mathrm{Kr}}(\boldsymbol{a})$ denotes the Kronecker δ-function equal to 0 if $\boldsymbol{a} \neq 0$ and equal to 1 if $\boldsymbol{a} = 0$; on the other hand, we have put:

$$V_{|l|} = V^{-\frac{1}{2}} \int V_{ij}(|\boldsymbol{r}|) e^{-i(\boldsymbol{l}\cdot\boldsymbol{r})} d^3\boldsymbol{r}, \tag{V.192}$$

the Fourier component $V_{|l|}$ depending only on the modulus of $\boldsymbol{l}$, since the potential V_{ij} is a function of the distance $|\boldsymbol{r}|$ alone (central force hypothesis). Thus, we see that the effect of the interaction represented by $\hat{L}'$ is expressed by a sum of terms of the type (V.191) in which only two "wave vectors" are changed. This is a consequence of the binary nature of the interaction. Moreover, we prove *the law of conservation of wave vectors*, since we must have in every transition:

$$\boldsymbol{k}'_i + \boldsymbol{k}'_j = \boldsymbol{k}_i + \boldsymbol{k}_j, \tag{V.193}$$

which, in fact, expresses the translational invariance of the properties of the system (central force hypothesis). Although these properties obviously depend on the form of the Hamiltonian, they play an important role in the theory later on.

We shall consider now the case of a homogeneous system; by definition we must have:

$$\varrho(\boldsymbol{p}_1, \boldsymbol{p}_2, \ldots, \boldsymbol{p}_N; \boldsymbol{r}_1 + \boldsymbol{a}, \ldots, \boldsymbol{r}_N + \boldsymbol{a}; t) = \varrho(\boldsymbol{p}_1, \boldsymbol{p}_2, \ldots, \boldsymbol{p}_N; \boldsymbol{r}_1, \ldots, \boldsymbol{r}_N; t).$$

If we substitute this relation in the expansion (V.186), we can prove that all the coefficients $\varrho_{\{k\}}$, whose sum of the wave vectors is non-vanishing, must be zero. Because of the conservation law (V.193), this condition is not changed with time and we can see therefore that the Fourier coefficients, the sum of whose wave vectors is non-vanishing, are related to deviations of the system from the state of spatial homogeneity.

For example, the velocity distribution function of N particles is given by:

$$\varrho(\boldsymbol{p}_1, \boldsymbol{p}_2, \ldots, \boldsymbol{p}_N; t) = \int \varrho(P, t)\, d^3\boldsymbol{r}_1\, d^3\boldsymbol{r}_2 \ldots d^3\boldsymbol{r}_N = V^{N/2}\varrho_{\{0\}}; \quad \text{(V.194)}$$

it can be expressed in terms of the Fourier coefficients $\varrho_{\{0\}}$ having all its wave numbers zero. Similarly, the deviation from spatial homogeneity for a particle is related to the Fourier coefficients having a single vector $\boldsymbol{k}$ which is non-vanishing (we shall denote them by $\varrho_{k_1\{0\}}$). Actually, by using definition (V.74) and denoting by $\varphi_1(\boldsymbol{p}_1, t)$ the velocity distribution function for a single particle, we have (with $c = N/V$)

$$f_1(\boldsymbol{p}_1, \boldsymbol{r}_1; t) - c\varphi_1(\boldsymbol{p}_1, t) = cV^{N/2} \sum_{k_1 \neq 0} \int \varrho_{k_1\{0\}}\, e^{i(k_1 \cdot r)} d^3\boldsymbol{p}_2 \ldots d^3\boldsymbol{p}_N. \quad \text{(V.195)}$$

On the other hand, it can be shown that the Fourier coefficients, whose wave vector sum is equal to zero, are related to different correlation functions of the system in the uniform state. Thus we have for the pair correlation function $g_2(\boldsymbol{r}_1 - \boldsymbol{r}_2; t)$:

$$g_2(\boldsymbol{r}_1 - \boldsymbol{r}_2; t) = V^2 \int \varrho(P, t)\, d^3\boldsymbol{p}_1 \ldots d^3\boldsymbol{p}_N\, d^3\boldsymbol{r}_3 \ldots d^3\boldsymbol{r}_N - 1$$

$$= V^{N/2} \sum_{k \neq 0} \int \varrho_{k,-k\{0\}} e^{i(k \cdot [r_1 - r_2])} d^3\boldsymbol{p}_1\, d^3\boldsymbol{p}_2 \ldots d^3\boldsymbol{p}_N; \quad \text{(V.196)}$$

in this case, the integral over the moments of the coefficient $\varrho_{k,-k\{0\}}$ is simply the Fourier coefficient of the function g_2. It can be proved similarly that the coefficients $\varrho_{kk'k''\{0\}}$, with $\boldsymbol{k}+\boldsymbol{k}'+\boldsymbol{k}''=0$, correspond to conditions in which three particles are correlated together; we could obtain a similar interpretation for the Fourier coefficients which have a larger number of non-zero wave vectors.

Since the distribution functions f_s (defined by (V.74)) have Fourier coefficients containing at most s non-zero vectors $\boldsymbol{k}$, it can be seen that all the physical quantities can be expressed directly as a function of Fourier coefficients having a very small number of non-zero wave vectors. Amongst these coefficients, those whose $\boldsymbol{k}$-vector sum is zero correspond to correlations of the system in the uniform state, the others describing deviations from the state of spatial homogeneity. On the other hand, the interactions included in the operator $\hat{L}'$ have the effect of inducing transitions between the different Fourier coefficients; thus, we obtain a "correlation dynamics", the equations of which are given by equation (V.187). Moreover, because of the conservation relations (V.193), the time evolution of the coefficients with $\sum \boldsymbol{k}=0$ is completely independent of that of the coefficients with $\sum \boldsymbol{k}\neq 0$; in a non-uniform system these two types of coefficients have a completely independent time behaviour. The study of this correlation dynamics will enable us to establish kinetic equations which are valid for sufficiently long times.

Before dealing with this problem, we must draw attention to an important property of the Fourier coefficients in the limit as $N, V\to\infty$, with $N/V=c$. We note first of all that according to the foregoing results it is suggestive to rearrange the expansion (V.186) according to increasing number of non-zero wave vectors $\boldsymbol{k}$ which occur in the $\varrho_{\{k\}}$. On the other hand, in order to study the limit of infinite systems, it is convenient to introduce the coefficients $\tilde{\varrho}_{\{k\}}$

$$\tilde{\varrho}_{\{k\}} = V^{N/2}\Omega_r\varrho_{\{k\}}, \tag{V.197}$$

in which we have put

$$\Omega=\frac{V}{(2\pi)^3}, \tag{V.198}$$

and where r denotes the number of non-zero and independent $\boldsymbol{k}$-vectors which occur in $\varrho_{\{k\}}$ (that is to say, the number of non-zero wave vectors less the number of relations of the type $\sum\limits_i \boldsymbol{k}_i=0$).

Under these conditions, the expansion (V.186) now assumes the following form (Prigogine, 1962):

$$\varrho(P, t) = V^{-N}[\tilde{\varrho}_0 + \Omega^{-1} \sum_k{}' \sum_j \tilde{\varrho}(j_k)\, e^{i(k\cdot r_j)}$$

$$+\Omega^{-1} \sum_k{}' \sum_{j,l} \tilde{\varrho}(j_k, l_{-k})\, e^{i(k\cdot[r_j - r_l])} + \Omega^{-2} \sum_{kk'}{}' \sum_{j,l} \tilde{\varrho}(j_k, l_{k'})\, e^{i(k\cdot r_j)+i(k'\cdot r_l)}$$

$$+ \ldots], \tag{V.199}$$

where we have denoted by $\tilde{\varrho}(j_k, l_{k'}, \ldots)$ a Fourier coefficient such that the vector $\boldsymbol{k}$ corresponds to the particle j, the vector $\boldsymbol{k}'$ to the particle l, ...† If, now, we proceed to the limit $\Omega \to \infty$ we can replace the sums over the wave vectors in equation (V.199) by integrals. Since, because of equation (V.177), the density of states is equal to Ω, we have the relationship $\Omega^{-1} \sum_k \to \int d^3\boldsymbol{k}$ and formula (V.199) becomes:

$$\varrho(P, t) = V^{-N}[\tilde{\varrho}_0 + \sum_j \int d^3\boldsymbol{k}\, \tilde{\varrho}(j_k)\, e^{i(k\cdot r_j)} \cdot$$

$$+ \sum_{j,l} \int d^3\boldsymbol{k}\, \tilde{\varrho}(j_k, l_{-k})\, e^{i(k\cdot[r_j - r_l])}$$

$$+ \sum_{j,l} \iint d^3\boldsymbol{k}\, d^3\boldsymbol{k}'\, \tilde{\varrho}(j_k, l'_k)\, e^{i(k'\cdot r_j)+i(k'\cdot r_l)} + \ldots]. \tag{V.200}$$

However, if we consider the distribution functions f_s defined by equation (V.74), we must accept that they tend towards a finite limit value which depends only on $c = N/V$, when N and V approach infinity with c constant: *this is the hypothesis relating to the finite nature of the distribution functions*, which is necessary for defining intensive physical quantities. If, therefore, we introduce the expansion (V.200) in Equation (V.74), important restrictions concerning the Fourier coefficients $\tilde{\varrho}_{\{k\}}$ result from this hypothesis: it can be seen that the coefficients $\tilde{\varrho}_0$, $\tilde{\varrho}(j_k)$, $\tilde{\varrho}(j_k, l_{-k})$, ... need not depend explicitly on N or V but only on the ratio N/V. This restriction, which is equivalent to the hypothesis concerning the finite nature of the distribution functions, can also be expressed in another form; actually, if we compare the Fourier transform of $\varrho(P, t)$ with the supposedly valid expansion (V.200), we see that it is necessary that this Fourier transform can be expressed as a sum of terms, each containing one singularity of the type $\delta(\boldsymbol{k}_1)\, \delta(\boldsymbol{k}_2) \ldots \delta(\boldsymbol{k}_j)$ ....

† $\sum_k{}'$ denotes that we are summing over all the wave vectors $\boldsymbol{k}$, except $\boldsymbol{k} = 0$; we have here written $\tilde{\varrho}_0$ in place of $\tilde{\varrho}_{\{0\}}$.

In conclusion, we point out that the expansion (V.200) bears a close relation to an expansion of the f_s functions in clusters generalising a similar expansion used for the equilibrium state. This expansion leads to the introduction of the cluster functions $u_s(\boldsymbol{p}_1, \ldots, \boldsymbol{p}_s; \boldsymbol{r}_1, \ldots, \boldsymbol{r}_s; t) \equiv u_s(1, 2, \ldots, s)$ defined by ($F_s = c^{-s} f_s$):

$$
\begin{aligned}
F_1 &= u_1 + \Phi_1, \\
F_2 &= u_2(1, 2) + u_1(1)\,\Phi_1(2) + u_1(2)\,\Phi_1(1) + \Phi_2(1, 2), \qquad \text{(V.201)} \\
F_s &= \sum_{r=0}^{s} \sum_{(P_r)} u_r \Phi_{s-r},
\end{aligned}
$$

where $\Phi_s(\boldsymbol{p}_1, \ldots, \boldsymbol{p}_s; t)$ is the velocity distribution function of s particles. The $\sum_{(P_r)}$ represents the sum over all the permutations associated with the different ways of assigning the r variables of the functions u_r (but not over the permutations of these variables themselves, once they are fixed). It can be seen that the Fourier coefficients of the functions u_r, $\langle\{k\}|u_r\rangle$, are non-vanishing only if the set of the non-zero vectors $\boldsymbol{k}$ occurring in $\{\boldsymbol{k}\}$ corresponds to the r variables of u_r. Moreover, if we accept the factorisation property† of the functions Φ_s, $\Phi_s(1, 2, \ldots, s) = \Phi_1(1)\,\Phi_1(2)\ldots\Phi_1(s)$, it can be shown that the functions u_s can be expressed as sums of Fourier coefficients. More precisely, the functions u_s are in this case the Fourier transforms of the sums of all the coefficients $\tilde{\varrho}(1_{k_1}, 2_{k_2}, \ldots s_{k_s})$, with non-zero wave vectors corresponding to the particles $(1,2,\ldots,s)$; for example, we have

$$
\begin{aligned}
u_1(1) &= \Omega^{-1} \sum_{k}{}' \tilde{\varrho}(1_k) e^{i(\boldsymbol{k}\cdot\boldsymbol{r}_1)}, \qquad \text{(V.202)} \\
u_2(1, 2) &= \Omega^{-2} \sum_{kk'}{}' \,[\tilde{\varrho}(1_k, 2_{k'})\, e^{i(\boldsymbol{k}\cdot\boldsymbol{r}_1) + i(\boldsymbol{k}'\cdot\boldsymbol{r}_2)} \\
&\quad + \Omega\, \delta^{\mathrm{Kr}}_{k+k'}\, \tilde{\varrho}(1_k, 2_{-k})\, e^{i(\boldsymbol{k}\cdot[\boldsymbol{r}_1 - \boldsymbol{r}_2])}]
\end{aligned}
$$

We note that these relations are simplified again in the uniform case; in the previous example, it can be seen that $u_1(1) = 0$ and $u_2(1, 2)$ is reduced to the second sum of the bracket.

4. *Generalised kinetic equations*

We proceed now to a study of the time-evolution of the coefficients $\varrho_{\{k\}}(t)$, for which we shall use the resolvant formalism ex-

† It corresponds to the hypothesis according to which the correlations have a finite range (cf. Prigogine, 1962).

plained in Section VI.2. If we put the Fourier expansion (V.186) into equation (V.171), we obtain the fundamental equation:

$$\varrho_{\{k\}}(t) = -\frac{1}{2\pi i}\oint dz e^{-itz} \sum_{\{k'\}} \times$$
$$\times \sum_{n=0}^{\infty} (-\lambda)^n \langle\{\boldsymbol{k}\}|\hat{R}_0(z)\,[\hat{L}', \hat{R}_0(z)]^n|\{\boldsymbol{k}'\}\rangle\, \varrho_{\{k'\}}(0), \quad \text{(V.203)}$$

where the resolvant operator $\hat{R}_0(z)$ of the unperturbed system is diagonal in the Fourier representation, which can be written as:

$$\langle\{\boldsymbol{k}\}\,|\hat{R}_0(z)|\,\{\boldsymbol{k}'\}\rangle = \frac{1}{\sum_{j=1}^{N}(\boldsymbol{k}_j\cdot\boldsymbol{v}_j) - z}\,\delta_{\{k\},\{k'\}}. \quad \text{(V.204)}$$

Although it is obviously out of the question to calculate explicitly the sum of all the terms of the expansion (V.203) which would amount to supplying the exact solution of an N-body problem, nevertheless we can obtain the equations describing the time-evolution of the various $\varrho_{\{k\}}(t)$ by rearranging and regrouping the terms in equation (V.203). We point out first of all that each operator $\hat{L}'$ introduces a sum of $N(N-1)/2$ operators of the type $\hat{L}'_{ij}$ corresponding to the binary interaction of the two particles i and j. To each of these $\hat{L}'_{ij}$ there corresponds a matrix element of the type $\langle\{\boldsymbol{k}\}|\hat{L}'_{ij}|\{\boldsymbol{k}'\}\rangle$ which we have calculated in (V.191): it has the effect of modifying the "initial state" characterised by the set of wave vectors $\{\boldsymbol{k}'\}$, by inducing a "transition" from this state to the state represented by the system of vectors $\{\boldsymbol{k}\}$; we note that because of equation (V.193), the sum of all the wave vectors must be conserved during these transitions. Thus, we can speak of an "initial" state of correlations $\{\boldsymbol{k}'\}$ which are propagated from right to left, in undergoing "transitions" induced by the matrix elements of the type $\langle\{\boldsymbol{k}\}|\hat{L}'_{ij}|\{\boldsymbol{k}'\}\rangle$.

Thus, we are led to representing this situation by the system of diagrams (or graphs) invented by Prigogine and his co-workers, in which the state of the system corresponding to the coefficient $\varrho_{\{k'\}}(0)$ is described by a set of superimposed horizontal lines, the number of lines being equal to the number of non-zero wave vectors of $\{\boldsymbol{k}'\}$; each line is denoted by the suffix of the particle associated with the corresponding wave vector. These states are propagated freely from right to left by the action of the operator $\hat{R}_0(z)$ but are modified by the interaction terms $\hat{L}'_{ij}$ which induce the transitions represented by "vertices"; these are the points of meeting or intersection of lines which are associated with the matrix element $\langle\{k\}|L'_{ij}|\{k'\}\rangle$. Because of the conservation relations (V.193), it can be seen easily that only the six fundamental vertices of Fig. 1 can be encountered.

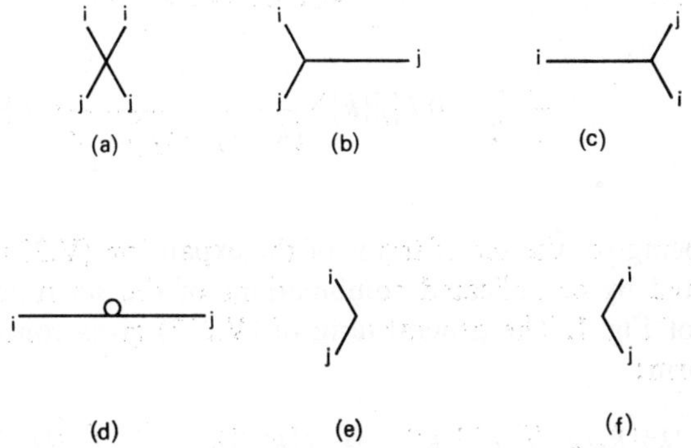

Figure 1

We note that the six basic diagrams can be arranged in three categories:

(α) those which correspond to a *propagation of correlations* in which the number of lines is not modified by the vertex (diagrams (a) and (d));

(β) those which describe the *creation of correlations* where the number of lines is larger to the left than to the right of the peak (diagrams (b) and (e));

(γ) finally, those corresponding to the *destruction of correlations*, where the number of lines is smaller to the left than to the right of the peak (diagrams (c) and (f)).

Amongst the various combinations of these elementary diagrams, we must point out the "cycle" represented by Fig. 2,

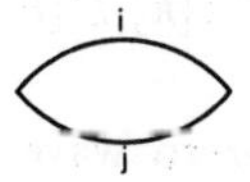

Figure 2

which, after the creation and the destruction of a correlation between the particles i and j, leads the system from an initial state without correlations to a similar final state. This diagram, which plays a fundamental role in the study of the velocity distribution function, corresponds to the operator: $\langle 0|\hat{R}_{ij}(z)|0\rangle$ which, according to (V.203) and (V.204), can be written as:

$$\langle 0|\hat{R}_{ij}(z)|0\rangle = -\frac{1}{z} + \lambda^2 \langle 0|\frac{1}{z}\hat{L}'_{ij}[\hat{R}_0(z)\,\hat{L}'_{ij}]\frac{1}{z}|0\rangle = -\frac{1}{z}$$

$$+\frac{\lambda^2}{z^2}\sum_{\{k\}}\langle|0\,\hat{L}'_{ij}|\{\boldsymbol{k}\}\rangle \frac{1}{(\boldsymbol{k}\cdot(\boldsymbol{v}_i-\boldsymbol{v}_j))-z}\langle\{\boldsymbol{k}\}|\hat{L}'_{ij}|0\rangle. \tag{V.205}$$

This being so, the set of terms of the expansion (V.203) can be represented by complicated combinations of the six fundamental vertices of Fig. 1. The general term of (V.203) gives contributions of the form:

$$\sum_{\{k'\}}\sum_{\{k''\}}\langle\{\boldsymbol{k}\}|\hat{R}_0(z)\sum_{i<j}\hat{L}'_{ij}|\{\boldsymbol{k}''\}\rangle\sum_{\{k'''\}}\langle\{\boldsymbol{k}''\}|\hat{R}_0(z)\sum_{i<j}\hat{L}'_{ij}|\{\boldsymbol{k}'''\}\rangle \ldots \times$$

$$\times \sum_{i<j}\hat{L}'_{ij}\hat{R}_0(z)|\{\boldsymbol{k}'\}\rangle\,\varrho_{\{k'\}}(0).$$

These different terms can be rearranged and expressed by means of one of the three operators defined by the following formulae (Stecki and Taylor, 1965):

—The general operator of correlation creation $\hat{\mathscr{C}}(\{\boldsymbol{k}\},\{\boldsymbol{k}'\};z)$:

$$\hat{\mathscr{C}}(\{\boldsymbol{k}\},\{\boldsymbol{k}'\};z) = \sum_{n=1}^{\infty}(-\lambda)^n\langle\{\boldsymbol{k}\}|[\hat{R}_0(z),\hat{L}']^n|\{\boldsymbol{k}'\}\rangle, \tag{V.206}$$

in which the number of non-zero wave vectors is larger in $\{\boldsymbol{k}\}$ than in $\{\boldsymbol{k}'\}$.

—The general diagonal operator $\hat{\Psi}(\{\boldsymbol{k}\};z)$:

$$\hat{\Psi}(\{\boldsymbol{k}\};z) = \sum_{n=2}^{\infty}(-\lambda)^n\langle\{\boldsymbol{k}\}|\hat{L}'[\hat{R}_0(z),\hat{L}']^{n-1}|\{\boldsymbol{k}\}\rangle, \tag{V.207}$$

where the number of non-zero wave vectors is the same in the initial and final states.

—The general destruction operator $\hat{\mathscr{D}}(\{k\},\{k'\};z)$:

$$\hat{\mathscr{D}}(\{k\},\{k'\};z) = \sum_{n=1}^{\infty} (-\lambda)^n \langle\{k\}|\hat{L}', \hat{R}_0(z)]^n|\{k'\}\rangle, \quad (V.208)$$

where the number of non-zero k vectors is smaller in $\{k\}$ than in $\{k'\}$; in the foregoing definitions, it is assumed implicitly that no intermediate, state $\{k''\}, \{k'''\}, \ldots$ is identical with the initial state $\{k'\}$. Thus, we obtain three functions of z whose analytic properties are similar to those of the resolvant operator $\hat{R}(z)$. We shall assume that these are analytic functions of z in the upper half-plane Im $z \geq 0$, and that they have analytic continuations into the lower half-plane Im $z < 0$; we shall assume, in addition, that these continuations have poles located in the lower half-plane at a finite distance from the real axis.† The functions which we have just defined, therefore, will be denoted in future by $\hat{\mathscr{C}}^+$, $\hat{\Psi}^+$ and $\hat{\mathscr{D}}^+$; in addition, we note that the function $\hat{\Psi}^+$ is identical with the sum of all the "diagonal fragments" represented by Prigogine by the diagram , whilst the functions $\hat{\mathscr{C}}^+$ and $\hat{\mathscr{D}}^+$ correspond to the diagrams and respectively. Finally, these functions have inverse Laplace transforms which we shall denote in the following by $\hat{\mathscr{C}}(\{k\},\{k'\};\tau)$, $\hat{\mathscr{G}}(\{k\};\tau)$ and $\hat{\mathscr{D}}(\{k\},\{k'\};\tau)$, where τ is the auxiliary time variable.

Following the method proposed by Prigogine and Résibois (1961), it is then possible to obtain generalised equations satisfied by the various Fourier coefficients of the phase density $\varrho(P,t)$. It can be seen immediately that the velocity distribution function corresponding to the Fourier coefficient $\varrho_0(t)$ (with all the k-vectors zero) satisfies an equation which can be written symbolically as:

$$\varrho_0(t) = \sum_{m=0}^{\infty} \bigcirc^m [\varrho_0(0) + \sum_{k'} \blacktriangleleft \varrho_{k'}(0)], \quad (V.209)$$

where $\bigcirc^m$ represents a sequence of m arbitrary "diagonal fragments". With the previous notation, equation (V.209) can be

† We note that the nature and position of these poles depend on the intermolecular potential force; for certain potentials, we can have other types of singularity (branch points or essential singularities).

written analytically as:

$$\varrho_0(t) = -\frac{1}{2\pi i}\oint_C dz e^{-izt}\frac{1}{z}\sum_{m=0}^{\infty}(-1)^m\left(\frac{\hat{\Psi}^+(z)}{z}\right)^m[\varrho_0(0)+\hat{\mathscr{D}}^+(z)]; \tag{V.210}$$

by taking the derivative with respect to t and by applying the convolution theorem for Laplace transforms, we have finally

$$\frac{\partial\varrho_0}{\partial t} = \int_0^t \hat{\mathscr{G}}_0^+(\tau)\,\varrho_0(t-\tau)\,d\tau + \hat{\mathscr{D}}_0^+(t); \tag{V.211}$$

this is the general kinetic equation satisfied by the velocity distribution function. It can be seen that this equation is non-Markovian and that the effect of initial correlations is described by the inhomogeneous term $\hat{\mathscr{D}}_0^+(t)$ which represents the sum of all the "destruction" terms leading to a final state with $\{k\} = \{0\}$.

This method can be applied easily to the study of the equations describing the time-evolution of correlations. We note first of all that, for every $\varrho_{\{k\}}(t)$, the terms of the expansion (V.186) divide themselves into two groups representing an entirely different time behaviour. Thus, it is quite in order to split up each of the $\varrho_{\{k\}}(t)$ according to the formula

$$\varrho_{\{k\}}(t) = \varrho'_{\{k\}}(t) + \varrho''_{\{k\}}(t); \tag{V.212}$$

the primed part corresponds to terms which result from the propagation and the destruction of the initial correlations, whilst $\varrho''_{\{k\}}(t)$ includes the other terms and represents that part of the correlations which is "created", starting from a correlation state of lower order. These two parts, ϱ' and ϱ'', correspond also to different types of equations. It can be shown easily that $\varrho'_{\{k\}}(t)$ satisfies an equation similar to (V.211) and, proceeding as for $\varrho_0(t)$, we obtain:

$$\frac{\partial\varrho'_{\{k\}}(t)}{\partial t} + i\left[\sum_{j=1}^{N}(\boldsymbol{k}_j\cdot\boldsymbol{v}_j)\right]\varrho'_{\{k\}}(t) = \int_0^t \hat{\mathscr{G}}^+(\{k\},\tau)\,\varrho'_{\{k\}}(t-\tau)\,d\tau + \hat{\mathscr{D}}^+_{\{k\}}(t), \tag{V.213}$$

which differs from (V.211) only by the presence of the flow term $i[\sum_j (\boldsymbol{k}_j \cdot \boldsymbol{v}_j)]\, \varrho'_{\{k\}}$. With regards to $\varrho''_{\{k\}}(t)$, it can be established, still by the same method, that it satisfies the equation

$$\varrho''_{\{k\}}(t) = \sum_{\{k'\}<\{k\}} \int_0^t \hat{\mathscr{C}}^{+}(\{\boldsymbol{k}\}, \{\boldsymbol{k}'\}; \tau)\, \varrho'_{\{k'\}}(t-\tau)\, d\tau, \qquad \text{(V.214)}$$

which states that the $\varrho''_{\{k\}}$ part of the correlations is obtained by the action of the operators $\hat{\mathscr{C}}(\tau)$ acting on the primed part of the Fourier coefficients corresponding to a correlation state $\{\boldsymbol{k}'\}$ of lower order than $\{\boldsymbol{k}\}$. Equations (V.211), (V.213) and (V.214) are the generalised kinetic equations satisfied by a classical fluid; we shall see when discussing Zwanzig's method in Chapter VI that we can obtain similar equations in quantum theory. It can be proved that these equations are all non-Markovian; this non-Markovian character results from the finite duration t_{int} of collision processes; the expression of $\varrho''_{\{k\}}$ and of the derivatives $\partial\varrho_0/\partial t$ and $\partial\varrho'_{\{k\}}/\partial t$ at time t depend on collisions which began at an earlier time of order $t - t_{\text{int}}$, and which finish at the time t; hence, the evolution integrals which occur in the second terms. Thus, the system has a "memory" which extends over a duration of the order of t_{int}, and this effect is governed by the time-dependence of the functions $\hat{\mathscr{C}}(\tau)$ and $\hat{\mathscr{G}}(\tau)$.

It can be shown, by following the method of Prigogine and Résibois, that the equations (V.211) and (V.213) reduce to Markov equations for times $t \gg t_{\text{int}}$. We note first of all that if the $\hat{G}(\tau)$ operators vanish practically for $\tau > t_{\text{int}}$, it is possible to extend the limit of integration to infinity without affecting the results. On the other hand, we can neglect the inhomogeneous terms $\hat{\mathscr{D}}_0^{+}(t)$ or $\hat{\mathscr{D}}_{\{k\}}^{+}(t)$ for sufficiently long times t. Elementary reasoning then consists in expanding $\varrho_0(t-\tau)$ or $\varrho'_{\{k\}}$ in Taylor series around $\varrho_0(t)$ and expressing the moments of $\hat{\mathscr{G}}_0^{+}(\tau)$ or of $\hat{\mathscr{G}}^{+}(\{\boldsymbol{k}\}, \tau)$ as functions of the derivatives at the point $z = 0$ of the Laplace transform $\hat{\Psi}_0(z)$ or $\hat{\Psi}(\{\boldsymbol{k}\}, z)$. We obtain finally for the velocity distribution function, the Markov equation:

$$\frac{\partial\varrho_0}{\partial t} = \hat{\Omega} i \hat{\Psi}_0^{+}(0)\, \varrho_0(t) = \hat{\Omega}$$  $$\varrho_0(t), \qquad \text{(V.215)}$$

where $\hat{\Omega}$ is defined by

$$\hat{\Omega} = 1 + \hat{\Psi}_0^{(1)+}(0) + \hat{\Psi}_0^{(1)+}(0)\,\hat{\Psi}_0^{(1)+}(0) + \tfrac{1}{2}\hat{\Psi}_0^{(2)+}(0)\,\hat{\Psi}_0^{+}(0) + \dots$$

with $\hat{\Psi}_0^{(\nu)+}(0)$ denoting the νth order derivative of $\hat{\Psi}_0^{+}(z)$ at the point $z = 0$. Similarly, equation (V.213) is replaced by the Markov equation

$$\left[\frac{\partial}{\partial t} + i\sum_{j=1}^{N}(\boldsymbol{k}_j \cdot \boldsymbol{v}_j)\right]\varrho'_{\{k\}}(t) = \hat{O}_k\varrho'_{\{k\}}(t); \qquad \text{(V.216)}$$

equation (V.214) remains, for its part, unchanged, except that the $\varrho'_{\{k\}}$ now satisfies the Markov equation (V.216). We note that these equations have been proved more rigorously by Prigogine and Résibois (1961).

The Markovian equations which we have just established are irreversible, in contrast to the fundamental equations (V.211–214) which, in principle, provide exact solutions of Liouville's equation. Equations (V.215–216) describe the return to equilibrium of the system and a generalised H-theorem can be established which includes the following three stages:

(a) By making use of equation (V.215) and with a power expansion of the concentration, we prove that $\varrho_0(t)$ tends towards the velocity distribution function at equilibrium, ϱ_0^{eq}.

(b) In the same way, starting from (V.216) we show that

$$\lim_{t\to\infty} \varrho'_{\{k\}}(t) = 0, \qquad \text{(V.217)}$$

which expresses the "destruction" of the correlations contained in ϱ' after a sufficiently long time t.

(c) By substituting the previous result in equation (V.214), correlations of the type $\varrho''_{\{k\}}(t)$ are given at the end of a very long time, by:

$$\varrho''_{\{k\}}(t) = \int_0^t \hat{\mathscr{C}}^{+}(\{\boldsymbol{k}\},\{0\};\tau)\,\varrho_0(t-\tau)\,d\tau, \qquad \text{(V.218)}$$

which shows that the correlations are then functionals of the velocity distribution function ϱ_0. In the limit as $t \to \infty$, $\varrho_0(t) \to \varrho_0^{\text{eq}}$, and we have:

$$\lim_{t\to\infty} \varrho''_{\{k\}}(t) = \int_0^\infty \hat{\mathscr{C}}^{+}(\{\boldsymbol{k}\},\{0\};\tau)\,\varrho_0^{\text{eq}}\,d\tau. \qquad \text{(V.219)}$$

We then show, and this is the third part of the generalised *H*-theorem, that these correlations correspond to those calculated starting from a canonical ensemble by equilibrium statistical mechanics.

Thus, we prove that the Markov equations (V.215)–(V.216) and (V.218) provide a sharp interpretation of the mechanism of irreversibility. What distinguishes these equations from the corresponding reversible equations, is that it suffices for calculating $\varrho_0(t)$, or $\varrho'_{\{k\}}(t)$, to know $\varrho_0(0)$ or $\varrho_{\{k\}}(0)$. On the other hand, it can be seen from equation (V.214) that the calculation of $\varrho''_{\{k\}}(t)$ only involves the correlations $\varrho_{\{k\}}$ (0) of a lower order $\{k'\}$; consequently, there are no terms existing in the Markov cycle which express a correlation of lower order as a function of a higher-order correlation. These terms, on the other hand, exist in the reversible equations (V.211) to (V.214) and they establish a link between the corresponding equations for various $\varrho_{\{k\}}(t)$. Transition to Markov equations thus has the effect of decoupling the kinetic equations which determine the evolution of the correlations and, in consequence, of decoupling the equations of the B.B.G.K.Y. hierarchy. In the limit as $t \to \infty$, the $\varrho'_{\{k\}}(t)$ vanish, which expresses the "destruction" of the initial correlations, and the state of the system is described simply by the velocity distribution $\varrho_0(t)$, satisfying equation (V.215), and by the correlations $\varrho''_{\{k\}}(t)$ which are expressed according to equation (V.218) as functionals of $\varrho_0(t)$.

In conclusion we point out that we find again, in a different form, Bogolyubov's ideas of expressing higher-order correlations as functionals of the distribution function of one particle. In fact, it can be proved that the Markov equations which we have just discussed contain Bogolyubov's kinetic equations, at least in the lowest order of the expansion parameter. Nevertheless, a complete comparison of the two methods is not easy, since the explicit form of Bogolyubov's kinetic equations does not include terms of higher order (see Stecki and Taylor, 1965).

CHAPTER VI

H-Theorems and Kinetic Equations in Quantum Statistical Mechanics

The study of the evolution in time of macroscopic systems develops in quantum theory by considerations similar to those followed for classical systems. We shall begin by establishing an $\bar{H}$-theorem for ensembles of quantum systems by introducing (as in the classical case) coarse-grained statistical densities in the sense of Ehrenfest: we shall see that the special properties of quantum observables make the introduction of macroscopic operators indispensible for defining the quantum parallel of finite cells of extension in phase (or stars in Γ-space) in classical statistical mechanics and that the results obtained with these definitions are open to the same interpretations and to the same criticisms as those of classical theory.

In order to obtain a more detailed description of the evolution of a macroscopic quantum system, we shall prove a quantum kinetic equation by making use of time-proportional transition probabilities: in this way, we shall find without difficulty all the results, especially those concerning the quantal H-theorems, showing moreover the significance of the perturbation potential V which causes the transitions. As in classical statistical mechanics, we shall prove that the formulae obtained are valid only for the short interval of time that follows the instant of macroscopic observation; their extension to any instant, which alone can enable us to make a quantitative study of irreversible phenomena, thus requires a supplementary assumption of the same kind as that introduced in classical theory: it is necessary that the quantum parallel of the property of molecular chaos be conserved in time.

Above all, in this chapter, we shall discuss the various assumptions which are essential for describing the irreversible evolution of a macroscopic quantum system, by using the essential quantum theoretical relations (such as Klein's lemma) and by comparing at each stage of reasoning the classical and quantum aspects of the foundations of statistical mechanics of systems in evolution. We shall show, particularly, in the last section that the considerations of Sections IV and V of the preceding chapter can be extended to quantum theory by means of certain adaptions which enable us to obtain, for example, a chain of equations similar to the B.B.G.K.Y. equations, a quantum Boltzmann equation and a "Master Equation", whose justification poses in final analysis problems similar to those encountered in classical theory.

I. Fine- and Coarse-grained Densities in Quantum Mechanics

1. *Properties of the fine-grained densities* $\hat{\varrho}$

As in classical mechanics, the fine-grained densities defined in Chapter II by the density matrices [equation (II.20)] cannot describe the irreversible evolution of the system; in fact, they obey an equation similar to that of Liouville:

$$\frac{\partial \hat{\varrho}}{\partial t} = \frac{i}{\hbar}[\hat{\varrho}, \hat{H}]_- . \qquad \text{(VI.1)}$$

which can be derived from Schrödinger's equation

$$\hat{H}\Psi(t) = i\hbar\frac{\partial \Psi(t)}{\partial t}, \qquad \text{(VI.2)}$$

with which is associated the evolution operator

$$\hat{U}(t) = \exp[-i\hat{H}t/\hbar]. \qquad \text{(VI.3)}$$

However, it is easy to show that the transformation (VI.3) is reversible and that the same is true for (VI.1), the formal solution of which can be written as

$$\hat{\varrho}(t) = \hat{U}(t)\,\hat{\varrho}(0)\,\hat{U}^*(t); \qquad \text{(VI.1')}$$

there will thus be no privileged direction of time for ensembles of systems described by a fine-grained density $\hat{\varrho}$. In order to establish this property of reversibility, we suppose first of all that in terms of some base φ_i we have $\Psi = \sum_i c_i\varphi_i$, or $\Psi^* = \sum_i c_i^*\varphi_i^*$. If the φ_i

are the eigenfunctions of an operator which is a real analytic expression, φ_i^* is also an eigenfunction and corresponds to the same eigenvalue (the case of position operators q_k); on the contrary, if φ_i is an eigenfunction of an operator corresponding to a pure imaginary expression, φ_i^* is an eigenfunction corresponding to the same eigenvalue with its sign changed $\left(\text{the case of momenta } p_k = \frac{h}{2\pi i}\frac{\partial}{\partial q_k}\right)$. Thus, the functions Ψ and Ψ^* lead to the same predictions for the measurement of observables with real operators and to opposite eigenvalues of purely imaginary operators. In order to prove the reversibility of (VI.2), it is sufficient then to show that the wave equation describing the evolution of a system for a reverse order of time $\theta = -t$ is obtained by taking the complex conjugate of (VI.2). Assuming that the Hamiltonian $\hat{H}$ is a real analytic expression (as is always the case in normal wave mechanics), we obtain from (VI.2):

$$\hat{H}\Psi^*(t) = -i\hbar\frac{\partial\Psi^*(t)}{\partial t} \tag{VI.2'}$$

which can be written also as

$$\hat{H}\Psi^*(t) = i\hbar\frac{\partial\Psi^*(t)}{\partial(-t)}. \tag{VI.2''}$$

Putting $-t = \theta$ and $\Psi^*(-\theta) = \chi(\theta)$, (VI.2') becomes

$$\hat{H}\chi(\theta) = i\hbar\frac{\partial\chi(\theta)}{\partial\theta},$$

which is equivalent to (VI.2) with a reversed order of time: the wave function $\chi(\theta)$ takes a value at the instant t which is the complex conjugate of Ψ at the instant $-t$.†

In addition, we can express the reversibility of evolution by introducing a quantity similar to that used in classical theory [equation (V.2)]; let

$$\sigma = \mathrm{Tr}(\hat{\varrho}\ln\hat{\varrho}) \tag{VI.4}$$

† This property of microscopic reversibility appears to be completely general; the foregoing proof, in fact, can be applied to other equations such as those of Dirac, quantum electrodynamics (see in particular Watanabe, 1951, 1955a, b, c).

be this quantity. Because of the properties of the trace of a matrix under unitary transformations, it can be seen that σ remains constant with time [since the evolution is described by the unitary operator (VI.3)]; in this case, we can write

$$\frac{d\sigma}{dt} = 0, \tag{VI.5}$$

which shows the need for introducing, as in classical statistical mechanics, coarse-grained densities in order to establish an *H*-theorem.

We note also that σ possesses the same minimum properties as its classical counterpart. If we are dealing with an isolated system whose energy is contained in a shell δE, σ is a minimum if $\hat{\bar{\varrho}}$ is the microcanonical ensemble.

$$\bar{\varrho}_{ij} = \begin{cases} \bar{\varrho}_0\, \delta_{ij} & \text{if} \quad E \leqq E_i < E + \delta E, \\ 0 & \text{in other cases.} \end{cases} \tag{VI.6}$$

Similarly, if only the average energy of the system is known, σ is minimum if $\hat{\bar{\varrho}}$ is the canonical ensemble

$$\hat{\bar{\varrho}} = e^{\beta(\Psi - \hat{H})}; \tag{VI.7}$$

and the grand-canonical ensemble

$$\hat{\bar{\varrho}} = e^{\beta(\Omega + \mu\hat{N} - \hat{H})} \tag{VI.7'}$$

makes σ minimum if only the average energy and the average number of particles of the system are known. We obtain these results by following the same route as in classical theory: we look for the minimum of equation (VI.4), connected by the relationships

$$\operatorname{Tr} \hat{\bar{\varrho}} = 1, \quad \operatorname{Tr}(\hat{\bar{\varrho}}\hat{H}) = \bar{E}, \quad \operatorname{Tr}(\hat{\bar{\varrho}}\hat{N}) = \bar{n},$$

and we use the method of Lagrangian multipliers.

Besides the property of reversibility, we can also, as in classical mechanics, establish the quasi-periodic nature of the evolution by proving (Bocchieri and Loinger, 1957) the quantum parallel of Poincaré's recurrence theorem. Suppose, in fact, that we have a system whose energy spectrum is discrete; we can show that there is always at least one time, τ, such that $\|\Psi(\tau) - \Psi(0)\| < \varepsilon$, where $\Psi(t)$ is the state vector in the Schrödinger representation. With the notation of Chapter III, § II, we have

$$\Psi(t) = \sum_i r_i e^{i(\alpha_i - E_i t/\hbar)} \psi_i$$

whence

$$\|\Psi(\tau) - \Psi(0)\| = 2 \sum_i r_i^2 (1 - \cos E_i \tau/\hbar) .$$

Since we can always choose N such that:

$$\sum_{i=N}^{\infty} r_i^2 (1 - \cos E_i \tau/\hbar) < \frac{\varepsilon}{2},$$

it is sufficient to prove that there is a value of τ such that

$$\sum_{i=0}^{N-1} (1 - \cos E_i \tau/\hbar) < \frac{\varepsilon}{2};$$

however, this is a normal result of the theory of quasi-periodic functions. Thus, we have proved a quantum recurrence theorem, in which the assumption concerning the discrete nature of the H-spectrum (bounded system) plays a similar role to the assumption relating to the finite size of the Σ hypersurfaces in classical mechanics. Mechanical reversibility and periodicity, therefore, in quantum theory as in classical theory, contradict the irreversibility of macroscopic evolution. This involves the same consequences and poses the same problems in both theories: in particular, we must define coarse-grained densities in quantum theory. Before dealing with these problems, we must still study a specific quantum property of fine-grained densities.

2. *Klein's lemma*

The quantity σ is not involved directly in the calculations leading to the H-theorem; it is the quantity

$$\sum_k \bar{\varrho}_{kk} \ln \bar{\varrho}_{kk} \tag{VI.8}$$

which plays a predominant role. We shall now prove an essential property of (VI.8), known by the name of Klein's lemma (1931): during evolution, expression (VI.8) is decreasing continually (or, at best, it is constant in the case where $\hat{\bar{\varrho}}$ is a diagonal matrix). For this purpose, we must calculate first of all the expression for the density matrix $\hat{\bar{\varrho}}(t)$ at t, starting from the expression for this matrix at $t_0 = 0$.

Since the evolution of a quantum system is described by the unitary operator (VI.3), the $c_i(t)$ coefficients of the expansion of $\Psi(t) = \sum_i c_i(t)\,\varphi_i$ satisfy the equations

$$c_i(t) = \sum_j U_{ij}(t)\, c_j(0). \tag{VI.9}$$

Taking the complex conjugate of (VI.9) and multiplying term by term, we obtain

$$c_k^*(t)\, c_i(t) = \sum_{j,l} U_{ij} U_{kl}^* c_j(0)\, c_l^*(0) \tag{VI.10}$$

which gives us the time-evolution of the density matrix of a pure case. Taking mean values (as we showed in Chapter II), we proceed from elementary density matrices to density matrices $\hat{\bar{\varrho}}$, whose time evolution is determined according to (VI.10) by

$$\bar{\varrho}_{ik}(t) = \sum_{j,l} U_{ij} U_{kl}^* \bar{\varrho}_{jl}(0). \tag{VI.11}$$

This equation, which we shall use frequently in what follows, can be derived also directly from (VI.1). We shall deal often with the case where $\bar{\varrho}_{jl}(0)$ is diagonal (assumption of a uniform distribution of the initial phases); equation (VI.11) can then be written as

$$\bar{\varrho}_{ik}(t) = \sum_j U_{ij} U_{kj}^* \bar{\varrho}_{jj}(0), \tag{VI.12}$$

which, for the diagonal terms $\bar{\varrho}_{ii}(t)$, becomes:

$$\bar{\varrho}_{ii}(t) = \sum_j |U_{ij}|^2\, \bar{\varrho}_{jj}(0). \tag{VI.13}$$

Having done this, we are now able to prove Klein's lemma; for this purpose, we consider two instants t' and t'', with $t'' > t'$, to which correspond the values $\hat{\bar{\varrho}}'$ and $\hat{\bar{\varrho}}''$ of the matrix $\hat{\bar{\varrho}}$: we shall suppose that $\hat{\bar{\varrho}}'$ is diagonal, so that we have

$$\bar{\varrho}'_{kl} = \bar{\varrho}'_{kk}\, \delta_{kl}. \tag{VI.14}$$

Applying formula (VI.13), we get

$$\bar{\varrho}''_{ii} = \sum_j |U_{ij}|^2 \varrho'_{jj}, \tag{VI.15}$$

where U_{ij} denotes here the matrix element of the operator $\hat{U}(t'' - t')$.

We introduce now the following expression:

$$Q_{ji} = \bar{\varrho}'_{jj}[\ln \bar{\varrho}'_{jj} - \ln \bar{\varrho}''_{ii} - 1] + \bar{\varrho}''_{ii} \qquad \text{(VI.16)}$$

which is always non-negative; actually, it is of the form $L(x, y) = x(\ln x - \ln y) - x + y$: if x and $y > 0$, this quantity is positive or zero according as $x \neq y$ or $x = y$ (as we saw in Chapter III); the elements $\bar{\varrho}'_{jj}$ and $\bar{\varrho}''_{ii}$ are of course positive (since they are probabilities) and they are generally different if the observable chosen for the expansion of $\Psi(t)$ is not an integral of motion (in other words, if it does not commute with the Hamiltonian). We multiply Q_{ji} by the positive quantity $|U_{ij}|^2$ and we sum over the indices i and j; we obtain

$$\sum_{i,j} |U_{ij}|^2 \bar{\varrho}'_{jj} \ln \bar{\varrho}'_{jj} - \sum_{i,j} |U_{ij}|^2 \bar{\varrho}'_{jj} \ln \bar{\varrho}''_{ii} - \sum_{i,j} |U_{ij}|^2 \bar{\varrho}'_{jj} + \sum_{i,j} |U_{ij}|^2 \bar{\varrho}''_{ii} \geqq 0. \qquad \text{(VI.17)}$$

Since the matrix U_{ij} is unitary, we have, in addition

$$\sum_i U^*_{ik} U_{il} = \delta_{kl}. \qquad \text{(VI.18)}$$

Putting (VI.18) and (VI.15) in equation (VI.17), we obtain

$$\sum_j \bar{\varrho}'_{jj} \ln \bar{\varrho}'_{jj} - \sum_i \bar{\varrho}''_{ii} \ln \bar{\varrho}''_{ii} - \sum_j \bar{\varrho}'_{jj} + \sum_i \bar{\varrho}''_{ii} \geqq 0. \qquad \text{(VI.19)}$$

Since the last two terms of the preceding expression represent the trace of the density matrix $\hat{\varrho}$ at t' and t'', which is always equal to 1, equation (VI.19) can be written

$$\sum_j \bar{\varrho}'_{jj} \ln \bar{\varrho}'_{jj} - \sum_i \bar{\varrho}''_{ii} \ln \bar{\varrho}''_{ii} \geqq 0. \qquad \text{(VI.20)}$$

This is Klein's lemma: the equality sign is valid only if the matrix $\hat{\bar{\varrho}}''$ (at t'') is itself diagonal, which is not generally the case. Thus, it is the existence of the non-diagonal terms $\hat{\bar{\varrho}}''_{ij} \neq 0$ [calculated using (VI.12)] which leads to the inequality (VI.20): this inequality is contrary to the invariance of the trace expressed by (VI.5) which corresponds to the classical relationship (V.3); it expresses a certain spreading of the fine-grained density over the different states, which has no classical equivalent. We note that the preceding result is valid only if *the matrix* $\hat{\bar{\varrho}}$ *is diagonal at* t'. We can also put (VI.20) in a different form, by associating with $\hat{\bar{\varrho}}$ a diagonalised matrix $\hat{\varrho}_d$, defined by

$$(\hat{\bar{\varrho}}_d)_{ij} = \bar{\varrho}_{ij}\delta_{ij}, \tag{VI.21}$$

to which a new quantity σ_d corresponds:

$$\sigma_d = \mathrm{Tr}\,(\hat{\bar{\varrho}}_d \ln \hat{\bar{\varrho}}_d). \tag{VI.22}$$

Klein's lemma then takes the form

$$\sigma'_d \geqq \sigma''_d; \tag{VI.23}$$

this relation expresses the fact that the transformation making the transformation from $\hat{\bar{\varrho}}'_d$ to $\hat{\bar{\varrho}}''$ is not unitary.

3. *Coarse-grained densities in quantum mechanics*

In order to define the coarse-grained densities necessary for describing the irreversible evolution of a macroscopic system, we must define first of all the macroscopic state of a quantum system. The usual method (employed in particular by Pauli and Tolman) was as follows: suppose that we have a quantum-mechanical macroscopic system represented by a matrix $\bar{\varrho}_{ij}(t)$, whose elements are written in the system with base $\{\varphi_i\}$ corresponding to the observable $\hat{A}$; the spectrum $\{\alpha_i\}$ of $\hat{A}$ is assumed to be discrete (spatially bounded system) but very dense, so that the calculations can be carried out as if we were dealing with a continuous spectrum (time-proportional transition probabilities). A macroscopic observation of A does not allow us to assign to the system a specified eigenvalue α_i but allows us only to distinguish between various groups of these eigenvalues; thus, macroscopic observation leads us to divide the set $\{\alpha_i\}$ into groups denoted by the suffix ν and each including G_ν eigenvalues.

The probability that a macroscopic observer finds that the observable A of the system has a value contained in a group ν is then given by

$$P_\nu(t) = \sum_{i=1}^{G_\nu} \bar{\varrho}_{ii}(t) \tag{VI.24}$$

and we define, starting from $P_\nu(t)$, a coarse-grained density $P_{ij}(t)$ by the relation

$$P_{ij}(t) = \frac{P_\nu(t)}{G_\nu}\delta_{ij} = \frac{\delta_{ij}}{G_\nu}\sum_{i=1}^{G_\nu} \bar{\varrho}_{ii}(t). \tag{VI.25}$$

The coarse-grained matrix $\hat{P}(t)$ is normalised like $\hat{\varrho}$; according to (VI.24) we have

$$\operatorname{Tr} \hat{P}(t) = 1. \tag{VI.26}$$

Under a change of coordinates, the matrix $\hat{P}$ is transformed in the usual way; if $\hat{S}$ is a unitary transformation of Hilbert space, we have

$$\hat{P}' = \hat{S}^{-1}\hat{P}\hat{S},$$

or

$$P'_{kl} = \sum_{i,j} S_{ki}^{-1} P_{ij} S_{jl} = \sum_{i} S_{ki}^{-1} P_{ii} S_{il}$$

which, using definition (VI.25), can be written as

$$P'_{kl} = \sum_{\nu} \frac{P_{\nu}}{G_{\nu}} \sum_{i=1}^{G_{\nu}} S_{ki}^{-1} S_{il}. \tag{VI.27}$$

We can then use this matrix $\hat{P}$ as we used Ehrenfest's coarse-grained densities in the classical theory; in this way, we prove an $\bar{H}$-theorem for an ensemble of systems (Pauli, 1928; Tolman, 1938).

We must point out, however, that the definition of the matrix $\hat{P}$ depends on the choice of the base $\{\varphi_i\}$, i.e. on the observable $\hat{A}$ which is being considered: in particular, the matrix $\hat{P}$ will be diagonal only in the system $\{\varphi_i\}$, as equation (VI.27) shows. On the other hand, the quantity $\bar{H}$ is defined by the relation

$$\bar{H} = \operatorname{Tr}(\hat{P} \ln \hat{P}) = \sum_{k} P_{kk}(t) \ln P_{kk}(t), \tag{VI.28}$$

and it assumes the form of the last term only if the matrix $\hat{P}$ is diagonal. Since this happens only for the base $\{\varphi_i\}$, the definition (VI.25) of coarse-grained densities has the disadvantage of making the macroscopic properties of the system depend on the nature of the observed quantities, and thus on the observation itself; thus, we find a difficulty which is inherent to quantum statistical mechanics, due to the non-commutability of microscopic observables.

We can avoid this difficulty here by using the definition of the macroscopic observable which we discussed in Chapter II. We recall that this method consisted in constructing a set of simultaneously measurable macroscopic observables which define, in the complete Hilbert space, a set of sub-spaces (similar to the classical stars of Γ-space) subtended by the eigenvectors $\Omega_i^{(\alpha)}$ and grouping together $s_\nu^{(\alpha)}$ macroscopic eigenvalues equal to one another (macroscopic degeneracy). The eigenvectors $\Omega_i^{(\alpha)}$ are defined by the change of variables (II.55)

$$\Omega_i^{(\alpha)} = \sum_{k=1}^{S(\alpha)} C_{ik}^{(\alpha)} \psi_k, \tag{VI.29}$$

where ψ_k are the eigenfunctions of the microscopic Hamiltonian. These operators will enable us to define the macroscopic state of a quantum system and to study its evolution.

We start by defining in this case the coarse-grained density matrices. Let $\Psi(t)$ be the wave function of the system considered. We can put

$$\Psi(t) = \sum_\alpha \sum_{i=1}^{S(\alpha)} c_i^{(\alpha)}(t)\, \Omega_i^{(\alpha)}, \tag{VI.30}$$

since the $\Omega_i^{(\alpha)}$ constitute an orthonormal base in Hilbert space. We point out first of all that there is no exchange of probability between the various shells $e^{(\alpha)}$ during evolution. Consider, in fact, the matrix elements $(\Omega_i^{(\alpha)}, \hat{U}(t)\, \Omega_j^{(\beta)})$; we can write them, according to (VI.29) and (VI.3) as:

$$\left(\sum_{k=1}^{S(\alpha)} C_{ik}^{(\alpha)} \psi_k ,\, e^{-i\hat{H}t/\hbar} \sum_{l=1}^{S(\beta)} C_{jl}^{(\beta)} \psi_l \right)$$

and, by expanding the operator $e^{-i\hat{H}t/\hbar}$, we can prove that non-zero contributions could originate only from terms corresponding to $k = l$; however, according to (VI.29), k and l are necessarily different if $\alpha \neq \beta$; we have, therefore,

$$(\Omega_i^{(\alpha)}, \hat{U}(t)\, \Omega_j^{(\beta)}) = 0 \quad \text{if} \quad \alpha \neq \beta, \tag{VI.31}$$

which is a consequence of the approximations made in Section II of Chapter II in order to obtain (II.25) or (VI.29). We can then write

$$\sum_{i=1}^{S(\alpha)} |c_i^{(\alpha)}(t)|^2 = \sum_{i=1}^{S(\beta)} |c_i^{(\alpha)}(0)|^2, \tag{VI.32}$$

which reduces, however, to stating the principle of conservation of macroscopic energy which is an integral of motion.

Thus, it is sufficient to limit ourselves in (VI.30) to the expansion of $\Psi(t)$ in terms of the functions $\Omega_i^{(\alpha)}$ of a single energy shell; in what follows, we shall always use the reduced expansion

$$\Psi(t) = \sum_{i=1}^{S(\alpha)} c_i^{(\alpha)}(t)\Omega_i^{(\alpha)}. \tag{VI.33}$$

An elementary density matrix

$$\varrho_{ij}^{(\alpha)}(t) = c_j^{(\alpha)*}(t)c_i^{(\alpha)}(t), \tag{VI.34}$$

corresponds to this pure case. The point of view of statistical mechanics consists then in representing the macroscopic system by an ensemble of systems (Jancel, 1956) defined by a mean density matrix

$$\overline{\varrho_{ij}^{(\alpha)}}(t) = \overline{c_j^{(\alpha)*}(t)\, c_i^{(\alpha)}(t)}, \tag{VI.35}$$

where the averages are taken with a suitable probability distribution as we have seen in Chapter II.

Having done this, a macroscopic observation will enable us only to know the statistical weight of a cell ν (according to its definition). If $P_\nu^{(\alpha)}(t)$ is this statistical weight, we have

$$P_\nu^{(\alpha)}(t) = \sum_{i=1}^{s_\nu(\alpha)} \overline{\varrho_{ii}^{(\alpha)}}(t). \tag{VI.36}$$

These are the quantities $P_\nu^{(\alpha)}(t)$ which describe the macroscopic state of the system; by starting from these, we can define a coarse-grained matrix by putting

$$P_{ij}^{(\alpha)}(t) = \frac{\delta_{ij}}{s_\nu^{(\alpha)}} P_\nu^{(\alpha)}(t) = \frac{\delta_{ij}}{s_\nu^{(\alpha)}} \sum_{i=1}^{s_\nu(\alpha)} \overline{\varrho_{ii}^{(\alpha)}}(t). \tag{VI.37}$$

This matrix $\hat{P}^{(\alpha)}$ possesses the same properties as the matrix $\hat{P}$ defined in (VI.25). In particular, we have always the relation

$$\mathrm{Tr}\, \hat{P}^{(\alpha)}(t) = 1. \tag{VI.38}$$

But $\hat{P}^{(\alpha)}(t)$ no longer gives rise to the difficulties encountered previously, since it has been defined in a system of coordinates where all the macroscopic observables are diagonal; it is related to the quantities $P_\nu^{(\alpha)}(t)$, which have a physical meaning, by the relation

$$P_\nu^{(\alpha)}(t) = s_\nu^{(\alpha)} P_{ii}^{(\alpha)}(t). \tag{VI.39}$$

II. The *H*-Theorem for an Ensemble of Quantum Systems

We are now able to use a reasoning similar to that of Chapter V in order to prove a generalised $\bar{H}$-theorem (Jancel, 1956). Starting from the coarse-grained matrix $\hat{P}^{(\alpha)}$ of (VI.37), we define first of all a quantity $\bar{H}^{(\alpha)}(t)$ given by

$$\bar{H}^{(\alpha)}(t) = \mathrm{Tr}\,(\hat{P}^{(\alpha)} \ln \hat{P}^{(\alpha)}) = \sum_{i=1}^{S^{(\alpha)}} P_{ii}^{(\alpha)}(t) \ln P_{ii}^{(\alpha)}(t); \qquad \text{(VI.40)}$$

$\bar{H}^{(\alpha)}(t)$ has the same properties as its classical parallel. Actually, according to (VI.37) we have

$$\bar{H}^{(\alpha)}(t) = \sum_{i=1}^{S^{(\alpha)}} \overline{\varrho_{ii}^{(\alpha)}} \ln P_{ii}^{(\alpha)} = \overline{\mathrm{Tr}\,(\ln \hat{P}^{(\alpha)})}, \qquad \text{(VI.41)}$$

where the average is taken over the statistical ensemble defined by $\overline{\varrho_{ij}^{(\alpha)}}$; furthermore, $\bar{H}^{(\alpha)}(t)$ has the same minimum properties as the quantity σ of (VI.4). Contrary to the case for σ, however, we shall show that $\bar{H}^{(\alpha)}(t)$ is not constant in time, because the transformation determining the evolution of $\hat{P}^{(\alpha)}(t)$ is not unitary.

Suppose, in fact, that at $t_0 = 0$ the macroscopic state of the system be completely defined, i.e. that the quantities $P_{\nu}^{(\alpha)}(0)$ are known; we must then represent the macroscopic system by a statistical ensemble of systems defined by a matrix $\hat{\bar{\varrho}}^{(\alpha)}(0)$ in accordance with the fundamental postulates of quantum statistical mechanics. Thus initially† we have

$$\overline{\varrho_{ij}^{(\alpha)}}(0) = \frac{P_{\nu}^{(\alpha)}(0)}{s_{\nu}^{(\alpha)}} \delta_{ij} = P_{ij}^{(\alpha)}(0). \qquad \text{(VI.42)}$$

The fine-grained and coarse-grained density matrices are then equal; the quantity $\bar{H}^{(\alpha)}$ assumes thus, at the initial time, the form

$$\bar{H}^{(\alpha)}(0) = \sum_i P_{ii}^{(\alpha)}(0) \ln P_{ii}^{(\alpha)}(0) = \sum_i \overline{\varrho_{ii}^{(\alpha)}}(0) \ln \overline{\varrho_{ii}^{(\alpha)}}(0). \quad \text{(VI.43)}$$

At a later time t, the value of the coarse-grained matrix will be calculated by starting from (VI.37), which requires a knowledge of

† Note that we define in this way a statistical ensemble which is not a stationary one; see also § III.2 and the discussion of equation (VI.56).

$\overline{\varrho_{ii}^{(\alpha)}}(t)$; by using (VI.13) and (VI.42) we obtain

$$\overline{\varrho_{ii}^{(\alpha)}}(t) = \sum_{j=1}^{S(\alpha)} |U_{ij}|^2 \overline{\varrho_{jj}^{(\alpha)}}0) = \sum_{\nu'} \frac{P_{\nu'}(0)}{S_{\nu'}^{(\alpha)}} \sum_{j=1}^{s_{\nu'}(\alpha)} |U_{ij}|^2 \varrho_{ii}^{(\alpha)}(t) \quad \text{(VI.44)}$$

without difficulty. We can see that the value of $\overline{\varrho_{ii}^{(\alpha)}}(t)$ depends on the contributions originating from all the cells ν' of the shell $e^{(\alpha)}$: these various contributions are determined by the evolution of the vector $\Psi(t)$ of (VI.33), which describes in time a trajectory over the unit hypersphere of the Hilbert sub-space corresponding to the shell $e^{(\alpha)}$. Initially, $\Psi(0)$ is determined by the coefficients $c_i^{(\alpha)}(0)$ which satisfy the relationships $P_\nu^{(\alpha)}(0) = \sum_{i=1}^{s_\nu(\alpha)} |c_i^{(\alpha)}(0)|^2$ for each cell ν; the initial statistical ensemble is then obtained by taking, in each of these cells, a uniform distribution over the hypersphere of $s_\nu^{(\alpha)}$ dimensions of radius $[P_\nu^{(\alpha)}(0)]^{\frac{1}{2}}$, which corresponds to relation (VI.42): the $P_\nu^{(\alpha)}(t)$ are calculated by starting from this initial statistical ensemble. If, now, we consider the trajectory of $\Psi(t)$, we can see from (VI.12) and (VI.44) that it is such that the quantities $\sum_{i=1}^{s_\nu(\alpha)} |c_i^{(\alpha)}(t)|^2$ are not invariant; the result is that at t, we have

$$\bar{H}^{\alpha}(t) = \sum_{i=1}^{S(\alpha)} \overline{\varrho_{ii}^{(\alpha)}}(t) \ln P_{ii}^{(\alpha)}(t), \quad \text{with} \quad \overline{\varrho_{ii}^{(\alpha)}}(t) \neq P_{ii}^{(\alpha)}(t). \qquad \text{(VI.45)}$$

If, now,we form the difference $\bar{H}^{(\alpha)}(0) - \bar{H}^{(\alpha)}(t)$ we obtain

$$\bar{H}^{(\alpha)}(0) - \bar{H}^{(\alpha)}(t) = \sum_{i=1}^{S(\alpha)} \left(\overline{\varrho_{ii}^{(\alpha)}}(0) \ln \overline{\varrho_{ii}^{(\alpha)}}(0) - \overline{\varrho_{ii}^{(\alpha)}}(t) \ln P_{ii}^{(\alpha)}(t)\right); \qquad \text{(VI.46)}$$

applying Klein's lemma (VI.20), it becomes

$$\bar{H}^{(\alpha)}(0) - \bar{H}^{(\alpha)}(t) \geqq \sum_{i=1}^{S(\alpha)} \left(\overline{\varrho_{ii}^{(\alpha)}}(t) \ln \overline{\varrho_{ii}^{(\alpha)}}(t) - \overline{\varrho_{ii}^{(\alpha)}}(t) \ln P_{ii}^{(\alpha)}(t)\right). \quad \text{(VI.47)}$$

As before, we can add to the right-hand side a term of the form $\sum_i (-\overline{\varrho_{ii}^{(\alpha)}}(t) + P_{ii}^{(\alpha)}(t))$, with the result that (VI.47) can be written as

$$\bar{H}^{(\alpha)}(0) - \bar{H}^{(\alpha)}(t) \geqq$$
$$\sum_{i=1}^{S(\alpha)} \left(\overline{\varrho_{ii}^{(\alpha)}}(t) \ln \overline{\varrho_{ii}^{(\alpha)}}(t) - \overline{\varrho_{ii}^{(\alpha)}}(t) \ln P_{ii}^{(\alpha)}(t) - \overline{\varrho_{ii}^{(\alpha)}}(t) + P_{ii}^{(\alpha)}(t)\right). \quad \text{(VI.48)}$$

Again, the right-hand side of (VI.48) is essentially positive since we have $\overline{\varrho_{ii}^{(\alpha)}}(t)$ and $P_{ii}^{(\alpha)}(t) > 0$; thus we have

$$\bar{H}^{(\alpha)}(0) - \bar{H}^{(\alpha)}(t) \geqq 0. \qquad \text{(VI.49)}$$

Hence, we confirm that the quantity $\bar{H}^{(\alpha)}(t)$ decreases, starting from its initial value $\bar{H}^{(\alpha)}(0)$, and we can show, by considerations which are comparable point by point with those of Chapter V,† that $\bar{H}^{(\alpha)}(t)$ probably will continue to decrease with time, until it has reached its minimum value corresponding to a stationary state of the macroscopic system. When this state is reached, $\hat{P}^{(\alpha)}$ is given by one of the microcanonical, canonical or grand-canonical distributions defined by (VI.6), (VI.7) or (VI.7′): starting from this instant, we have again

$$\overline{\varrho_{ii}^{(\alpha)}}(t) = P_{ii}^{(\alpha)}(t). \qquad \text{(VI.50)}$$

The same remarks and the same reservations as in classical theory (cf. Chapter V, § II.4) can be made concerning the qualitative nature of this result and we can establish also a parallel between the ergodic theorem in Hopf's form (studied in Chapter III) and the generalised $\bar{H}$-theorem (see also § III.2). Finally, if we analyse the causes of the decrease of $\bar{H}$, we can see that they are two in number:

(a) The first is due to equations (VI.42) and (VI.45) which express the initial equality of the fine-grained and coarse-grained matrices and their inequality at t: it corresponds to the difference between the fine-grained and coarse-grained density in classical mechanics;

(b) The second concerns the property of fine-grained matrices $\hat{\varrho}$ expressed by Klein's lemma: there is no parallel in classical mechanics since $\sum_i \overline{\varrho_{ii}^{(\alpha)}}(t) \ln \overline{\varrho_{ii}^{(\alpha)}}(t)$ is not the trace of a quantum operator; since it expresses the fine-grained density spread over different states, certain authors make the proof of the H-theorem depend on this single property. We shall return to this subject in the

† The fine-grained density spread over the very numerous stars is replaced here by the spread of the probabilities $\overline{\varrho_{ii}^{(\alpha)}}(t)$ over the different subspaces (corresponding to macroscopic cells) of Hilbert space: we find again the process of "mixing" of the initial conditions connected with formulae (VI.12) and (VI.44).

next chapter and we shall see that this property can be related to the irreversible perturbation due to quantum observations on the system: its role, therefore, will be of little importance in all cases where the perturbation of the system by the observer can be neglected; since this is the case of macroscopic observations, we can consider this second cause as not playing an essential role in the decrease of $\bar{H}(t)$.

III. The Kinetic Equation and Irreversible Processes

1. "*Complementarity*" *between microscopic and macroscopic descriptions*

The above proof of the generalised $\bar{H}$-theorem for a statistical ensemble of systems is formally analogous to the usual proofs of Pauli and Tolman, although it uses a different definition of coarse-grained matrices; we shall see now that this definition enables us also to avoid certain difficulties in the considerations of isolated systems. In fact, if we apply the usual method to such a system, we must expand its wave function in terms of the energy eigenfunctions; since the energy is in practice known only approximately within a range δE, the expansion will involve only $n(= S^{(\alpha)})$ terms and we shall have

$$\Psi(t) = \sum_{i=1}^{n} c_i(t)\psi_i. \tag{VI.51}$$

with

$$\sum_{i=1}^{n} |c_i(t)|^2 = 1.$$

In the absence of other data about the system, the initial statistical ensemble will be determined by

$$\overline{|c_i(0)|^2} = \frac{1}{n}; \tag{VI.52}$$

this is the microcanonical distribution. Since the energy is an integral of motion, it is not possible to prove that the system evolves: we find a difficulty which has been encountered already in ergodic theory, namely the absence of evolution of probability amplitudes due to the particular time-dependence of the integrals of motion in wave mechanics.

We avoid this difficulty by introducing a perturbation potential $\hat{V}$ representing the effect of the surroundings (usually we justify this potential by a discussion of the concept of an isolated system). The total Hamiltonian can then be written as

$$\hat{H} = \hat{H}_0 + \hat{V}, \tag{VI.53}$$

where $\hat{H}_0$ represents the unperturbed Hamiltonian. The observed macroscopic quantities correspond to groups of eigenvalues of $\hat{H}_0$, so that there are transitions between these states induced by the operator $\hat{V}$. Thus, we obtain evolution because the operator $\hat{H}_0$ does not commute with $\hat{H}$; we have, in fact,

$$[\hat{H}, \hat{H}_0]_- = [\hat{V}, \hat{H}_0]_- \neq 0. \tag{VI.54}$$

This is the same as saying that macroscopic quantities do not commute with the corresponding microscopic quantities. In the probabilistic language of wave mechanics, we can say that there exists a "complementarity" between the microscopic and macroscopic descriptions of physical phenomena. The term "complementarity", however, is used here in a different sense from that taken, in quantum theory, in the discussion of the Heisenberg uncertainty relations satisfied by canonically conjugate quantities. It means simply that any macroscopic description necessarily involves a certain inaccuracy of our knowledge of the subjacent microscopic state and, likewise, that a precise microscopic description does not enable us to restore, without a restrictive assumption, the properties of the macroscopic evolution. This complementarity is thus not specifically quantal, since it does not involve Planck's constant; it involves the existence of fluctuations† and it is connected, in fact, with the introduction of macroscopic observables with which we associate a coarse-grained density (in the sense of Ehrenfest) for describing the irreversible evolution of a physical system.

We can prove without difficulty, by referring to the construction

† An example of these "complementary" quantities is provided by the energy and temperature of an isolated system. If the energy is known with precision, we must expect to find temperature fluctuations; conversely, if the temperature is fixed, it is the energy which fluctuates.

of macroscopic operators discussed in Chapter II, that these operators do not commute with the corresponding microscopic operators. The only exception to this rule is provided by the macroscopic energy: actually, the operator $\hat{\mathscr{H}}$ possesses by construction the same eigenfunctions ψ_k as the microscopic Hamiltonian $\hat{H}$; we have in this case,

$$[\hat{\mathscr{H}}, \hat{H}]_- = 0, \tag{VI.55}$$

which ensures conservation of macroscopic energy and the validity of (VI.31). On the contrary, we will have for other operators $\hat{\mathscr{A}}$ corresponding to microscopic observables $\hat{A}$ (non-commuting with $\hat{H}$)

$$[\hat{\mathscr{A}}, \hat{A}]_- \neq 0, \quad [\hat{\mathscr{A}}, \hat{H}]_- \neq 0\dagger. \tag{VI.56}$$

However, macroscopic observation has been defined with respect to the operators $\hat{\mathscr{A}}$ (since it allows us to distinguish by hypothesis between the $N^{(\alpha)}$ groups of eigenvalues of $\hat{\mathscr{A}}$ in the shell $e^{(\alpha)}$): the macroscopic state is thus described in the representation of eigenfunctions of $\hat{\mathscr{A}}$ and, according to (VI.56), there will necessarily occur an evolution starting from the initial state.‡

Thus, we see that the evolution of an isolated system can be obtained in wave mechanics by the introduction of macroscopic operators which enable us to avoid (or to justify) the use of a perturbation potential external to the system. In what follows, we shall again define more precisely the relations existing between the two methods.

Since wave mechanics always allows us to integrate formally the equations of motion, we can calculate the coarse-grained density starting from the fine-grained density; according to (VI.36) and (VI.44) we have:

$$P_\nu^{(\alpha)}(t) = \sum_{i=1}^{s_\nu(\alpha)} \overline{\varrho_{ii}^{(\alpha)}}(t) = \sum_{\nu'=1}^{N(\alpha)} \frac{P_{\nu'}^{(\alpha)}(0)}{S_{\nu'}^{(\alpha)}} \sum_{i=1}^{s_\nu(\alpha)} \sum_{j=1}^{s_\nu'(\alpha)} |U_{ij}(t)|^2; \tag{VI.57}$$

† The fact that we might have at the same time

$$[\hat{\mathscr{A}}, \hat{H}]_- \neq 0, \quad \text{with} \quad [\hat{\mathscr{H}}, \hat{H}]_- = [\hat{\mathscr{H}}, \hat{\mathscr{A}}]_- = 0,$$

originates, obviously, from the fact that $\hat{H}$ is, in the shell $e^{(\alpha)}$, a whole multiple of the unit operator because of its definition.

‡ We note that due to the non-commuting of $\hat{\mathscr{A}}$ and of $\hat{H}$ the ensembles introduced in § II and § III.2 are not stationary.

this relation can be rewritten as

$$P_\nu^{(\alpha)}(t) = \sum_{\nu'} \mathscr{T}_{\nu\nu'}^{(\alpha)}(t)\, P_{\nu'}^{(\alpha)}(0), \tag{VI.58}$$

with

$$\mathscr{T}_{\nu\nu'}^{(\alpha)}(t) = \frac{1}{s_{\nu'}^{(\alpha)}} \sum_{i=1}^{s_\nu^{(\alpha)}} \sum_{j=1}^{s_{\nu'}^{(\alpha)}} |U_{ij}(t)|^2. \tag{VI.59}$$

The time-evolution of the quantities $P_\nu^{(\alpha)}(t)$ depends, through the intermediary of $\mathscr{T}_{\nu\nu'}^{(\alpha)}(t)$, on the matrix elements $U_{ij}(t)$; according to (VI.3), these can be written as

$$\begin{aligned} U_{ij}(t) &= (\Omega_i^{(\alpha)}, \hat{U}\Omega_j^{(\alpha)}) \\ &= \delta_{ij} - \frac{i}{\hbar} t (\Omega_i^{(\alpha)}, \hat{H}\Omega_j^{(\alpha)}) - \frac{1}{\hbar^2}\frac{t^2}{2!}(\Omega_i^{(\alpha)}, \hat{H}^2\Omega_j^{(\alpha)}) + \ldots; \end{aligned} \tag{VI.60}$$

from (VI.29) we then derive

$$\begin{aligned} (\Omega_i^{(\alpha)}, \hat{H}\Omega_j^{(\alpha)}) &= \Big(\sum_k C_{ik}^{(\alpha)}\psi_k, \hat{H}\sum_l C_{jl}^{(\alpha)}\psi_l\Big) \\ &= \Big(\sum_k C_{ik}^{(\alpha)}\psi_k, \sum_l C_{jl}^{(\alpha)}E_l\psi_l\Big) = \sum_k C_{ik}^{(\alpha)*}C_{jk}^{(\alpha)}E_k \end{aligned}$$

and similarly

$$(\Omega_i^{(\alpha)}, \hat{H}^2\Omega_j^{(\alpha)}) = \sum_k C_{ik}^{(\alpha)*}C_{jk}^{(\alpha)}E_k^2, \ldots .$$

Substituting into (VI.60), we obtain

$$\begin{aligned} U_{ij}(t) &= \delta_{ij} - \frac{i}{\hbar} t \sum_k C_{ik}^{(\alpha)*}C_{jk}^{(\alpha)}E_k - \frac{i}{\hbar^2}\frac{t^2}{2!}\sum_k C_{ik}^{(\alpha)*}C_{jk}^{(\alpha)}E_k^2 + \ldots \\ &= \delta_{ij} + \sum_k C_{ik}^{(\alpha)*}C_{jk}^{(\alpha)}\left(-\frac{i}{\hbar}tE_k - \frac{1}{\hbar^2}\frac{t^2}{2!}E_k^2 + \ldots\right) \\ &= \delta_{ij} + \sum_k C_{ik}^{(\alpha)*}C_{jk}^{(\alpha)}(e^{-iE_k t/\hbar} - 1), \end{aligned}$$

whence, because of the unitarity of the matrix $C_{ik}^{(\alpha)}$, we derive:

$$U_{ij}(t) = \sum_k C_{ik}^{(\alpha)*}C_{jk}^{(\alpha)}e^{-iE_k t/\hbar}; \tag{VI.61}$$

the corresponding expression for $\mathscr{T}_{\nu\nu'}^{(\alpha)}(t)$ is derived immediately.

We are now able to compare the preceding results, obtained by means of macroscopic observables, with the usual formulae derived from (VI.53). In fact, we suppose in this formalism that the states actually observed correspond to groups of eigenvalues of the non-perturbed Hamiltonian $\hat{H}_0$, the perturbing potential $\hat{V}$ enabling us to define the transition probabilities from one group to another in the unperturbed spectrum. It is interesting, therefore, to expand the wave function of the system in terms of the eigen-functions of $\hat{H}_0$; it then becomes

$$\psi(t) = \sum_i c_i(t)\, \psi_i^{(0)}, \tag{VI.62}$$

where the $c_i(t)$ obey the relation

$$c_i(t) = \sum_j U_{ij}(t)\, c_j(0). \tag{VI.63}$$

If we introduce the eigenfunctions ψ_n of the total Hamiltonian $\hat{H} = \hat{H}_0 + \hat{V}$, we can expand the unperturbed functions $\psi_i^{(0)}$ in terms of the ψ_n, or

$$\psi_i^{(0)} = \sum_n d_{in}\psi_n \quad \text{and} \quad \psi_n = \sum_i d_{ni}^{-1}\psi_i^{(0)}, \tag{VI.64}$$

with

$$d_{in}^{-1} = d_{ni}^*$$

since the matrix d_{in} is unitary. If, now, we calculate the matrix elements $U_{ij}(t)$ in terms of the eigenfunctions and eigenvalues of the total Hamiltonian $\hat{H}$, we obtain (by a calculation which is similar at all points to the preceding calculation):

$$U_{ij}(t) = \sum_n d_{in}^* d_{jn} e^{-iE_n t/\hbar}. \tag{VI.65}$$

Thus, we see that expression (VI.65), which determines the evolution in terms of the eigenstates of $\hat{H}_0$, is similar to (VI.61) which defines the evolution in terms of the eigenfunctions of the macroscopic operators: the operator $\hat{H}_0$ plays here the role of the macroscopic operator $\hat{\mathscr{H}}$ and the change of variables (VI.64) is similar to (VI.29). This provides a possible interpretation of the usual separation of the Hamiltonian of a system into two parts, the second of which can be considered as a perturbation. [The only difference arises from the fact that $\hat{H}_0$ does not commute with $\hat{H}$ whilst $\hat{\mathscr{H}}$ commutes with $\hat{H}$ according to (VI.55); however, this is of no fundamental importance since we know, from the calculation of transition probabilities, that the conservation of energy within an interval ΔE, compatible with the duration of the observation, is assured, provided that the perturbation potential $\hat{V}$ is small.]

In particular, if we stick to the first approximation, we have for the eigenfunctions of the total Hamiltonian:

$$\psi_i = \psi_i^{(0)} + \psi_i^{(1)}$$

and, by expressing $\psi_i^{(1)}$ as a function of the $\psi_i^{(0)}$ we can calculate the coefficients d_{in} of (VI.64):

$$d_{in} = \begin{cases} 1 & \text{if } i = n, \\ \dfrac{V_{in}^*}{E_n^{(0)} - E_i^{(0)}} & \text{if } i \neq n, \end{cases}$$

$$d_{ni}^{-1} = \begin{cases} 1 & \text{if } i = n, \\ \dfrac{V_{in}}{E_n^{(0)} - E_i^{(0)}} & \text{if } i \neq n, \end{cases} \qquad \text{(VI.66)}$$

We have by substitution in (VI.65) and neglecting terms in $\hat{V}^2$ (first approximation)

$$U_{ii}(t) = e^{-iE_it/\hbar},$$

$$U_{ij}(t) = d_{ji}e^{-iE_it/\hbar} + d_{ij}^* e^{-iE_it/\hbar}$$

$$= \frac{V_{ji}^*}{E_i^{(0)} - E_j^{(0)}} e^{-iE_it/\hbar} + \frac{V_{ij}}{E_j^{(0)} - E_i^{(0)}} e^{-iE_jt/\hbar} \quad (i \neq j). \qquad \text{(VI.67)}$$

Using the property of Hermiticity of $\hat{V}(V_{ji}^* = V_{ij})$ and the expression for the total energy, which can to a first approximation be written as $E_i = E_i^{(0)} + V_{ii}$, we can write

$$U_{ij}(t) = \frac{V_{ij}}{E_i^{(0)} - E_j^{(0)}} [e^{-iE_i^{(0)}t/\hbar} e^{-iV_{ii}t/\hbar} - e^{-iE_j^{(0)}t/\hbar} e^{-iV_{jj}t/\hbar}] \, (i \neq j).$$

Substituting this result in (VI.63) and putting

$$c_i(t) = a_i(t)\, e^{-iE_i^{(0)}t/\hbar},$$

it becomes

$$a_i(t) = a_i(0)\, e^{-iV_{ii}t/\hbar}$$
$$+ \sum_{j\neq i} \frac{V_{ij} e^{iE_i^{(0)} t/\hbar}}{(E_i^0 - E_j^0)} [e^{-i(E_i^{(0)}+V_{ii})t/\hbar} - e^{-i(E_j^{(0)}+V_{jj})t/\hbar}]\, a_j(0), \qquad \text{(VI.68)}$$

whence, by confining ourselves to a sufficiently small interval of time t (we neglect the products $V_{ii}t$), we obtain

$$a_i(t) = a_i(0) + \sum_{j\neq i} \frac{V_{ij}}{E_j^{(0)} - E_i^{(0)}} [e^{i(E_i^{(0)}-E_j^{(0)})t/\hbar} - 1]\, a_j(0). \quad \text{(VI.68')}$$

Starting from this formula, the transition probabilities between the various eigenstates of a system can be calculated; in the case where the spectrum is very dense, we see that these probabilities are proportional to the time: thus, we have proved the relationship between the method based on the definition of macroscopic operators and those normally used in quantum statistical mechanics.

If we use the scheme provided by macroscopic operators the more general equations connected with the evolution of macroscopic quantities are given by (VI.58) and (VI.59). They enable us to determine, for the time interval which follows the macroscopic observation, the quantities $P_\nu^{(\alpha)}(t)$ which represent the physically observable quantities. The form of this equation depends essentially on the matrix $\mathscr{T}_{\nu\nu'}^{(\alpha)}(t)$ which determines the transition probabilities from a cell ν' to a cell ν; this matrix depends in its turn on the microscopic nature of the system observed and on the way in which the macroscopic observables are defined [on the one hand by the coefficients $C_{ik}^{(\alpha)}$ according to (VI.61) and, on the other hand, by the quantities $s_\nu^{(\alpha)}$]: therefore, the form of equations (VI.58) can only be defined precisely by choosing special examples and by adapting each time the formalism to the conditions of the problem considered.

2. *Return to ergodicity conditions*

It should be noted that the introduction of averages over the macroscopic cells ν (implied, for example, in (VI.42)) and the definition of a coarse-grained density $P_\nu^{(\alpha)}(t)$ enables us to show the evolution of a quantum system and to obtain new conditions of ergodicity for ensuring that a macroscopic system tends towards an equilibrium state. Actually, we suppose that the initial state of the system is such that only the cell ν' is occupied; in this case: $P_{\nu'}^{(\alpha)}(0) = 1$ and, according to (VI.58), $P_\nu^{(\alpha)}(t) = \mathscr{T}_{\nu\nu'}^{(\alpha)}(t)$. Ergodicity will exist if we have

$$\overline{P_\nu^{(\alpha)}(t)}^{\infty} = \frac{s_\nu^{(\alpha)}}{S^{(\alpha)}},$$

whence, according to (VI.59), we have for the condition of ergodicity

$$\frac{1}{s_{\nu'}^{(\alpha)}} \sum_{i=1}^{s_\nu^{(\alpha)}} \sum_{j=1}^{s_{\nu'}^{(\alpha)}} \overline{|U_{ij}(t)|^2}^{\infty} = \frac{s_\nu^{(\alpha)}}{S^{(\alpha)}}. \tag{E$_1$}$$

We emphasise again that the evolution of the system is shown to exist because of the introduction of macroscopic operators and cells and by the definition of a statistical ensemble of systems which does not remain stationary during the evolution (contrary to the microcanonical ensemble used in the first part). In the present case, this statistical ensemble is defined initially as the uniform ensemble over the cell ν'; according to the notation of Chapter II, an initial density matrix

$$\hat{\varrho}(0) = \frac{1}{s_{\nu'}^{(\alpha)}} \sum_{j=1}^{s_{\nu'}^{(\alpha)}} P_{\Omega_j}.$$

corresponds to it.

We can obtain less restrictive conditions of ergodicity by imposing only a convergence in quadratic mean (Prosperi and Scotti, 1960); the statistical weight of a cell ν can be written (for a system located initially in the cell ν') as

$$w_\nu^{(\alpha)}(t) = \sum_{i=1}^{s_\nu^{(\alpha)}} |c_i^{(\alpha)}(t)|^2 = \sum_i \sum_{j,k=1}^{s_{\nu'}^{(\alpha)}} U_{ij} U_{ik}^* c_j^{(\alpha)}(0)\, c_k^{(\alpha)*}(0),$$

and we have the following time-average

$$\frac{\overline{\left(w_\nu^{(\alpha)}(t) - \dfrac{s_\nu^{(\alpha)}}{S^{(\alpha)}}\right)^2}^{\infty}}{\left(\dfrac{s_\nu^{(\alpha)}}{S^{(\alpha)}}\right)^2}$$

$$= \sum_{j,k=1}^{s_{\nu'}^{(\alpha)}} \sum_{j',k'=1}^{s_{\nu'}^{(\alpha)}} E_{jk}^{(\nu)} E_{j'k'}^{(\nu)} c_j^{(\alpha)}(0)\, c_k^{(\alpha)*}(0)\, c_{j'}^{(\alpha)}(0)\, c_{k'}^{(\alpha)*}(0),$$

where we have put

$$E_{jk}^{(\nu)} = \frac{S^{(\alpha)}}{s_\nu^{(\alpha)}} \left\{ \sum_{i=1}^{s_\nu^{(\alpha)}} \overline{U_{ij} U_{ik}^*}^{\infty} - \frac{s_\nu^{(\alpha)}}{S^{(\alpha)}} \delta_{jk} \right\}.$$

If, now, we take the average over an ensemble of systems uniformly distributed over the cell ν', we have first (denoting this average by $\overline{\cdots}^{\nu'}$) according to the results of Appendix II:

$$\overline{c_j^{(\alpha)}(0)\, c_k^{(\alpha)*}(0)\, c_{j'}^{(\alpha)}(0)\, c_{k'}^{(\alpha)*}(0)}^{\nu'} = \frac{1}{s_{\nu'}^{(\alpha)}(s_{\nu'}^{(\alpha)}+1)} (\delta_{jk}\,\delta_{j'k'} + \delta_{jk'}\,\delta_{kj'}),$$

and we obtain a condition of ergodicity by writing:

$$\sum_{\nu=1}^{N(\alpha)} \frac{S^{(\alpha)2}}{s_\nu^{(\alpha)2}} \overline{\left(\overline{w_\nu^{(\alpha)}(t)}^{\infty} - \frac{s_\nu^{(\alpha)}}{S^{(\alpha)}} \right)^2}^{\nu'}$$

$$\simeq \sum_{\nu=1}^{N(\alpha)} \frac{1}{(s_{\nu'}^{(\alpha)})^2} \left[\left(\sum_{j=1}^{s_{\nu'}^{(\alpha)}} E_{jj}^{(\nu)} \right)^2 + \sum_{j,k=1}^{s_{\nu'}^{(\alpha)}} |E_{jk}^{(\nu)}|^2 \right] \ll 1. \tag{E$_2$}$$

This condition can be satisfied by assuming, for example, that

$$\left| \frac{1}{s_{\nu'}^{(\alpha)}} \sum_{j=1}^{s_{\nu'}^{(\alpha)}} E_{jj}^{(\nu)} \right| \ll \frac{1}{(N^{(\alpha)})^{\frac{1}{2}}}, \quad \frac{1}{(s_{\nu'}^{(\alpha)})^2} \sum_{j,k=1}^{s_{\nu'}^{(\alpha)}} |E_{jk}^{(\nu)}|^2 \ll \frac{1}{N^{(\alpha)}}. \tag{E$'_2$}$$

The inequalities (E$'_2$) provide the most general conditions of ergodicity for a convergence in quadratic mean; if they are satisfied, we have an ergodic theorem which is a generalised form of Hopf's quantum theorem. A much more restrictive way of ensuring this convergence would be to assume the condition

$$\sum_{i=1}^{s_\nu^{(\alpha)}} \overline{U_{ij} U_{ik}^*}^{\infty} = \frac{s_\nu^{(\alpha)}}{S^{(\alpha)}} \delta_{jk}, \tag{E$_3$}$$

in which case the equality $\overline{w_\nu^{(\alpha)}(t)}^{\infty} = s_\nu^{(\alpha)}/S^{(\alpha)}$ would be satisfied, whatever the initial conditions.

For a better understanding of the nature of these various conditions of ergodicity, we must express the $U_{ij}(t)$ as functions of the $C_{ki}^{(\alpha)}$ according to (VI.61): we then verify that the conditions (E) bear at the same time on the definition of macroscopic cells and on the structure of the microscopic Hamiltonian of the system. Their analysis is obviously very complicated and a new difficulty appears—that of proving that a physical system actually possesses such properties. By way of example we have, in the case where there is no degeneracy of the microscopic energy,

$$\sum_{i=1}^{s_\nu^{(\alpha)}} \overline{U_{ij}U_{ik}^*}^{\infty} = \sum_{l=1}^{S^{(\alpha)}} C_{kl}^{(\alpha)*}C_{jl}^{(\alpha)} \sum_{i=1}^{s_\nu^{(\alpha)}} |C_{il}^{(\alpha)}|^2,$$

and condition (E_3) can then be written as

$$\sum_{i=1}^{s_\nu^{(\alpha)}} |C_{il}^{(\alpha)}|^2 = \frac{s_\nu^{(\alpha)}}{S^{(\alpha)}};$$

this condition is similar to the one we have already encountered when discussing Fierz's method (cf. Chapter III, § IV.3).

3. *The kinetic equation in quantum mechanics. Pauli's equation*

For the purpose of arriving at a quantum kinetic equation, we shall make an assumption concerning the matrix $\mathscr{T}_{\nu\nu'}^{(\alpha)}(t)$, suggested by the result of first-order perturbation theory; we shall assume that, given the large number of degrees of freedom of the system, the transition probabilities are proportional to the time: this enables us to write (knowing that we must justify this assumption for each problem studied):

$$\mathscr{T}_{\nu\nu'}^{(\alpha)}(t) = \delta_{\nu\nu'} + W_{\nu\nu'}^{(\alpha)}\Delta t, \tag{VI.69}$$

whence we derive, according to (VI.58):

$$P_\nu^{(\alpha)}(t) = P_\nu^{(\alpha)}(0) + \Delta t \sum_{\nu'} W_{\nu\nu'}^{(\alpha)} P_{\nu'}^{(\alpha)}(0). \tag{VI.70}$$

Since Δt is macroscopically small, we can conclude from this that

$$\left(\frac{dP_\nu^{(\alpha)}(t)}{dt}\right)_{t=0} = \frac{P_\nu^{(\alpha)}(t) - P_\nu^{(\alpha)}(0)}{\Delta t} = \sum_{\nu'} W_{\nu\nu'}^{(\alpha)} P_{\nu'}^{(\alpha)}(0). \tag{VI.71}$$

In order to obtain a kinetic equation at t, we must repeat at each instant the reasoning which we have pursued for (VI.71), that is to say, we must construct new statistical ensembles compatible with our information at t. If we accept that the fundamental assumption of quantum statistical mechanics expressed by (VI.42) can be applied at every t (we shall return to this point later), we have

$$\frac{dP_\nu^{(\alpha)}(t)}{dt} = \sum_{\nu'} W_{\nu\nu'}^{(\alpha)} P_{\nu'}^{(\alpha)}(t). \qquad \text{(VI.72)}$$

This is the most general form of the kinetic equation in quantum statistical mechanics which could be derived from the formalism of macroscopic observables. It enables us to describe the irreversible evolution of a macroscopic system, as we shall see at the end of this section, by calculating $d\bar{H}/dt$ with time-proportional transition probabilities.

We shall mention briefly a few properties of the matrix $W_{\nu\nu'}^{(\alpha)}$; from the relation

$$\sum_{\nu=1}^{N^{(\alpha)}} P_\nu^{(\alpha)}(t) = 1,$$

we derive

$$\sum_{\nu=1}^{N^{(\alpha)}} \frac{dP_\nu^{(\alpha)}}{dt} = 0,$$

whence, according to (VI.72),

$$\sum_{\nu=1}^{N^{(\alpha)}} W_{\nu\nu'}^{(\alpha)} = 0.$$

Equation (VI.72) can then be written as

$$\frac{dP_\nu^{(\alpha)}}{dt} = \sum_{\nu' \neq \nu} \left(W_{\nu\nu'}^{(\alpha)} P_{\nu'}^{(\alpha)} - W_{\nu'\nu}^{(\alpha)} P_\nu^{(\alpha)}\right). \qquad \text{(VI.73)}$$

On the other hand, if we suppose that $P_{\nu'}(0)$ is initially of the form $\delta_{\nu'\nu_0}$ it becomes, according to (VI.58) and (VI.69),

$$P_\nu(t) = \mathscr{T}_{\nu\nu_0} = W_{\nu\nu_0}\Delta t, \quad \text{when} \quad \nu \neq \nu_0 .$$

It follows that the $W_{\nu\nu'}$ are necessarily positive and that they can be interpreted as transition probabilities. Thus, equation (VI.73) is a probabilistic equation of the Master Equation type, satisfied by a Markovian process.

Finally, the $W_{\nu\nu'}$ satisfy also important symmetry properties; according to (VI.61) we have $U_{ji}(t) = U_{ij}^*(-t)$. If we take the case in which the operators $\hat{H}$ and $\hat{A}$ are real, the same is true for their eigenfunctions; we can then choose the functions $\Omega_i^{(\alpha)}$ to be real, so that the coefficients $C_{ki}^{(\alpha)}$ are real in their turn and the matrix U_{ij}

is symmetrical. Thus, we have $U_{ij}(t) = U_{ji}(t)$ whence, by substituting in (VI.59), we obtain:

$$s_\nu^{(\alpha)} \mathscr{T}_{\nu'\nu}^{(\alpha)} = s_{\nu'}^{(\alpha)} \mathscr{T}_{\nu\nu'}^{(\alpha)} \tag{VI.74}$$

and, according to (VI.69),

$$s_\nu^{(\alpha)} W_{\nu'\nu}^{(\alpha)} = s_{\nu'}^{(\alpha)} W_{\nu\nu'}^{(\alpha)}. \tag{VI.75}$$

We can then easily prove, starting from these general relations (and by introducing slightly different variables), the symmetry relationships which correspond to Onsager's reciprocity formulae in the theory of irreversible phenomena [see especially van Kampen, 1954].†

The foregoing results are valid only if the transition probabilities are time-proportional; we have seen that it is necessary to verify this assumption for each particular case and we know already that it is valid if we accept the perturbed Hamiltonian model. It was with this model that Pauli proved for the first time a Master Equation of the type (VI.73), which is usually called Pauli's equation. It does not deal with a coarse-grained distribution but with the probabilities $|c_i(t)|^2 = p_i$ for the unperturbed quantum states of the system which we are assuming to lie sufficiently densely so that we can consider the spectrum of $\hat{H}_0$ to be continuous. This equation can be written as

$$\frac{dp_i}{dt} = \sum_\nu (w_{ii'}p_{i'} - w_{i'i}p_i) \tag{VI.73'}$$

where the $w_{ii'}$, which are given by

$$w_{ii'} = 2\pi\lambda^2 \delta(E_{i'}^{(0)} - E_i^{(0)}) |V_{ii'}|^2, \quad w_{ii'} = w_{i'i},$$

† We obtain symmetry relationships which generalise (VI.75) by taking complex operators $\hat{A}$ and $\hat{H}$ (presence of a magnetic field). We note also van Kampen's point of view according to which it would not be necessary to use the concept of ensembles of quantum systems in order to obtain the Master Equation. This point of view depends, on the one hand, on the form of the average quantum value for a pure case of a macroscopic observable $\hat{\mathscr{A}}$ [which, according to (III.37) can always be compared with an average over a certain classical ensemble, with the weights $w_\nu^{(\alpha)}(t)$] and, on the other hand, it depends on an assumption of irregularity of the phases of the coefficients $c_i^{(\alpha)}$ for the pure case (VI.33), recurring at every instant t.

Nevertheless, it seems difficult not to adopt a statistical interpretation for this, since it corresponds in fact to the conservation with time of the property of molecular chaos.

are obtained by starting from (VI.68′) and the assumption of uniformly and randomly distributed phases at every time t (random phase approximation). As for λ, this is a small parameter which characterises the order of magnitude of the perturbation $\hat{V}$ and which enables us to define more precisely the conditions necessary for the validity of the Born approximation.

Thus, the kinetic equations (VI.73) have been proved only at the price of a supplementary assumption, namely that we can use for any time the condition (VI.42) which was valid for the initial statistical ensembles; it can be expressed as

$$\overline{\varrho_{ij}^{(\alpha)}(t)} = 0 \quad \text{if} \quad i \neq j. \tag{VI.76}$$

This is the hypothesis of "molecular disorganisation" (Pauli, 1928) which amounts to neglecting the off-diagonal elements of the density matrix. It is necessary for enabling us to calculate $P_\nu^{(\alpha)}(t)$ by a formula similar to (VI.44); we note, however, that it is incompatible with equation (VI.12), according to which the off-diagonal terms $\overline{\varrho_{ij}^{(\alpha)}(t)}$ are generally non-vanishing, even if $\overline{\varrho_{ij}^{(\alpha)}(0)} = 0$. This assumption can thus be contradictory to the laws of wave mechanics: we encounter the same difficulty as in classical theory, where we must assume that the property of molecular chaos is conserved with time; this leads to restriction of the possible movements of the representative point in Γ-space. Here, it is condition (VI.76) which corresponds to this extension of the property of molecular chaos: moreover, we can develop in quantum theory a discussion which is similar to that in Section III of Chapter V, by making the trajectory of $\Psi(t)$ over the $S^{(\alpha)}$-dimensional hypersphere $\sum_{i=1}^{S(\alpha)} |c_i^{(\alpha)}(t)|^2 = 1$ correspond to the movement of the point $\mathscr{P}(t)$ in Γ-space, and the cells with $s_\nu^{(\alpha)}$ dimensions to the stars in Γ-space.

Since this condition is realised after a quantum measurement on the system, certain authors have assumed that one could not speak about the irreversible evolution of a quantum system without carrying out a series of measurements (see for example, Delbrück and Molière, 1936) on this system: this point of view is unsatisfactory, since the quantities considered are macroscopic and must be independent of the observation actually carried out on the system. (We shall return, however, to this point in the next chapter when stressing the difference between macroscopic observation and quantum measurement.)

In this way we are led, as in classical mechanics, to regard (VI.76) as a supplementary assumption of a stochastic nature which introduces a certain random element into the evolution of a macroscopic system. As in classical theory there are two routes for research: either to try to show that there are no actual discrepancies between the microscopic mechanical evolution of the system and the statistical description adopted for macroscopic phenomena, or to attempt to construct models of particular stochastic processes which are suitable for describing the physical phenomena under consideration (we shall give an example later on in this chapter).

In the first class of ideas, it would be necessary to take account of the large number of degrees of freedom of the system (the $s_\nu^{(\alpha)}$ will be very large) in order to justify the models used; in this connection we mention Van Hove's method (1955) who attempts to determine what conditions must be satisfied by a perturbation potential $\hat{V}$ in order that the property of molecular chaos is conserved with time: this is a problem which we have encountered already in classical statistical mechanics. Van Hove obtains, in this way, a Pauli type kinetic equation for a coarse-grained distribution, by making the assumption of molecular chaos only at the initial instant t_0. In a general way, the proof of a kinetic equation of the type (VI.72) rests in final analysis, as in classical theory, on the study of the time-evolution of the fine-grained densities $\hat{\varrho}$ described by the exact equation (VI.1). We shall show in the next section that the classical methods described in Section V of Chapter V can be extended in a certain way to quantum theory and that we can in this way obtain a quantum Boltzmann equation.

We shall end this section now by showing the form taken by the generalised $\bar{H}$-theorem, with the assumption of time-proportional transition probabilities. For this purpose we shall adopt definitions (VI.24) and (VI.25) of coarse-grained densities, according to which $\bar{H}$ can be written as

$$\bar{H}(t) = \sum_i P_{ii}(t) \ln P_{ii}(t) = \sum_\nu P_\nu \ln \frac{P_\nu}{G_\nu}, \qquad \text{(VI.77)}$$

whence we derive

$$\frac{d\bar{H}}{dt} = \sum_{\nu} (\ln P_\nu - \ln G_\nu) \frac{dP_\nu}{dt}. \tag{VI.78}$$

In order to evaluate $d\bar{H}/dt$, we must know the quantities dP_ν/dt; we calculate them by using equations (VI.68′) which give the first approximation of perturbation theory (with an external perturbation $\hat{V}$). We assume (as we have already pointed out previously), that the unperturbed energy spectrum of $\hat{H}_0$ is sufficiently dense to be able to treat it as a continuous spectrum: with these conditions the transition probabilities are proportional to the time. If the system exists at $t = 0$ in the group of states $G_\varkappa$, we can write at t

$$P_\nu(t) = \frac{1}{G_\varkappa} T_{\varkappa\nu} t \tag{VI.79}$$

and, if the system is at $t = 0$ in the group G_ν, we have also

$$P_\varkappa(t) = \frac{1}{G_\nu} T_{\nu\varkappa} t, \tag{VI.80}$$

with

$$T_{\nu\varkappa} = T_{\varkappa\nu}. \tag{VI.81}$$

More generally, if the system is distributed initially in many groups of states $G_\varkappa$, G_ν, with the respective weights $P_\varkappa$ and P_ν, the transition probabilities per unit time between the states $\varkappa$ and ν are given by

$$Z_{\varkappa\nu} = A_{\varkappa\nu} G_\nu P_\varkappa, \quad Z_{\nu\varkappa} = A_{\nu\varkappa} G_\varkappa P_\nu, \tag{VI.82}$$

where (because of the Hermiticity of the perturbation operators and because of our statistical assumptions)

$$A_{\varkappa\nu} = A_{\nu\varkappa} \quad \text{and} \quad A_{\varkappa\nu} = \frac{1}{G_\varkappa G_\nu} T_{\varkappa\nu}. \tag{VI.83}$$

The quantities dP_ν/dt can then be written as

$$\frac{dP_\nu}{dt} = \sum_{\varkappa} (Z_{\varkappa\nu} - Z_{\nu\varkappa}) = \sum_{\varkappa} A_{\nu\varkappa}(G_\nu P_\varkappa - G_\varkappa P_\nu), \tag{VI.84}$$

whence, by substituting in (VI.78):

$$\frac{d\bar{H}}{dt} = \frac{1}{2} \sum_{\nu,\varkappa} A_{\nu\varkappa}(G_\nu P_\varkappa - G_\varkappa P_\nu) \ln \frac{G_\varkappa P_\nu}{G_\nu P_\varkappa}$$

$$= -\frac{1}{2} \sum_{\nu,\varkappa} A_{\nu\varkappa} G_\varkappa G_\nu \left(\frac{P_\varkappa}{G_\varkappa} - \frac{P_\nu}{G_\nu}\right)\left(\ln \frac{P_\varkappa}{G_\varkappa} - \ln \frac{P_\nu}{G_\nu}\right). \quad \text{(VI.85)}$$

We see that $d\bar{H}/dt$ is always negative, with the result that $\bar{H}$ decreases to the point where we have

$$\frac{P_\nu}{G_\nu} = \frac{P_\varkappa}{G_\varkappa}, \quad \text{(VI.86)}$$

or

$$Z_{\varkappa\nu} = Z_{\nu\varkappa}. \quad \text{(VI.86')}$$

Equation (VI.86) shows that equilibrium is attained with a uniform coarse-grained density (microcanonical ensemble) and relation (VI.86′) expresses the principle of detailed balancing of quantum statistical mechanics.

In the formalism of macroscopic operators, we put similarly

$$\bar{H}(t) = \sum_\nu P_\nu^{(\alpha)} \ln \frac{P_\nu^{(\alpha)}}{s_\nu^{(\alpha)}} \quad \text{(VI.77')}$$

and we see that, because of assumption (VI.69), $d\bar{H}/dt \leq 0$ and that statistical equilibrium is achieved when we have

$$\frac{P_{\nu'}^{(\alpha)}}{s_{\nu'}^{(\alpha)}} = \frac{P_\nu^{(\alpha)}}{s_\nu^{(\alpha)}}. \quad \text{(VI.86'')}$$

We note that the foregoing results are open to the same discussions as in classical theory: the statistical interpretation which we must give is precisely similar to that which we developed in Section III of Chapter V. Their range of validity is limited to a macroscopically small interval of time after the instant of observation and their extension to any instant necessitates an additional assumption: this amounts to assuming that the mechanical evolution of a macroscopic system (with a large number of degrees of freedom) is such that the initial property of molecular chaos is conserved with time. We encounter, therefore, in quantum theory difficulties of the same nature as in classical mechanics in order to reconcile mechanical reversibility and quasi-periodicity with macroscopic irreversibility. We could try specially, as in Section IV of Chapter V, to define the equations which govern the evolution of the distribution function $P(n_1, ..., n_i, ...; t)$ of a Markovian process whose transition probabilities are provided by the hypothesis of molecular chaos, i.e. by formulae (VI.93) of Section IV. We shall see at the end of the next section that equations similar to (V.56) can be obtained which open up a field of interesting research for a more thorough study of irreversible phenomena in quantum physics.

IV. Boltzmann's Equation and Stochastic Processes in Quantum Theory

We shall now show how the reasoning of Sections IV and V of the preceding chapter can be extended to quantum theory, with particular emphasis on Boltzmann's quantum equation and on the derivation of the Master Equation.

1. *The individual H-theorem*

Since we have been interested up to now only in the generalised $\bar{H}$-theorem, we shall start by proving an H-theorem for a single quantum system, similar to that of Boltzmann's H-theorem in classical theory. In fact, if we are interested in a single system we can, exactly as in classical mechanics, make a quantity H correspond to it which is defined along Boltzmann's lines (by making all the P_ν equal to zero, except one) by

$$H = -\ln G, \tag{VI.87}$$

where G represents the number of microscopic complexions of the system compatible with the macroscopic state observed: this is analogous to the volume of a star in Γ-space. We calculate G by enumerating the various states occupied by the particles which make up the system (the method is similar to the transition from Γ-space to μ-space). If we have divided the energy spectrum of a particle into groups containing g_i eigenstates with energy ε_i and if n_i is the number of particles occupying the group with index i, G will assume different values according as the particles obey Boltzmann, Bose–Einstein or Fermi–Dirac statistics. Using an obvious notation, we have

$$\left.\begin{aligned} G_{\text{Bo}} &= \prod_i \frac{g_i^{n_i}}{n_i!}, \\ G_{\text{B.E.}} &= \prod_i \frac{(n_i + g_i - 1)!}{n_i!(g_i - 1)!} \\ G_{\text{F.D.}} &= \prod_i \frac{g_i!}{(g_i - n_i)!\, n_i!}. \end{aligned}\right\} \tag{VI.88}$$

From these expressions we derive

$$H = \sum_i \left[\alpha g_i \ln \frac{g_i}{n_i} - (n_i + \alpha g_i) \ln \left(\frac{g_i}{n_i} + \alpha \right) \right] \tag{VI.89}$$

(where $\alpha = 0, +1$ or -1 for Boltzmann, Bose–Einstein and Fermi–Dirac statistics, respectively) and

$$\frac{dH}{dt} = \sum_i \frac{dn_i}{dt} \ln \frac{n_i}{g_i + \alpha n_i} \tag{VI.90}$$

which is the quantum parallel of (V.37). As in classical mechanics, we cannot calculate dH/dt without knowing dn_i/dt: however, for this it would be necessary to integrate the equations of motion. Thus, we can only evaluate the average values of dn_i/dt, $(dn_i/dt)_m$, and of dH/dt, $(dH/dt)_m$, by the method of transition probabilities. We can use this method by supposing that the Hamiltonian of the system can be written in the form $\hat{H} = \hat{H}_0 + \hat{V}$, where

$$\hat{H}_0 = \sum_i \hat{H}_i \tag{VI.91}$$

is the Hamiltonian of the free particles. We suppose further that the interaction $\hat{V}$ can be put into the form

$$\hat{V} = \sum_{i \neq j} \hat{V}_{ij}, \tag{VI.92}$$

where $\hat{V}_{ij}$ denotes the interaction potential between two particles i and j; because of the various approximations used and of the fact, especially, that we are dealing with the first approximation in perturbation theory, this latter assumption amounts to considering only binary collisions as in Boltzmann's equation (case of dilute gases). We point out that the perturbation potential $\hat{V}$ has a precise physical meaning here, corresponding to the internal structure of the system and that the splitting of $\hat{H}$ into two terms falls exactly within the framework of the theory involving macroscopic operators; the results that we derive can provide, therefore, an acceptable interpretation of the irreversible phenomena.

If we denote by $Z_{ij}^{i'j'}$ the average number of transitions per unit time which take particles from the initial cells (i, j) to the final cells (i', j') we have (always to a first approximation)

$$Z_{ij}^{i'j'} = A_{ij}^{i'j'} n_i n_j (g_{i'} + \alpha n_{i'}) (g_{j'} + \alpha n_{j'}),$$

$$Z_{i'j'}^{ij} = A_{i'j'}^{ij} n_{i'} n_{j'} (g_i + \alpha n_i) (g_j + \alpha n_j), \tag{VI.93}$$

with $A_{ij}^{i'j'} = A_{i'j'}^{ij}$.

These expressions are calculated by means of the fundamental assumption of statistical mechanics which is, at the initial instant, the same as the hypothesis of molecular chaos expressed by (VI.76) (for details of the calculations see Tolman, 1938, pp. 436–450); thus, equation (VI.93) is the quantum parallel of formulae (V.38)–(V.38′) of classical theory. In this case we have

$$\left(\frac{dn_i}{dt}\right)_m = \sum_{j(i'j)} (Z_{i'j'}^{ij} - Z_{ij}^{i'j'})$$

$$= \sum_{j(i'j')} A_{ij}^{i'j'} \{n_{i'} n_{j'} (g_i + \alpha n_i) (g_j + \alpha n_j) - n_i n_j (g_{i'} + \alpha n_{i'}) (g_{j'} + \alpha n_{j'})\}, \tag{VI.94}$$

whence we derive

$$\left(\frac{dH}{dt}\right)_m = \sum_{(ij),(i'j')} A_{ij}^{i'j'} \{n_{i'} n_{j'} (g_i + \alpha n_i) (g_j + \alpha n_j) - n_i n_j (g_{i'} + \alpha n_{i'}) (g_{j'} + \alpha n_{j'})\} \ln \frac{n_i}{g_i + \alpha n_i}. \tag{VI.95}$$

From this we see easily that $\left(\frac{dH}{dt}\right)_m \leqq 0$, the equality sign occurring when the n_i are distributed according to the Maxwell–Boltzmann, Bose–Einstein or Fermi–Dirac laws depending on the values of α. Equation (VI.94) is thus the quantum parallel of (V.49) and its interpretation gives rise to the same remarks as in classical theory, especially concerning the role of the hypothesis of molecular disorganisation (cf. § 3 in Section III of Chapter V); we shall see now under what conditions a quantum Boltzmann equation can be obtained.

2. *A quantum Boltzmann equation*

As we have emphasised already, the quantum kinetic equations of the form (VI.72) can be justified in the final analysis only by studying the fine-grained density represented by the statistical operator $\hat{\varrho}$; this includes the maximum amount of data that we can collect concerning a macroscopic quantum system. As in classical theory, we shall see that the fundamental equation (VI.1) of the Liouville type can be replaced by a chain of equations obtained by successive "contractions" and that we end up in this way with a quantum kinetic equation of the Boltzmann type, by means of certain approximations.

For this purpose we adopt the coordinate-space representation of the statistical operator $\hat{\bar{\varrho}}$ (q-representation, where the base functions are the functions $\delta(\boldsymbol{r})$) defined by†

$$\bar{\varrho}_n(\boldsymbol{r}^{(n)}, \boldsymbol{r}^{(n)\prime}; t) = \overline{\Psi^*(\boldsymbol{r}^{(n)\prime}, t)\, \Psi(\boldsymbol{r}^{(n)}, t)} = \sum_{i,j} \bar{\varrho}_{ij} \varphi_j^*(\boldsymbol{r}^{(n)\prime})\, \varphi_i(\boldsymbol{r}^{(n)}). \tag{VI.96}$$

It can be seen that, in this representation, the density matrix $\hat{\bar{\varrho}}_n$ depends on the $6n$ space coordinates $\boldsymbol{r}^{(n)}$ and $\boldsymbol{r}^{(n)\prime}$, just as in classical theory the phase density ϱ_n depends on the n space vectors $\boldsymbol{r}^{(n)}$ and on the n momenta $\boldsymbol{p}^{(n)}$. To the operation which consists of integration with respect to the variables $\boldsymbol{p}^{(n)}$, which leads for example to $\varrho_n(\boldsymbol{r}^{(n)}) = \int \varrho_n(\boldsymbol{r}^{(n)}, \boldsymbol{p}^{(n)})\, d^{3n}\boldsymbol{p}^{(n)}$, there corresponds in quantum theory the operation which consists in taking the diagonal element, for example, $\hat{\varrho}_n(\boldsymbol{r}^{(n)}, \boldsymbol{r}^{(n)})$. The relation Tr $\hat{\bar{\varrho}} = 1$ can in this notation be written as

$$\int \bar{\varrho}_n(\boldsymbol{r}^{(n)}, \boldsymbol{r}^{(n)\prime})\, \delta(\boldsymbol{r}^{(n)} - \boldsymbol{r}^{(n)\prime})\, d^{3n}\boldsymbol{r}^{(n)\prime}\, d^{3n}\boldsymbol{r}^{(n)} = \int \bar{\varrho}_n(\boldsymbol{r}^{(n)}, \boldsymbol{r}^{(n)})\, d^{3n}\boldsymbol{r}^{(n)} = 1; \tag{VI.97}$$

† Once again we use the notation $\boldsymbol{r}^{(n)}$ for the vector $(\boldsymbol{r}_1, \ldots, \boldsymbol{r}_n)$ of the configuration space of the system and $\hat{\varrho}_n$ for the density matrix of a system with n particles.

also, for the average value of an observable A, we have

$$\bar{A} = \text{Tr}(\hat{\bar{\varrho}}_n \hat{A}) = \int \{[\hat{\bar{\varrho}}_n, \hat{A}]_- (\boldsymbol{r}^{(n)}, \boldsymbol{r}^{(n)})\} \, d^{3n}\boldsymbol{r}^{(n)}. \qquad \text{(VI.98)}$$

The calculation of the average value of a quantity reduces thus to integrating the diagonal element of $\{[\hat{\varrho}_n, \hat{A}]_- (\boldsymbol{r}^{(n)}, \boldsymbol{r}^{(n)})\}$ over $\boldsymbol{r}^{(n)}$; this clearly resembles the classical integration $\int \varrho_n(\boldsymbol{r}^{(n)}, \boldsymbol{p}^{(n)}) \, A(\boldsymbol{r}^{(n)}, \boldsymbol{p}^{(n)}) \, d^{3n}\boldsymbol{p}^{(n)} \, d^{3n}\boldsymbol{r}^{(n)}$; we can say, therefore, that the quantum operation $\int \ldots \delta(\boldsymbol{r}^{(n)} - \boldsymbol{r}^{(n)\prime}) \, d^{3n}\boldsymbol{r}^{(n)\prime}$, corresponds to the classical operation $\int \ldots d^{3n}\boldsymbol{p}^{(n)}$.

Having done this, we can make reduced density matrices, (introduced for the first time by Husimi, 1940) $\bar{\varrho}_s(\boldsymbol{r}^{(s)}, \boldsymbol{r}^{(s)\prime}; t)$ correspond to the classical probability densities $\bar{\varrho}_s(\boldsymbol{r}_1, \ldots, \boldsymbol{r}_s, \boldsymbol{p}_1, \ldots, \boldsymbol{p}_s; t)$. We obtain these reduced density matrices by taking the diagonal elements for the coordinates of particles which do not belong to the group (s) considered and by integrating over these variables; we have, therefore,

$$\bar{\varrho}_s(\boldsymbol{r}^{(s)}, \boldsymbol{r}^{(s)\prime}; t) = \int \bar{\varrho}_n(\boldsymbol{r}^{(s)}\boldsymbol{r}^{(n-s)}, \boldsymbol{r}^{(s)\prime}\boldsymbol{r}^{(n-s)}) \, d^{3n}\boldsymbol{r}^{(n-s)}. \qquad \text{(VI.99)}$$

The diagonal terms $\bar{\varrho}_s(\boldsymbol{r}^{(s)}, \boldsymbol{r}^{(s)}; t)$ are the quantum parallels of the classical distribution functions $\varrho_s(\boldsymbol{r}^{(s)}, t) = \int \varrho_s(\boldsymbol{r}^{(s)}, \boldsymbol{p}^{(s)}, t) \, d^{3s}\boldsymbol{p}^{(s)}$; similarly, the generic probability densities $f_s(\boldsymbol{r}^{(s)}, \boldsymbol{p}^{(s)}, t)$ have the reduced matrices:

$$\mathscr{F}_s(\boldsymbol{r}^{(s)}, \boldsymbol{r}^{(s)\prime}; t) = \frac{n!}{(n-s)!} \bar{\varrho}_s(\boldsymbol{r}^{(s)}, \boldsymbol{r}^{(s)\prime}; t). \qquad \text{(VI.100)}$$

for their quantum analogues.

By relying on these definitions, we can according to Born and Green derive from (VI.1), a chain of quantum equations which is similar to the B.B.G.K.Y. equations. By introducing the operation Tr_s (reduced trace) which, applied to an operator $\hat{A}_s(\boldsymbol{r}^{(s)}, \boldsymbol{r}^{(s)\prime})$, has the form

$$\text{Tr}_s(\hat{A}_s) = \iint A_s(\boldsymbol{r}^{(s)}, \boldsymbol{r}^{(s)\prime}) \, \delta(\boldsymbol{r}_s - \boldsymbol{r}_s') \, d^{3s}\boldsymbol{r}_s \, d^{3s}\boldsymbol{r}_s', \qquad \text{(VI.101)}$$

we obtain in fact—starting from the equation $\partial \hat{\varrho}_n / \partial t = i[\hat{\bar{\varrho}}_n, \hat{H}_n]_- / \hbar$ the reduced equation

$$\frac{\partial \hat{\bar{\varrho}}_{n-1}}{\partial t} = \frac{i}{\hbar} [\hat{\bar{\varrho}}_{n-1}, \hat{H}_{n-1}]_- + \frac{i}{\hbar} \sum_{i=1}^{n-1} \text{Tr}_n [\hat{\bar{\varrho}}_n, \hat{\Phi}_{in}]_-, \qquad \text{(VI.102)}$$

where $\hat{H}_n$ is the quantum Hamiltonian corresponding to the classical definition (V.80). By repeating the contraction $(n - s)$ times, we have finally

$$\frac{\partial \hat{\bar{\varrho}}_s}{\partial t} = \frac{i}{\hbar} [\hat{\bar{\varrho}}_s, \hat{H}_s]_- + \frac{i}{\hbar} \sum_{i=1}^{s} \mathrm{Tr}_{s+1} [\hat{\bar{\varrho}}_{s+1}, \hat{\Phi}_{i,s+1}]_- \quad \text{(VI.103)}$$

which is the quantum parallel of equation (V.79). Similar equations can be written for the generic matrices (VI.100) and, in the case $s = 1$, an equation similar to (V.82) is obtained.

It is not possible to proceed directly from (VI.103) for $s = 1$ to an equation of the Boltzmann type, since it is not possible, in wave mechanics, to define precisely distribution functions in the Γ-space of the system. However, if we take the limit as $h \to 0$, it can be shown that equations (VI.103) are equivalent to the corresponding classical equations (correspondence principle): it is then possible to find a Boltzmann equation with certain characteristic modifications of the quantum theory (Kikuchi and Nordheim, 1930; Nordheim, 1928a, b; Uehling and Uhlenbeck, 1933).

Before showing the form of this equation, we shall analyse briefly another method which is suitable for founding quantum statistical mechanics, namely the Wigner function method (Wigner, 1932). We know that the concept of the distribution function cannot be applied directly to quantum theory, because of Heisenberg's uncertainty principle which does not allow a precise localisation of the representative point of the state of the system in phase space; this leads us to define the statistical operators or density matrices that we have used throughout this book. Wigner's method consists in constructing precisely a distribution function $\mathscr{W}_n(\boldsymbol{r}^{(n)}, \boldsymbol{p}^{(n)}, t)$ which has no simple interpretation in terms of probability concepts (for example, it can become negative) but which can be used directly for calculating average values in a manner similar to the corresponding classical formulae.†

The Wigner distribution function can be defined by starting from the density matrix in the coordinate or momentum re-

† There is yet a third method for dealing with quantum statistical mechanics: this is the method of second quantisation. It is not our intention to deal with it here; a complete discussion of it will be found in the book by Massignon (1957).

presentation; we have (with $\mathscr{F}_n = n!\,\varrho_n$):

$$\mathscr{W}_n(\boldsymbol{r}^{(n)}, \boldsymbol{p}^{(n)}, t) = \frac{1}{h^{3n}} \int e^{(\boldsymbol{p}(u)\cdot\boldsymbol{r}^{(n)\prime})/\hbar} \times$$

$$\times \mathscr{F}_n(\boldsymbol{r}^{(n)} - \tfrac{1}{2}\boldsymbol{r}^{(n)\prime}, \boldsymbol{r}^{(n)} + \tfrac{1}{2}\boldsymbol{r}^{(n)\prime}, t)\, d^{3n}\boldsymbol{r}^{(n)\prime}$$

$$= \frac{1}{h^{3n}} \int e^{i(\boldsymbol{p}^{(n)\prime}\cdot\boldsymbol{r}^{(n)})/\hbar} \times$$

$$\times \mathscr{F}_n(\boldsymbol{p}^{(n)} - \tfrac{1}{2}\boldsymbol{p}^{(n)\prime}, \boldsymbol{p}^{(n)} + \tfrac{1}{2}\boldsymbol{p}^{(n)\prime}, t)\, d^{3n}\boldsymbol{p}^{(n)\prime}, \tag{VI.104}$$

where $\mathscr{W}_n$ is real but not necessarily positive. Integration of this function over all momenta gives the diagonal elements of the density matrix in coordinate space; similarly, integration over the coordinates gives the diagonal elements of the density matrix in the momentum representation. The interest in Wigner's function is that it enables us to calculate average values by formulae which are identical to the classical formulae; in fact, it can be shown that because of (VI.104) we have

$$\bar{G} = \iint \mathscr{W}_n(\boldsymbol{r}^{(n)}, \boldsymbol{p}^{(n)}, t)\, G(\boldsymbol{r}^{(n)}, \boldsymbol{p}^{(n)})\, d^{3n}\boldsymbol{r}^{(n)}\, d^{3n}\boldsymbol{p}^{(n)}, \tag{VI.105}$$

which avoids the operational technique of the density matrix; similarly, numerous results obtained with $\hat{\varrho}_n$ remain unchanged with the function $\mathscr{W}_n$.

The $\mathscr{W}_n$ functions satisfy a differential equation which is derived from the fundamental equation (VI.1) by taking account of definition (VI.104); similarly, we can define successive "contractions" of $\mathscr{W}_n$. Thus, it has recently been possible to show (Mori and Ono, 1952) that it would be possible to obtain a quantum Boltzmann equation for dilute gases, starting from the differential equation which is satisfied by Wigner's function $\mathscr{W}_n$. If we introduce again the distribution function $f(\boldsymbol{r}, \boldsymbol{v}, t)$, this quantum Boltzmann equation assumes the form

$$\frac{\partial f}{\partial t} + (\boldsymbol{v}\cdot\nabla_r f) + \left(\frac{\boldsymbol{X}}{m}\cdot\nabla_v\right) f = \iiint [f_1' f'(1 + \theta f_1)(1 + \theta f)$$

$$- f_1 f(1 + \theta f_1')(1 + \theta f')]\, \alpha(g, \chi) \sin\chi\, d\chi\, d\varepsilon\, d^3\boldsymbol{v}_1, \tag{VI.106}$$

where $\alpha(g, \chi)$ is the quantum cross-section of two particles moving with the relative velocity g and where $\theta = (h^3/m^3 G)\,\varepsilon$; G is the statistical weight corresponding to the particle considered and ε is equal to -1, 0 or 1, depending on whether we are using Fermi–Dirac, Boltzmann or Bose–Einstein statistics. It can be seen that the flow term is identical with the classical term (as was shown by Kikuchi and Nordheim, 1930) and that the collision term involves two essential differences: (a) we have replaced $gb\,db\,d\varepsilon$ by the quantum differential cross-section $\alpha(g, \chi) \sin \chi \, d\chi \, d\varepsilon$, which introduces quantum diffraction effects in the collision process of the two particles (the modifications obtained are particularly important for hydrogen and helium); (b) we see terms in $(1 + \theta f)$ appearing, which are particularly important at low temperatures and which express the effects of symmetry and take into account the behaviour of degenerate gases. Of course, Boltzmann's equation (VI.106), inasfar as it is valid from the quantum point of view, gives rise to the same criticism as we have made concerning the corresponding classical equation; we note, finally, that one can obtain an *H*-theorem which is identical with (VI.95), if we define *H* by:

$$H = \int \left[f \ln f - \left(\frac{1}{\theta} + f \right) \ln \left(\frac{1}{\theta} + f \right) \right] d^3\boldsymbol{v}. \qquad \text{(VI.107)}$$

3. *The Master Equation in quantum theory*

We shall end this section by giving a few brief observations on the formulation of the Master Equation in quantum theory (we shall follow a recent paper by Mathews, Shapiro and Falkoff, 1960). As in classical mechanics, we can try to describe the irreversible evolution of a macroscopic system by a Markovian stochastic process with the probabilities for transitions per unit time given by first order perturbation theory [see, for example, formulae (VI.66)–(VI.68)]. The probability distribution $P(n_1, \ldots, n_i, \ldots; t) = P(\langle n|; t)$ is then expressed as a function of the occupation numbers n_i of the different individual states of the particles and it satisfies the equation

$$P(\langle n|; t + dt) = \sum_{\langle m|} P(\langle m|; t) \, P(\langle m|; t \to \langle n|; t + dt) \qquad \text{(VI.108)}$$

[which is similar to (VI.55)], where $P(\langle m|; t \to \langle n|; t + dt)$ is the conditional probability of finding the occupation numbers $\langle n|$ at the instant $t + dt$, given the numbers $\langle m|$ at the instant t. If we restrict ourselves to stationary transition probabilities, where

$P(\langle m|; t \to \langle n|; t + dt)$ depends only on the interval dt and not on t, we can define a matrix $Q(\langle m| \to \langle n|)$ by

$$Q(\langle m| \to \langle n|) = \lim_{dt \to 0} \left[\frac{P(\langle m|; t \to \langle n|; t + dt) - \prod_i \delta_{n_i m_i}}{dt} \right], \tag{VI.109}$$

which, by virtue of the definition of transition probabilities, has the following properties:

$$Q(\langle m| \to \langle n|) \geqq 0 \quad \text{if} \quad \langle m| \neq \langle n|, \tag{VI.110}$$

$$Q(\langle m| \to \langle m|) = - \sum_{\langle n| \neq \langle m|} Q(\langle m| \to \langle n|).$$

If $\langle m| \neq \langle n|$, $Q(\langle m| \to \langle n|)$ represents the probability of the system passing from the state $\langle m|$ to the state $\langle n|$ per unit time. With these definitions, equation (VI.108) can be written as

$$\frac{\partial P(\langle n|; t)}{\partial t} = \sum_{\langle m|} P(\langle m|; t)\, Q(\langle m| \to \langle n|); \tag{VI.111}$$

this is the "Master Equation" in the quantum case [formally equivalent to (VI.73) and similar to equation (V.55$_2$)] the formal solution of which is:

$$\langle P(t)| = \langle P(0)| e^{Qt}, \tag{VI.112}$$

where $\langle P(0)|$ is the row vector corresponding to the initial distribution $P(\langle n|; 0)$. In the case where the number of states is infinite, the Master Equation (VI.111) is equivalent to an infinite set of differential equations; its solution can be made easier by introducing a generating function Φ defined by

$$\Phi(z_1, z_2, \ldots; t) \equiv \Phi(|z\rangle; t) = \sum_{\langle n|} z_1^{n_1} z_2^{n_2} \ldots P(\langle n|; t). \tag{VI.113}$$

and which corresponds to P.

The interest in this function Φ is due to the fact that we can deduce all the moments of P and its marginal distributions, by noting that the summation of P over all values of n_i for each value of i amounts to putting $z_i = 1$ in Φ. Thus, we obtain the relations

$$\Phi(|z\rangle; t)_{z_1 = z_2 = \ldots = 1} = 1;$$

$$\overline{n_i n_j \ldots n_l} = \left[z_i \frac{\partial}{\partial z_i} z_j \frac{\partial}{\partial z_j} \ldots z_l \frac{\partial}{\partial z_l} \Phi \right]_{z_1 = z_2 = \ldots = 1}. \tag{VI.114}$$

On the other hand, the marginal distribution $P(n_1, n_2; t)$, for example, is given by the set of coefficients of $z_1^{n_1} z_2^{n_2}$ in the series expansion of $\Phi(z_1, z_2; t)$ which is obtained, starting from (VI.113), by putting all z_i except z_1 and z_2 equal to 1. According to (VI.111) we can see is that Φ satisfies the partial differential equation

$$\frac{\partial \Phi}{\partial t}(|z\rangle; t) = \sum_{\langle n|} \sum_{\langle m|} P(\langle m|; t)\, Q(\langle m| \to \langle n|) \prod_i z_i^{n_i}$$

$$\equiv \Gamma\left(\left\langle z \frac{\partial}{\partial z}\right|; |z\rangle\right) \Phi(|z\rangle; t), \qquad \text{(VI.115)}$$

in which we have put

$$\Gamma(\langle m|; |z\rangle) \equiv \sum_{\langle n|} Q(\langle m| \to \langle n|) \prod_i z_i^{n_i - m_i}. \qquad \text{(VI.115')}$$

This equation is equivalent to the Master Equation (VI.111) and it is quite general; it assumes special forms once we determine the transition probabilities of the problem considered. If it be limited to the case of binary collisions (dilute monatomic gases), equation (VI.115) can be simplified and becomes:

$$\frac{\partial \Phi}{\partial t} = \frac{1}{4} \sum_{ijkl} A_{ij}^{kl} (z_k z_l - z_i z_j)\, [\partial_i \partial_j (1 + \theta z_k \partial_k)(1 + \theta z_l \partial_l)]\, \Phi \qquad \text{(VI.116)}$$

[equation (VI.116) is similar to the equations obtained by Moyal, 1949, and Siegert, 1949]; we note that even in this simple form, no general solution is known for (VI.116). However, we can deduce from it the equations satisfied by the moments of P of various orders; if we limit ourselves to the first order, we find the equation:

$$\frac{d\bar{n}_k}{dt} = \frac{1}{2} \sum_{jkl} \Big[A_{ij}^{kl} \overline{n_i n_j (1 + \theta n_k)(1 + \theta n_l)} - A_{kl}^{ij} \overline{n_k n_l (1 + \theta n_i)(1 + \theta n_j)} \Big]; \qquad \text{(VI.117)}$$

this is analogous to equation (VI.61) with $\theta = 0$ and it can only be reduced to a quantum Boltzmann equation if we make the additional assumption:

$$\overline{n_i n_j (1 + \theta n_k)(1 + \theta n_l)} = \bar{n}_i \bar{n}_j (1 + \theta \bar{n}_k)(1 + \theta \bar{n}_l), \qquad \text{(VI.118)}$$

which is equivalent to (V.62) with $\theta = 0$; thus, we meet again the hypothesis of molecular chaos which we have discussed already.

Of course, the question remains open of knowing to what extent we can deduce these probabilistic equations from the exact equations of wave mechanics. The foregoing involves, in fact, the implicit assumption that the occupation numbers remain "good" quantum numbers at every instant. Since they correspond to the diagonal elements of the density matrix, we encounter in fact Pauli's hypothesis according to which we can neglect the off-diagonal elements of the density matrix; this amounts to assuming that the property of molecular chaos is conserved during evolution. An important task therefore remains to be accomplished with a view to justifying the probabilistic equations of the form (VI.111), starting from equation (VI.1) which is satisfied by the fine-grained density $\hat{\varrho}$; thus, we have returned to the problems which we studied in section III of this chapter and here we recall once more the work of Van Hove (1955, 1957, 1959) who has succeeded, by means of the condition of molecular chaos at the initial instant, to obtain the Master Equation as a first approximation of the quantum equations of motion of a system composed of a large number of weakly coupled particles.

V. Zwanzig's Method

1. *Introduction*

Amongst the methods which have been proposed recently for attempting to resolve these problems, it seems appropriate to discuss Zwanzig's method (1960, 1964) which, at least in its early development, can be presented in a simple form. Without recourse to the complex formalism of graphs, it allows us to arrive by a direct route at a generalised "Master Equation" which is equivalent, apart from the notation, to the equations obtained previously by Nakajima (1958), by Montroll (1960, 1962) and by Prigogine and Résibois (1961; Résibois, 1963).

This method depends essentially on the following statement: in many problems of statistical physics, complete and precise knowledge of the statistical operator $\hat{\varrho}(t)$ is not generally necessary. This is due to the fact that, on the one hand, we are only interested in certain classes of observables and that, on the other hand, we only consider certain types of initial conditions; in this respect, it is sufficient to recall the important role played in the preceding sections by the diagonal elements $\overline{\varrho_{ii}(t)}$ of the density matrix and by the initial condition of random phases $\overline{\varrho_{ij}(0)} = 0$. We must therefore distinguish in the total operator $\hat{\varrho}(t)$ a "relevant" part denoted by $\hat{\varrho}_R(t)$ and an "irrelevant" part denoted by $\hat{\varrho}_I(t)$, which does not contribute to the calculation of the average values. More precisely, we assume that $\hat{\varrho}(t)$ can be split according to the formula

$$\hat{\varrho}(t) = \hat{\varrho}_R(t) + \hat{\varrho}_I(t), \tag{VI.119}$$

with the condition

$$\mathrm{Tr}\,(\hat{\varrho}_I \hat{A}) = 0. \tag{VI.120}$$

It follows that for the average value $\bar{A}(t)$ of an observable A, we have

$$\bar{A}(t) = \mathrm{Tr}(\hat{\varrho}_R \hat{A}). \tag{VI.121}$$

It is clear that this analysis will depend essentially on the nature of the problem studied. In a general way, it can be achieved by defining a Hermitian projection operator $\hat{P}$, such that

$$\hat{P}\hat{\varrho}(t) = \hat{\varrho}_R(t),\ (1 - \hat{P})\,\hat{\varrho}(t) = \hat{\varrho}_I(t), \tag{VI.122}$$

where $\hat{P}$ satisfies the relation

$$\hat{P}^2 = \hat{P}. \tag{VI.123}$$

For example, if we are interested only in the probabilities of the states i of a quantum system (as is the case in the kinetic equation), it is sufficient to know the diagonal part of $\hat{\varrho}$; therefore, we shall construct the projection operator $\hat{P}$ in such a way that its application would diagonalise the matrix $\hat{\varrho}$. Similarly, if we wish to introduce the reduced density matrices for s particles, $\hat{\varrho}_s(\boldsymbol{r}^{(s)}, \boldsymbol{r}^{(s)\prime}; t)$, the operator $\hat{P}$ will be defined by formula (VI.99); its application is a matter of taking the partial trace of $\hat{\varrho}$ over the $(n - s)$ particles which do not belong to the group considered. In addition, these considerations can be extended to the classical case where we define similarly distribution functions for s particles [cf. formula (V.74)]; the operator $\hat{P}$ reduces in this case to an integration over the $6(n - s)$-dimensional partial phase space.

This being so, Zwanzig's method consists in obtaining a precise equation of evolution for the "relevant" part $\hat{\varrho}_R(t)$, starting from Liouville's equation. We shall see that this equation, which is equivalent to Liouville's equation, can be put in the form of a generalised "Master Equation", although it does not yet involve any element of irreversibility. However, it constitutes an excellent point of departure for various approximations, according to the nature of the problem being studied, and we shall show that it leads, in the limit of "weak coupling", to an irreversible Master Equation of the type of Pauli's equation.

2. *The Liouville operator in quantum theory*

Before describing this method, it will be convenient to introduce Liouville's quantum operator $\hat{L}$, which is analogous to the operator which we have encountered already in the classical case.

We write the quantum parallel of Liouville's equation (frequently called von Neumann's equation) in the form:

$$\frac{\partial \hat{\varrho}}{\partial t} = -\frac{i}{\hbar}[\hat{H}, \hat{\varrho}]_- \tag{VI.124}$$

and we define a linear Hermitian operator $\hat{L}$ by

$$\hat{L}\hat{\varrho} \equiv \frac{1}{\hbar}[\hat{H}, \hat{\varrho}]_-. \tag{VI.125}$$

Equation (VI.124) then becomes

$$\frac{\partial \hat{\varrho}}{\partial t} = -i\hat{L}\hat{\varrho} \tag{VI.126}$$

and hence takes a form which is completely equivalent to the corresponding classical equation [see equation (V.155)].

This operator $\hat{L}$ [first introduced by Klein (1952), then studied by Fano (1957) and Kubo (1957)] acts on the set of operators in the space of quantum states and it has the effect, in a given representation, of transforming a matrix with two indices to another matrix with two indices. Liouville's quantum operator $\hat{L}$ is therefore a mathematical creation with four indices (called a tetrad), whose elements L_{ijkl} are defined by:

$$(\hat{L}\hat{\varrho})_{ij} = \sum_{kl} L_{ijkl}\varrho_{kl}; \tag{VI.127}$$

Since $\hat{L}$ is the commutator with $\hat{H}$, we can show that we have

$$L_{ijkl} = \frac{1}{\hbar}(H_{ik}\delta_{jl} - H_{lj}\delta_{ik}). \tag{VI.128}$$

In consequence of equation (VI.128) we have the following two relations:

$$L_{iikk} = 0, \quad \sum_k L_{ijkk} = 0. \tag{VI.129}$$

On the other hand, for the algebra of these tetrads we have the two obvious rules:

$$\left(\overset{\leftrightarrow}{1}\right)_{ijkl} = \delta_{ik}\delta_{jl}, \tag{VI.130}$$

which defines the unit-tetrad, and

$$(\hat{L}_1\hat{L}_2)_{ijkl} = \sum_m \sum_n (\hat{L}_1)_{ijmn}(\hat{L}_2)_{mnkl}, \tag{VI.131}$$

for the rule of multiplying two tetrads.

Once this operator $\hat{L}$ has been introduced, the formal solution of von Neumann's equation can be written as

$$\hat{\bar{\varrho}}(t) = e^{-i\hat{L}t}\hat{\bar{\varrho}}(0), \tag{VI.132}$$

where the unitary operator $e^{-i\hat{L}t}$ can be defined by an expansion of the exponential function; it is called the propagator, or the Green function, associated with von Neumann's equation, since it enables us to express the evolution of the system as a function of the initial conditions. Equation (VI.132) can be rewritten in matrix form:

$$\bar{\varrho}_{ij}(t) = \sum_{kl} (e^{-i\hat{L}t})_{ijkl}\, \bar{\varrho}_{kl}(0), \tag{VI.132'}$$

and, differentiating with respect to time and using equation (VI.128), we can prove the identity

$$(e^{-i\hat{L}t})_{ijkl} = (e^{-i\hat{H}t/\hbar})_{ik}\,(e^{i\hat{H}t/\hbar})_{jl}. \tag{VI.133}$$

Applying this last relation to equation (VI.132) we find the familiar solution:

$$\hat{\bar{\varrho}}(t) = e^{-i\hat{H}t/\hbar}\,\hat{\bar{\varrho}}(0)\, e^{i\hat{H}t/\hbar}, \tag{VI.134}$$

in which the unitary evolution operator $e^{-i\hat{H}t/\hbar}$ enters.

Similarly, an observable A which does not depend explicitly on the time satisfies, in the Heisenberg representation, the equation

$$\frac{d\hat{A}}{dt} = \frac{i}{\hbar}[\hat{H}, \hat{A}]_- = i\hat{L}\hat{A}, \tag{VI.135}$$

the formal solution of which can be written as

$$\hat{A}(t) = e^{i\hat{L}t}\hat{A}(0). \tag{VI.136}$$

It follows that for the average value of $\hat{A}(t)$ we have

$$\bar{A}(t) = \mathrm{Tr}[\hat{A}(t), \hat{\varrho}(0)]_{-} = \mathrm{Tr}[\hat{A}(0), \hat{\varrho}(t)]_{-}, \tag{VI.137}$$

because of the unitarity of the operator $e^{-i\hat{L}t}$; thus, we find for the average values a property similar to that which we have already established in classical theory [see equation (V.157)].

We note, in conclusion, that the introduction of the operator $\hat{L}$ has two advantages: on the one hand, it enables us to write in a more manageable form the perturbation expansions as a function of the coupling parameter; in fact, in the case where the Hamiltonian is given in the form $\hat{H} = \hat{H}_0 + \lambda\hat{V}$, we have similarly $\hat{L} = \hat{L}_0 + \lambda\hat{L}'$ and the perturbation operator $\hat{L}'$ always appears to the left of $\hat{\varrho}(0)$ in the expansion, whilst, in Hamiltonian formalism, the operator $\hat{V}$ appears both to left and to the right of $\hat{\varrho}(0)$. On the other hand, the operator $\hat{L}$ enables us also to arrive at a very simple expression for the resolvant equation associated with (VI.126). This can be verified by defining the Laplace transform $\hat{g}(s)$ associated with $\hat{\varrho}(t)$, through

$$\hat{g}(s) = \int_0^\infty dt\, e^{-st}\hat{\varrho}(t); \tag{VI.138}$$

the Laplace transform of equation (VI.126) then gives

$$s\hat{g}(s) - \hat{\varrho}(0) = -i\hat{L}\hat{g}(s), \tag{VI.139}$$

the formal solution of which can be written as

$$\hat{g}(s) = \frac{1}{s + i\hat{L}}\hat{\varrho}(0), \tag{VI.140}$$

where $(s + i\hat{L})^{-1}$ is the resolvant operator associated with the unitary operator $e^{-i\hat{L}t}$. (It is also called the propagator or the Green function of the problem.)

3. *Zwanzig's equation*

Let us suppose now that we have defined a projection operator $\hat{P}$ which is time-independent, with $\hat{P}^2 = \hat{P}$, such that:

$$\hat{P}\hat{\varrho}(t) = \hat{\varrho}_R(t), \quad (1 - \hat{P})\,\hat{\varrho}(t) = \hat{\varrho}_I(t),$$

whence

$$\hat{\hat{\varrho}}(t) = \hat{P}\hat{\hat{\varrho}}(t) + (1-\hat{P})\hat{\hat{\varrho}}(t) \equiv \hat{\varrho}_R(t) + \hat{\varrho}_I(t). \qquad \text{(VI.141)}$$

It can be seen from the above that this operator $\hat{P}$ is also a tetrad; for example, in the special case where $\hat{P}$ has the effect of diagonalising the matrix $\hat{\hat{\varrho}}$ (we denote this diagonalisation operator by $\hat{D}$, in what follows), we have

$$\hat{D}\hat{\hat{\varrho}} = \hat{\hat{\varrho}}_d \qquad \text{(VI.142)}$$

or, in matrix terms,

$$(\hat{D}\hat{\hat{\varrho}})_{ij} = \bar{\varrho}_{ii}\delta_{ij} \qquad \text{(VI.143)}$$

and

$$D_{ijkl} = \delta_{ij}\delta_{ik}\delta_{jl}. \qquad \text{(VI.144)}$$

This being so, we shall establish Zwanzig's equation in the general case, without discussing the special form of the operator $\hat{P}$. Let us consider Liouville's equation (VI.126) and let us apply to it successively the projectors $\hat{P}$ and $(1-\hat{P})$. Taking account of equation (VI.141), it becomes

$$\frac{\partial\hat{\varrho}_R(t)}{\partial t} = -i\hat{P}\hat{L}\hat{\varrho}_R(t) - i\hat{P}\hat{L}\hat{\varrho}_I(t), \qquad \text{(VI.145)}$$

$$\frac{\partial\hat{\varrho}_I(t)}{\partial t} = -i(1-\hat{P})\hat{L}\hat{\varrho}_R(t) - i(1-\hat{P})\hat{L}\hat{\varrho}_I(t), \qquad \text{(VI.146)}$$

or again, by noting that we have $\hat{P}\hat{\varrho}_R = \hat{\varrho}_R$ and $(1-\hat{P})\hat{\varrho}_I = \hat{\varrho}_I$,

$$\frac{\partial\hat{\varrho}_R(t)}{\partial t} = -i\hat{P}\hat{L}\hat{P}\hat{\varrho}_R(t) - i\hat{P}\hat{L}(1-\hat{P})\hat{\varrho}_I(t), \qquad \text{(VI.145')}$$

$$\frac{\partial\hat{\varrho}_I(t)}{\partial t} = -i(1-\hat{P})\hat{L}\hat{\varrho}_R(t) - i(1-\hat{P})\hat{L}(1-\hat{P})\hat{\varrho}_I(t). \qquad \text{(VI.146')}$$

Zwanzig's method consists in looking for an equation of evolution for the "relevant" part $\hat{\varrho}_R(t)$ only; we can do this by eliminating $\hat{\varrho}_I(t)$ from the two previous equations. For this purpose we write the formal solution of equation (VI.146) by considering it as a linear equation in $\hat{\varrho}_I(t)$ with the inhomogeneous term $-i(1-\hat{P})\hat{L}\hat{\varrho}_R$.

We find therefore:

$$\hat{\varrho}_I(t) = e^{-i(1-\hat{P})\hat{L}t}\,\hat{\varrho}_I(0) - i\int_0^t d\tau e^{-i(1-\hat{P})\hat{L}\tau}(1-\hat{P})\,\hat{L}\hat{\varrho}_R(t-\tau), \tag{VI.147}$$

where the operator $\hat{\mathscr{G}}(t) \equiv e^{-i(1-\hat{P})\hat{L}t}$ is the propagator associated with equation (VI.146). We obtain Zwanzig's equation by putting this formal solution into equation (VI.145). We have, finally,

$$\frac{\partial\hat{\varrho}_R(t)}{\partial t} = -i\hat{P}\hat{L}\hat{\varrho}_R(t) - i\hat{P}\hat{L}\hat{\mathscr{G}}(t)\,\hat{\varrho}_I(0) - \int_0^t d\tau\hat{G}(\tau)\,\hat{\varrho}_R(t-\tau), \tag{VI.148}$$

where

$$\hat{G}(\tau) = \hat{P}\hat{L}\hat{\mathscr{G}}(\tau)\,(1-\hat{P})\,\hat{L} = \hat{P}\hat{L}e^{-i(1-\hat{P})\hat{L}\tau}(1-\hat{P})\,\hat{L}. \tag{VI.149}$$

Equation (VI.148) is an exact equation satisfied by the "relevant" part $\hat{\varrho}_R(t)$ of $\hat{\varrho}(t)$; in this sense, it is equivalent to Liouville's equation and it results simply from the fact that we are interested only in one portion of the data contained in the total $\hat{\varrho}$. We note that it contains a convolution integral expressing the non-Markovian behaviour of $\hat{\varrho}_R(t)$ and that it still depends on the initial conditions $\hat{\varrho}_I(0)$.

A simpler form can be obtained for equation (VI.148) by particularising a little the problem being studied. We note first of all that we often have the relation:

$$\hat{P}\hat{L}\hat{P} = 0; \tag{VI.150}$$

this is the case, for example, with the operator $\hat{D}$ as we shall see in what follows. This relation has the effect of eliminating the first term on the right-hand side. On the other hand, we can restrict ourselves to the study of cases where the initial condition

$$\hat{\varrho}_I(0) = 0 \tag{VI.151}$$

is satisfied; if $\hat{P} \equiv \hat{D}$, this relation expresses the absence of initial correlations in the system: this is the hypothesis of "molecular chaos" at the initial instant. With the two conditions (VI.150) and (VI.151), Zwanzig's equation assumes the simple form:

$$\frac{\partial \hat{\varrho}_R(t)}{\partial t} = - \int_0^t d\tau \, \hat{G}(\tau) \, \hat{\varrho}_R(t - \tau), \qquad \text{(VI.152)}$$

which is always equivalent to Liouville's equation. If condition (VI.151) were not satisfied, we would have the supplementary term $-i\hat{P}\hat{L}\hat{\mathscr{G}}(t)\,\hat{\varrho}_I(0)$ which, in principle, enables us to study the effect of the initial conditions on the final evolution of the system.

4. *Derivation of a generalised "Master Equation"*

We shall see now how we can use equation (VI.152) to establish the generalised "Master Equation" obtained according to various methods by Nakajima, Zwanzig, Montroll, Prigogine and Résibois. Since the previous conditions are still too general, we must fix our model more precisely.

We shall suppose that, in what follows, the Hamiltonian is of the form

$$\hat{H} = \hat{H}_0 + \lambda \hat{V}, \qquad \text{(VI.153)}$$

where λ is a parameter characterising the order of magnitude of the perturbation $\hat{V}$; it follows that we have, similarly:

$$\hat{L} = \hat{L}_0 + \lambda \hat{L}'. \qquad \text{(VI.154)}$$

In addition, we shall use the representation of the eigenfunctions of $\hat{H}_0$, in which the matrix elements $(\hat{L}_0)_{ijkl}$ can be written as

$$(\hat{L}_0)_{ijkl} = (E_i - E_j)\, \delta_{ik}\delta_{jl}, \qquad \text{(VI.155)}$$

the E_i denoting the eigenvalues of $\hat{H}_0$. We shall assume further that $\hat{L}'$ has no diagonal elements, which means assuming that the perturbation $\hat{V}$ only has the effect of inducing transitions between the eigenstates of $\hat{H}_0$. Finally, we shall take as $\hat{\varrho}_R(t)$ the diagonal part $\hat{\varrho}_d(t)$ of the operator $\hat{\varrho}(t)$; in this case, the operator $\hat{P}$ is none other than the projector $\hat{D}$ defined by equations (VI.142), (VI.143) and (VI.144). In this case, we can prove the relation:

$$\hat{D}\hat{L}\hat{D} = 0, \qquad \text{(VI.156)}$$

since be definition we have

$$(\hat{D}\hat{L}\hat{D})_{ijkl} = \delta_{ij} L_{iikk} \delta_{kl},$$

which becomes zero because of equation (VI.129). Similarly, we can prove the formula

$$\hat{D}\hat{L}_0 = \hat{L}_0\hat{D}, \tag{VI.157}$$

as a consequence of equation (VI.155). By combining this latter equation with (VI.156), it can be shown that the operators $\hat{D}$ and $\hat{L}$ satisfy the following two relations:

$$\hat{D}\hat{L} = \lambda\hat{D}\hat{L}', \; (1 - \hat{D})\,\hat{L}\hat{D} = \lambda\hat{L}'\hat{D}. \tag{VI.158}$$

Under these conditions, the kernel $\hat{G}(\tau)$ of Zwanzig's equation beoomes

$$\hat{G}(\tau) = \lambda^2 \hat{D}\hat{L}'\hat{\mathscr{G}}(\tau)\,\hat{L}'\hat{D} \equiv \lambda^2 \hat{G}'(\tau), \tag{VI.159}$$

with the result that Zwanzig's equation can now be written as

$$\frac{\partial \hat{\varrho}_d(t)}{\partial t} = -\lambda^2 \int_0^t d\tau \hat{G}'(\tau)\,\hat{\varrho}_d(t - \tau). \tag{VI.160}$$

This equation is still exact, equivalent to Liouville's equation and, in consequence, reversible; it is strictly valid for our model in the case where the initial conditions $\hat{\varrho}_I(0) = 0$ are satisfied.

We write equation (VI.160) in the matrix form:

$$\frac{\partial \bar{\varrho}_{ii}}{\partial t} = \lambda^2 \sum_j \int_0^t W_{ij}(\tau)\,\bar{\varrho}_{jj}(t - \tau)\,d\tau \tag{VI.161}$$

with

$$W_{ij}(\tau) = -[\hat{L}'e^{-i\tau(1-\hat{D})\hat{L}}\,\hat{L}']_{iijj} = -[\hat{L}'\hat{\mathscr{G}}(\tau)\,\hat{L}']_{iijj}, \tag{VI.162}$$

using the property (VI.144) of the operator $\hat{D}$. In order to obtain a "Master Equation", it remains for us to show that the $W_{ij}(\tau)$ represent the transition probabilities of the state j to the state i. Since the trace of $\hat{\varrho}$ is equal to unity, it is sufficient to verify the relations

$$\sum_i W_{ij}(\tau) = 0, \quad \sum_j W_{ij}(\tau) = 0; \tag{VI.163}$$

these are obtained by writing the matrix element $[\hat{L}'\hat{\mathscr{G}}\hat{L}']_{iijj}$ and by using the property $\sum_i L'_{iikl} = 0$ [cf. equation (VI.129)]. In particular, we obtain from equation (VI.163):

$$W_{ii}(\tau) = -\sum_{j \neq i} W_{ij}(\tau),$$

with the result that equation (VI.161) assumes the form

$$\frac{\partial \bar{\varrho}_{ii}}{\partial t} = \lambda^2 \sum_{j \neq i} \int_0^t d\tau\, W_{ij}(\tau)\, [\bar{\varrho}_{jj}(t-\tau) - \bar{\varrho}_{ii}(t-\tau)] \quad \text{(VI.164)}$$

$$= \lambda^2 \sum_{j \neq i} \int_0^t d\tau [W_{ij}(\tau)\, \bar{\varrho}_{jj}(t-\tau) - W_{ji}(\tau)\, \bar{\varrho}_{ii}(t-\tau)].$$

This is Zwanzig's generalised "Master Equation"; it corresponds closely, in notation, to the equations obtained separately by the various authors that we have mentioned previously (cf. Zwanzig, 1960, 1964). Although this equation now has the familiar form of a "gains–losses" balance sheet, it is still equivalent to Liouville's equation for the model envisaged here. It is thus an exact and reversible equation and we can prove that it differs from a true irreversible "Master Equation" by the presence of the convolution integral. It should reduce to an equation of Pauli's type if it could be established that we have, under certain conditions, $W_{ij}(\tau) = W_{ij}\delta(\tau)$, with the W_{ij} constant and independent of τ. In order to distinguish the reversible equations such as equation (VI.164) from an irreversible "Master Equation", certain authors such as Mazur have proposed that they should be called "Pre-Master Equations".

We note, finally, that the integral on the right-hand side of equation (VI.164) is formally of second order in λ and that terms of higher order are contained in the exponential factor of the operator $\hat{\mathscr{G}}$. Zwanzig's equation is thus easily adapted to perturbation calculations, since the first-order term in λ^2 is obtained by replacing $\hat{L}$ by $\hat{L}_0$ in the index of $\hat{\mathscr{G}}$; the propagator associated with free particles $e^{-i\tau\hat{L}_0}$ is thus seen to appear.

5. *Time evolution; Pauli equation*

In order to discuss the conditions under which the behaviour of the system will be irreversible, we must study the time evolution which results from equation (VI.161) or (VI.164). In order to simplify this study, we shall suppose for the moment that $\hat{\bar{\varrho}}$ and $\hat{G}$ are scalar functions and that we have the initial condition $\hat{\bar{\varrho}}(0) = 1$. We shall consider the equation:

$$\frac{\partial \bar{\varrho}(t)}{\partial t} = -\lambda^2 \int_0^t G(\tau, \lambda)\, \bar{\varrho}(t-\tau)\, d\tau, \qquad \text{(VI.165)}$$

in which we have recalled the dependence of the function G on λ and we are interested in the case in which the intensity λ of the perturbation is small (weak coupling).

(a) The most elementary argument consists then in assuming that $\partial\bar{\varrho}/\partial t$ is small and that it is of the order of λ^2, which permits us to write

$$\bar{\varrho}(t-\tau) \simeq \bar{\varrho}(t) + O(\lambda^2). \qquad \text{(VI.166)}$$

Equation (VI.165) then reduces to

$$\frac{\partial \bar{\varrho}}{\partial t} = -\lambda^2 \left(\int_0^t G(\tau, \lambda)\, d\tau \right) \bar{\varrho}(t) + O(\lambda^4), \qquad \text{(VI.167)}$$

the solution of which is:

$$\bar{\varrho}(t) = \exp\left[-\lambda^2 \int_0^t d\tau \int_0^\tau d\tau' G(\tau', \lambda) + O(\lambda^4) \right], \qquad \text{(VI.168)}$$

since $\bar{\varrho}(0) = 1$. If we assume that the function $G(\tau, \lambda)$ is integrable from zero to infinity, a characteristic time exists for the system, τ_c, for which we have to a sufficient degree of accuracy

$$\int_0^t d\tau G(\tau, \lambda) \simeq \int_0^{\tau_c} d\tau G(\tau, \lambda) \qquad \text{(VI.169)}$$

for a sufficiently large value of t. Thus, we have

$$\int_0^t d\tau \int_0^\tau d\tau' G(\tau', \lambda) \simeq t \int_0^{\tau_c} d\tau G(\tau, \lambda),$$

with the result that for $t \gg \tau_c$ we have an exponential decrease of $\bar{\varrho}(t)$ given by:

$$\hat{\bar{\varrho}}(t) \simeq e^{-t/\tau_R} + O(\lambda^4), \qquad \text{(VI.170)}$$

with a relaxation time τ_R defined by:

$$\tau_R^{-1} = \lambda^2 \int_0^{\tau_c} G(\tau, \lambda)\, d\tau. \tag{VI.171}$$

(b) Although this reasoning gives interesting indications concerning the behaviour of the system with time and concerning the manner by which irreversible evolution can be obtained whilst λ is small, it does not have an exact character. We can attempt to improve this by resorting to the so-called "weak-coupling" limit introduced first by Van Hove. It consists in defining a new variable θ by

$$\theta = \lambda^2 t \tag{VI.172}$$

and making $\lambda \to 0$ and $t \to \infty$, whilst θ remains constant. By substituting equation (VI.172) in equation (VI.165), the function $\bar{\varrho}(t)$ is transformed to a new function $P(\theta)$ which satisfies the equation

$$\frac{\partial P(\theta)}{\partial \theta} = - \int_0^{\theta/\lambda^2} d\tau G(\tau, \lambda)\, P(\theta - \lambda^2 \tau). \tag{VI.173}$$

Taking the limit as $\lambda \to 0$, $t \to \infty$ with θ constant, we find

$$\frac{\partial P(\theta)}{\partial \theta} = - \left(\int_0^{\infty} d\tau G(\tau, 0) \right) P(\theta), \tag{VI.174}$$

the solution of which,

$$P(\theta) = \exp\left[-\theta \int_0^{\infty} d\tau G(\tau, 0) \right], \tag{VI.175}$$

is equivalent to equation (VI.170), with a relaxation time given on this occasion by

$$\tau_R^{-1} = \lambda^2 \int_0^{\infty} G(\tau, 0)\, d\tau. \tag{VI.176}$$

We note that we should obtain an identical result but in a more exact way by introducing the Fourier transform of the kernel $G(\tau, \lambda)$.

(c) Despite the elegance of this weak-coupling method, we may ask how it has been possible to derive an irreversible evolution from the fundamental equation (VI.165) and, in particular, what is the physical significance of the above arguments. Actually, we shall see that they cannot be applied to real physical situations because they depend on the hypothesis of integrability of the function $G(\tau,\lambda)$. This hypothesis, however, is not justified for systems with a *finite* number, N, of degrees of freedom: we can show, in fact, as we shall prove later when deriving Pauli's equation, that the operator $\hat{G}(\tau, \lambda)$ is generally a pseudo-periodic function of time, whose integral does not approach any definite limit.

In order to overcome this difficulty and to obtain an irreversible evolution, it is necessary to assume that the number of degrees of freedom increases indefinitely, that is to say, to introduce the limit $N \to \infty$. With a view to analysing the role played by this limiting process, we shall take once more a simple example given by Zwanzig, in which we suppose that $G(\tau, \lambda)$ has the form of a sum of cosines, which is a special case of an almost periodic function. We put

$$G(\tau, \lambda) = \frac{1}{N} \sum_{k=1}^{N} \cos\left(\frac{k\tau}{N}\right) \tag{VI.177}$$

and we consider the limit $N \to \infty$. The $\sum_k$ can then be replaced by an integral and we find:

$$\lim_{N \to \infty} G(\tau, \lambda) = \frac{\sin \tau}{\tau}, \tag{VI.178}$$

which, obviously, is no longer a periodic function of τ.

On the other hand, we know the precise form of $G(\tau, \lambda)$, since the kernel (VI.177) is a geometrical series. We have

$$G(\tau, \lambda) = \frac{1}{2N}(\cos \tau - 1) + \frac{\tau}{2N} \cot\left(\frac{\tau}{2N}\right) \frac{\sin \tau}{\tau}. \tag{VI.179}$$

Comparing expressions (VI.178) and (VI.179), it can be seen that in the case where N is *finite*, the limit (VI.178) is a good approximation only if the following two conditions are satisfied:

1. We consider only "small" times τ, such that $\tau \ll N$, i.e. small in comparison with the recurrence time of the system or the "return time", equal in our example to $2\pi N$.

2. Terms corresponding to microscopic fluctuations (or "noise"), of order of magnitude $1/N$, can be neglected.

In this way we see that the role of the limit $N \to \infty$ is essentially to make the period $T(N)$ of the Poincaré cycles of the system approach infinity: this is an indispensable condition for making irreversibility appear. In addition, it can be seen that the time-evolution of a real physical system is characterised by the three time scales τ_C, τ_R and $T(N)$. If the coupling λ is very small and if the number of degrees of freedom is very large, we shall always have $\tau_C \ll \tau_R \ll T(N)$, and for times τ of the order of τ_R, we have a Markovian evolution described by one of the formulae (VI.170) or (VI.175).

(d) It remains now for us to apply these arguments to the derivation of Pauli's equation. We start from equation (VI.164), which we shall rewrite by changing the notation of the quantum states. In a system with a large number of degrees of freedom, each quantum state i includes numerous quantum numbers; we denote by E the energy of the system and by α the set of other quantum numbers necessary for specifying completely the level i. Each state is then denoted by $(E\alpha)$ and we put:

$$\bar{\varrho}_{ii}(t) = \bar{\varrho}(E\alpha; t) \qquad W_{ij}(\tau, \lambda) = W(E\alpha, E'\alpha'; \tau, \lambda).$$

With this notation, equation (VI.164) can be written as

$$\frac{\partial \bar{\varrho}(E\alpha; t)}{\partial t} = \lambda^2 \sum_{E'} \sum_{\alpha'} \int_0^t d\tau \, W(E\alpha, E'\alpha'; \tau, \lambda) \, [\bar{\varrho}(E'\alpha'; t-\tau) - \bar{\varrho}(E\alpha, t-\tau)]. \qquad \text{(VI.180)}$$

We now give an explicit expression for the transition probabilities W by always using the model of § 4. To a first order in λ we have, according to equation (VI.162),

$$W_{ij}(\tau, \lambda) = -(\hat{L}' e^{-i\hat{L}_0\tau} \hat{L}')_{iijj} + O(\lambda). \qquad \text{(VI.181)}$$

Remembering that

$$(e^{-i\hat{L}_0\tau})_{klmn} = \delta_{kl}\delta_{mn} e^{-i(E_k - E_l)t/\hbar}$$

$$L'_{iikl} = \frac{1}{\hbar}[V_{ik}\delta_{il} - V_{li}\delta_{ik}],$$

we have

$$W_{ij}(\tau, \lambda) = \frac{2}{\hbar^2} |V_{ij}|^2 \cos \frac{E_i - E_j}{\hbar} \tau + O(\lambda)$$

$$= \frac{2}{\hbar^2} |V(E\alpha|E'\alpha')|^2 \cos \frac{E - E'}{\hbar} \tau + O(\lambda); \qquad \text{(VI.182)}$$

we note in passing that we find again the pseudo-periodic behaviour of the nucleus $G(\tau, \lambda)$. Under these conditions, equation (VI.180) becomes

$$\frac{\partial \bar{\varrho}(E\alpha; t)}{\partial t} = \lambda^2 \sum_{E'} \sum_{\alpha'} \frac{2}{\hbar^2} \int_0^t d\tau |V(E\alpha; E'\alpha')|^2 \cos \frac{E - E'}{\hbar} \tau \times$$

$$\times [\bar{\varrho}(E'\alpha'; t - \tau) - \bar{\varrho}(E\alpha; t - \tau)] + O(\lambda^3 \bar{\varrho}). \quad \text{(VI.183)}$$

We shall obtain Pauli's equation if we can remove the convolution integral on the right-hand side. We must take account here of the fact that the system contains a large number of degrees of freedom; in this case, the spectrum of the energy levels is sufficiently dense so that we can replace the $\sum_{E'}$ by an integral. If $N(E)$ is the density of levels, we have

$$\frac{\partial \bar{\varrho}(E\alpha; t)}{\partial t} = \lambda^2 \sum_{\alpha'} \frac{2}{\hbar^2} \int dE' N(E') \int_0^t d\tau |V(E\alpha; E'\alpha')|^2 \cos \frac{E - E'}{\hbar} \tau \times$$

$$\times [\bar{\varrho}(E'\alpha'; t - \tau) - \bar{\varrho}(E\alpha; t - \tau)] + O(\lambda^3 \bar{\varrho}). \quad \text{(VI.184)}$$

If we now carry out the transition to the "weak-coupling" limit (such as we defined under *b*), we see the appearance of a function $\delta(E - E')$ and we obtain finally:

$$\frac{\partial \bar{\varrho}(E\alpha; \theta)}{\partial \theta} = \sum_{\alpha'} W(\alpha, \alpha') [\bar{\varrho}(E\alpha'; \theta) - \bar{\varrho}(E\alpha; \theta)], \quad \text{(VI.185)}$$

where we have put

$$W(\alpha, \alpha') = \frac{2\pi}{\hbar} |V(E\alpha; E\alpha')|^2 N(E) = W(\alpha', \alpha).$$

This is Pauli's equation which we have obtained by assuming that the density matrix was initially diagonal (condition $\hat{\varrho}_I(0) = 0$)

and that the parameter λ is small (weak coupling). Nevertheless, in order to proceed from the exact equation (VI.183) to the irreversible equation (VI.185), we have had to make three additional assumptions:

1. The system has a large number of degrees of freedom.
2. The time t is large and of the order of $1/\lambda^2$.
3. The matrix elements $V(E\alpha; E'\alpha')$ are slowly varying functions of E'. Thus we prove once again the importance of transition to the limit $N \to \infty$, which must be carried out in the exact equations before taking the "weak-coupling" limit. If these three assumptions were not satisfied, it would then be necessary to return to equation (VI.183) which is still exact for small values of λ up to terms in $\lambda^3 \bar{\varrho}$ approximately.

If we wish to discuss the physical significance and the limitation of the foregoing results, we can apply to equation (VI.183) the elementary method described in (a). In this scheme, we replace $\hat{\bar{\varrho}}(t-\tau)$ by $\hat{\bar{\varrho}}(t) + O(\lambda^2 \bar{\varrho})$, with the result that equation (VI.183) becomes in this approximation:

$$\frac{\partial \bar{\varrho}(E\alpha; t)}{\partial t} = \lambda^2 \sum_{E'} \sum_{\alpha'} \frac{2}{\hbar^2} \int_0^t d\tau |V(E\alpha; E'\alpha')|^2 \cos \frac{E-E'}{\hbar} \tau \times$$

$$\times [\bar{\varrho}(E'\alpha'; t) - \bar{\varrho}(E\alpha; t] + O(\lambda^3 \bar{\varrho}). \qquad \text{(VI.186)}$$

We can then carry out integration over τ; we have immediately

$$\int_0^t d\tau W(E\alpha, E'\alpha'; \tau, \lambda) = \frac{2}{\hbar^2} |V(E\alpha; E'\alpha')|^2 \frac{\hbar}{E-E'} \sin \frac{E-E'}{\hbar} t. \qquad \text{(VI.187)}$$

If, now, we replace $\sum_{E'}$ by an integral of the form $\int N(E')\, dE'$, it can be seen that, if t is sufficiently large, the term $\frac{\hbar}{E-E'} \sin \frac{E-E'}{h} t$ approaches the function $\pi\hbar\delta(E-E')$, provided the inequality

$$\frac{t}{\hbar N(E)} \ll 1 \qquad \text{(VI.188)}$$

is satisfied.

We can deduce from this that the time interval for which Pauli's equation is valid depends on the number of degrees of freedom of the system. The higher the number N, the less restricted is condition (VI.188); thus, we eliminate the Poincaré cycles from the system.

In conclusion, we must emphasise the remarkable simplicity of the method proposed by Zwanzig which has enabled us to derive Pauli's "Master Equation" with the minimum of calculations. Moreover, Zwanzig has shown that his method could be applied also to classical theory and a "Master Equation" obtained, which was found previously by Brout and Prigogine. Whilst these results are not new, Pauli's equation having been obtained already with identical assumptions by Van Hove, it can be expected that this method will enable us to analyse in more detail the mechanisms which give rise to the irreversible evolution of macroscopic systems.

CHAPTER VII

General Conclusions. Macroscopic Observation and Quantum Measurement

In this book, we have studied the problem of the foundations of classical and quantum statistical mechanics and we have seen that the methods used are divided into two distinct groups: the ergodic theory and the *H*-theorems. In addition, these correspond to a different definition of macroscopic quantities starting from the microscopic description. It can be said that the ergodic theory (which justifies the replacement of time-averages by phase-averages over the microcanonical ensemble) is a particularly suitable method for studying the stationary states of isolated systems and that the *H*-theorems describe more particularly the time evolution of the macroscopic state of the system.

However, we have seen that the ergodic theory in itself involves the irreversible evolution of the system towards stationary states (although incapable of providing a quantitative analysis) and that the $\bar{H}$-theorems, in their generalised form, are comparable with Hopf's ergodic theorem and with its quantum parallel. We shall show now how the fundamental concepts of statistical mechanics are involved in the statistical interpretation of thermodynamic properties at equilibrium (statistical thermodynamics) and of the thermodynamics of irreversible processes.

I. Applications of Statistical Mechanics

1. *Statistical thermodynamics*

It is necessary here to justify the use of stationary ensembles and to deduce from the properties of these ensembles the laws of thermodynamics; for this, we can resort to two methods, used for defining macroscopic observables.

(a) In the first method, the observed physical quantities are the time averages of the functions $f(P_t)$ or of the quantum averages $A(t) = (\Psi_t, \hat{A}\Psi_t)$. The ergodic theory attempts to prove the equality of these averages with the statistical averages taken over a microcanonical ensemble (which describe the stationary state of an isolated system). This being so, the mechanical evolution of such systems is described either by the trajectory of the representative point $\mathscr{P}(t)$ over a hypersurface of Γ-space (or over an energy shell contained between two hypersurfaces), or by the trajectory of the affix $\Psi(t)$ over the unit hypersphere in the Hilbert subspace associated with the shell $e^{(\alpha)}$; the microcanonical ensembles are defined by invariant measures on the hypersurface or on the unit hypersphere. If we could prove an ergodic theorem which is valid for all trajectories, we should justify in this way—*starting from mechanical considerations*—the use of the microcanonical ensemble from which the fundamental theory of statistical mechanics would be deduced. We have seen, however, that this is not so and that the exact ergodic theorems always accept the possibility of *exceptional trajectories*; thus, the ergodic theory has an inevitable statistical aspect (almost certain convergence or convergence in quadratic mean) and can be applied, as we have seen, only to systems which satisfy very restrictive conditions (especially in quantum theory). This is why we have stressed particularly the probability ergodic theorem (which we developed in detail from the quantum point of view in Chapter IV), in which the statistical aspect of the theory is used with advantage for breaking away from the assumptions of metric transitivity in classical mechanics, and of the absence of degeneracy and of resonance frequencies in the energy spectrum in wave mechanics.

Thus, in classical statistical mechanics, as in quantum statistical mechanics (and without looking at the differences in detail between the two aspects of the theory which are mentioned throughout this book), the probability ergodic theory justifies the use of microcanonical ensembles and thus permits the founding of the atomistic interpretation of the laws of thermodynamics, by relying essentially on the very large number of degrees of freedom of a macroscopic system. The statistical nature of the ergodic theory determines, in addition, the limit of validity of the laws of classical thermodynamics, by leading us to expect *fluctuations* and to calculate them by means of the canonical and grand-canonical ensembles which we have defined, starting from the microcanonical ensemble.

We shall now show briefly how we can arrive at the usual results of statistical thermodynamics; our aim, obviously, is not to give a complete report on the calculation methods used in statistical mechanics [we can refer for that to specialised books; see for example Massignon, 1957], but simply to recall how the exclusively theoretical considerations of this book are involved in the practical applications.

(b) For this, we resort to the second more complete definition of the macroscopic state of a physical system. In classical theory, this macroscopic state is defined by covering μ-space (associated with a particle) by a network of finite cells of extension in phase ω_i, to which correspond the stars Ω_n of Γ-space. The more precise definition of the macroscopic state is obtained by the localisation of the point $\mathscr{P}(t)$ in a specified star in Γ-space. We see here that the results of the ergodic theory relative to a hypersurface of constant energy are inadequate (because of the natural inaccuracy of macroscopic observations), which proves the physical usefulness of Hopf's theorem. From this theorem we can then deduce the relation

$$\overline{f(P_t)}^{\infty} = \lim_{T\to\infty} \overline{f(P_t)}^{T} = \frac{\sum_{\Omega_n} f(\Omega_n)\, W(\Omega_n)}{\sum_n W(\Omega_n)}, \qquad \text{(VII.1)}$$

in which we assume that f is a function of the occupation numbers n_i of the cells ω_i and where $W(\Omega_n)$ represents the statistical "weight" of the stars Ω_n (which is taken to be proportional to the volume of these stars for the microcanonical ensemble). The calculation of the average values (VII.1) can be undertaken by means of the Darwin–Fowler method (1923) used in the fundamental books by Fowler (1936) and Fowler and Guggenheim (1939); the complicated formalism of this method, however, can be simplified considerably (as shown by Khinchin) by a systematic application of the central limit theorem of probability calculus (see Appendix I, § 4). On the other hand, we find Boltzmann's usual results by noting that the dimensions of the star $\Omega_{\max}$ (corresponding to the most probable state) are such that we can in all applications replace $W(Z)$ by $W(\Omega_{\max})$ (because of the large number of degrees of freedom of the system and of the insensitivity of Boltzmann's formula to the definition of the elementary probability). We can then write

$$\overline{f(P_t)}^{\infty} \cong f(\Omega_{\max}) \tag{VII.2}$$

and in this way we can obtain all the usual formulae (Maxwell–Boltzmann distribution, equipartition of energy, ...).

Quantum theory follows a similar route, by characterising the observed quantities by macroscopic operators: they define Hilbert sub-spaces with suffixes ν of $s_\nu^{(\alpha)}$ dimensions, which are analogous to the stars in Γ-space. We have seen that these operators, apart from the macroscopic energy, do not commute with the corresponding microscopic operators. In particular, we can describe a quantum gas by taking $\hat{H} = \hat{H}_0 + \hat{V}$ for the Hamiltonian, where $\hat{V}$ is the interaction energy of the particles whose free Hamiltonian is $\hat{H}_0$: macroscopic observable states are then represented by groups of eigenvalues of $\hat{H}_0$ between which quantum transitions occur, which are produced by the potential $\hat{V}$. We obtain the usual results by defining groups of eigenvalues of the energy of a single particle (by means of the well-known relationship $g_i = \omega_i/h^r$, where r is the number of internal degrees of freedom of the particles considered) and by calculating the number of complexions corresponding to the macroscopic state of the system according to the type of statistics considered.

The relation between the microscopic structure of the system and the thermodynamic functions is obtained by means of the sum over states (Planck's Zustandssumme), also called the partition function (Fowler); for one molecule it is defined by ($\beta = 1/kT$)

$$Z_\mu = \sum_i g_i \, e^{-\beta \varepsilon_i} \tag{VII.3}$$

where ε_i is the eigen-energy of one particle; the classical formula

$$Z_\mu = \int_\mu e^{-\beta \varepsilon} \, d\mu \tag{VII.3'}$$

corresponds to (VII.3). This partition function is related to the thermodynamic free energy F by the expression

$$Z_\mu = e^{-\beta F}. \tag{VII.4}$$

We can then evaluate, by means of formulae (VII.3) and (VII.4), all the usual thermodynamic functions (entropy, enthalpy, ...), from which we can deduce the thermodynamic quantities obtained by experiment.

In order to define the reversible transformations of thermodynamics, we must resort to the canonical ensemble which represents a system in thermal equilibrium with a thermostat. Using the property of adiabatic invariance of these ensembles (for this purpose the external parameters which occur in the Hamiltonian of the system are varied), we can work out a statistical interpretation of the isothermal and adiabatic reversible transformations of thermodynamics: these must be such that the statistical equilibrium state is conserved during transformation. The thermodynamic quantities are calculated in this case by means of the partition function in Γ-space:

$$Z_\Gamma = \sum_i \gamma_i \, e^{-\beta E_i}, \tag{VII.5}$$

where the γ_i are the weights of the states with total energy E_i; for a system composed of N independent particles, we have $Z_\Gamma = (Z_\mu)^N$. In the general case of real gases, the study of the thermodynamic properties of the gas involves a calculation of Z_Γ; this calculation has given rise to numerous works both in classical theory (the Ursell–Mayer method of cluster functions and Bogulyubov's method) as well as in quantum theory where the formalism of quantum field theory has recently been used with success (Matsubara, 1955; Bloch and de Dominicis, 1958; Bloch, 1960).

Finally, grand-canonical ensembles enable us to take up the study of the thermodynamics of physico-chemical processes. We point out, in addition, that the average values calculated by starting from grand-canonical ensembles are the same as those calculated by the Darwin–Fowler method; this result, which we have studied already from the theoretical point of view at the end of Chapter IV, is one of very great practical importance since it permits us to simplify the calculation of the average values (by dispensing with the restrictive conditions implied in the definition of microcanonical and canonical ensembles); this is of particular interest for studying real gases.

2. *Irreversible processes*

We have seen that the study of the irreversible evolution of macroscopic systems encounters difficulties which are greater than the justification of the method of statistical thermodynamics based on the use of stationary ensembles corresponding to states of macroscopic equilibrium. In this study, macroscopic observables must be introduced by means of the method of phase cells, which enables us to define coarse-grained statistical densities; in this way, we have shown that any initial distribution tends probably towards an equilibrium distribution because of the difference between coarse-grained and fine-grained densities; this is the generalised $\bar{H}$-theorem which we proved in quantum theory in Chapter VI. This method, however, as we have seen involves two serious disadvantages: on the one hand, the $\bar{H}$-theorem could not be proved exactly and, on the other hand, it can scarcely provide us with any more information about the evolution of the system than Hopf's ergodic theorem or its quantum equivalent; in particular, there is no means of obtaining practical quantitative results. Thus, in its generalised form the $\bar{H}$-theorem could only serve as a theoretical basis for the use of stationary ensembles in statistical thermodynamics.

In order to work out a quantitative theory which could serve as the basis of the thermodynamics of irreversible processes, we must try to analyse macroscopic evolution by comparing the statistical definitions of macroscopic quantities with the microscopic mechanical model of the system. From the point of view of mechanics, the evolution of the system is always described either in classical mechanics by the motion of the representative point $\mathscr{P}(t)$, which traverses successively different stars in Γ-space, or in wave mechanics by the trajectory of $\Psi(t)$ over the unit hypersphere of the Hilbert sub-space associated with the shell $e^{(\alpha)}$ (different probabilities for the sub-spaces with suffix ν correspond to different points of this trajectory). The macroscopic observer, however, can only ascertain successive transitions from one star to another in Γ-space, or the changes in probability associated with the quantum cells ν: *there is a "complementarity" between the macroscopic description and the microscopic description of the events*; we have emphasised at length this aspect of the problem in Section III of Chapter VI. The statistical method, in this case, consists in representing the evolution of macroscopic quantities by average values (or most probable values) calculated with statistical ensembles defined in accordance with the initial macroscopic observations.

We have seen that the results obtained by this method are only strictly valid for a (macroscopically) small interval of time after the instant of observation. If we require equations which are valid at any instant (such as the kinetic equations), we must assume that a statistical ensemble can be defined which corresponds to a macroscopic condition at any given instant, without taking account of the macroscopic states previously occupied by the system. This assumption is necessary for proving Boltzmann's equation and it reduces to accepting *the conservation in time of the property of molecular chaos*; moreover, we have found these conditions in both classical and quantum theory, when describing the irreversible evolution of the system by a Markovian statistical process, with probabilities given by the theory of molecular chaos. In both cases, the final justification of these theories depends ultimately on the approximate integration of Liouville's equation or of its quantum equivalent.

It is essential to note that the foregoing considerations are independent of whether the macroscopic observation has or has not been actually performed: the macroscopic state of a system is, in fact, objective and therefore is independent of the observation; in particular, this does not disturb the system in an uncontrollable way as in the case of the quantum measurement and its effect on the system can always be made as small as desired. These points are just as valid in classical statistical mechanics as in quantum statistical mechanics; in addition, they will allow us to study in more detail the difference encountered in the proof of the $\bar{H}$-theorem between the classical and quantum cases. In fact, we must bear in mind that the decrease of $\bar{H}$ was due in both cases to the difference between the fine-grained and coarse-grained densities and that it was intensified in the quantum case by Klein's lemma: we can show that this latter cause of the decrease of $\bar{H}$, which is without counterpart in classical theory, can be associated with the irreversible perturbation due to observations made on the system.

II. Quantum Measurement and Macroscopic Observation

1. *Quantal entropy*

Let us start by recalling the mechanism of a quantum measurement (by using the normal interpretation of wave mechanics). If we denote the system in which we wish to measure an observable by I, the procedure consists in coupling this system with another system II in such a manner that the observable being measured in I must correspond to an observable in II. If the two systems I and II are initially in a pure case represented by the wave functions φ and ψ respectively, the total system I + II also is in a pure case P_Φ, with $\Phi = \varphi\psi$. If φ_i are the eigenfunctions associated with the observable of I to be measured and if ψ_k are those associated with the corresponding observable of II, we can put

$$\varphi = \sum_i a_i \varphi_i \quad \text{and} \quad \psi = \sum_k d_k \psi_k, \tag{VII.6}$$

whence

$$\Phi = \sum_{i,k} c_{ik} \varphi_i \psi_k, \quad \text{with} \quad c_{ik} = a_i d_k. \tag{VII.7}$$

This being so, the first phase of the measuring process consists in suitably coupling systems I and II; this coupling leaves I + II in a pure case, but establishes a correspondence between the Φ_i and ψ_k such that Φ can be written as

$$\Phi = \sum_i a_i \varphi_i \psi_i. \tag{VII.8}$$

System I, which was described initially by a pure case with a density matrix

$$\varrho^{\text{I}}_{ij} = a_j^* a_i \tag{VII.9}$$

is described, after the coupling, by a mixture with the diagonal density matrix:

$$\varrho^{\text{I}}_{ij} = |a_i|^2 \, \delta_{ij}; \tag{VII.10}$$

the coupling between I and II thus has the effect of making the pure case (VII.9) change to the mixture (VII.10).

The second phase of the measuring process is the observation of the result of measuring ψ_i, which allows the wave packet to be reduced. After reading the measurement result, system I is again in a pure case, represented by φ_i. If we denote by σ^{I} the *quantal entropy* of I, defined by

$$\sigma^{\mathrm{I}} = \mathrm{Tr}\,(\hat{\varrho}^{\mathrm{I}} \ln \hat{\varrho}^{\mathrm{I}}), \tag{VII.11}$$

it is initially zero (pure case); then it increases sharply after coupling and becomes equal to $-\sum_i |a_i|^2 \ln |a_i|^2$ (mixture); the reduction of the wave packet returns it to 0. Thus, there are two quantum entropy steps of equal amplitude and with opposite sign: one on establishing the coupling between I and II and one on reading the measurement result; in particular, it can be seen that the increase of quantum entropy is linked to the coupling between the observed system and the measurement equipment, without reading the measurement result (Jancel and Kahan, 1955). Moreover, this is connected with the subjective nature of the wave function in wave mechanics (within the framework of the usual interpretation).

These results can be extended to a series of measurements undertaken on a series of observables of a quantum system: according as we actually proceed or not in reading the measurement results, there are two different descriptions of the system. In the first case, there are two quantum entropy steps for each measurement, corresponding, respectively, to the coupling with an intermediate system and to reading the measurement result; after each measurement the system returns to a pure case and the quantum entropy takes again a zero value until the next measurement. In the second case, where the result of the measurement is not read, we have a discontinuous increase of entropy at each coupling corresponding to a measurement; the quantum entropy can then be represented by a stepped curve which takes account of the irreversibility of successive interactions of the observed system with the intermediate systems. When we actually proceed to a maximal observation, the quantum entropy returns to zero: this behaviour shows actually that the quantum entropy expresses the state of our knowledge about the system.

2. *The role of Klein's lemma*

This being so, the proof of Klein's lemma depends essentially on the fact that the initial matrix $\overline{\varrho_{ij}^{(\alpha)}}\,(0)$ is diagonal. More precisely, if the initial statistical ensemble is determined by a macroscopic observation, corresponding to the quantities $P_\nu^{(\alpha)}(0)$, we can put

$$\overline{\varrho_{ij}^{(\alpha)}}(0) = \frac{P_\nu^{(\alpha)}(0)}{s_\nu^{(\alpha)}}\,\delta_{ij}, \tag{VII.12}$$

whence we obtain for the matrix elements $\overline{\varrho_{ij}^{(\alpha)}}(t)$:

$$\overline{\varrho_{ii}^{(\alpha)}}(t) = \sum_{\nu'} \frac{P_{\nu'}^{(\alpha)}(0)}{s_{\nu'}^{(\alpha)}} \sum_{j=1}^{s_{\nu'}^{(\alpha)}} |U_{ij}|^2 \tag{VII.13}$$

and

$$\overline{\varrho_{ik}^{(\alpha)}}(t) = \sum_{\nu'} \frac{P_{\nu'}^{(\alpha)}(0)}{s_{\nu'}^{(\alpha)}} \sum_{j=1}^{s_{\nu'}^{(\alpha)}} U_{ij}U_{kj}^*, \quad \text{with} \quad i \neq k. \tag{VII.14}$$

It can be seen from (VII.14) that the matrix $\overline{\hat{\varrho}^{(\alpha)}}(t)$, which was diagonal at time $t = 0$, does not generally continue to be diagonal and that it is the existence of off-diagonal terms which involve the inequality sign in Klein's lemma. On the other hand, as we saw when we derived the kinetic equation (see Chapter VI, § III.3), the calculation of the observable macroscopic quantities $P_\nu^{(\alpha)}(t)$, starting from (VII.13) [with $P_\nu^{(\alpha)}(t) = \sum_{i=1}^{s_\nu(\alpha)} \overline{\varrho_{ii}^{(\alpha)}}(t)$] is valid only for the interval of time which follows the instant of observation. If we wish to extend the foregoing results to any instant whatsoever, it must be assumed that:

(a) we have, after each macroscopic observation at time t,

$$\overline{\varrho_{ik}^{(\alpha)}}(t+0) = 0 \quad \text{if} \quad i \neq k; \tag{VII.15}$$

(b) it is possible to define a new statistical ensemble at $t + 0$ by

$$\overline{\varrho_{ij}^{(\alpha)}}(t+0) = \frac{1}{s_\nu^{(\alpha)}}\,P_\nu^{(\alpha)}(t)\,\delta_{ij}, \tag{VII.16}$$

where $P_\nu^{(\alpha)}(t)$ is calculated starting from (VII.13). We see that formula (VII.15) can be made to correspond to the diagonalisation [described by (VII.10)] obtained by coupling a quantum system with a measuring instrument. This comparison suggests that we interpret the decrease of $\overline{H}$ due to Klein's lemma as the irreversible perturbation produced by observation of the system: since condition (VII.15) is not conserved with time by the laws of wave mechanics, we are reduced to assuming a series of successive couplings between the system and the observer (or the surroundings of the system).

In addition, with Born and Green (Born, 1948, 1949; Born and Green, 1948), we can make the proof of the H-theorem depend on this single property of statistical matrices. By depending on the definition of $\hat{\bar{\varrho}}_d$ given in (VI.21) and by putting

$$H = \mathrm{Tr}\,(\hat{\bar{\varrho}}_d \ln \hat{\bar{\varrho}}_d) = \sigma_d, \qquad \text{(VII.17)}$$

the H-theorem is reduced to Klein's lemma; actually, according to (VI.23) we have

$$H' = \sigma'_d \geqq \sigma''_d = H''; \qquad \text{(VII.18)}$$

the equality sign is obtained if the matrix $\hat{\bar{\varrho}}$ is diagonal at t. However, (VII.18) gives us no information about the final behaviour of σ_d; for example, we cannot conclude that $\sigma'''_d < \sigma''_d$ for $t''' > t''$. This proof, therefore, is valid only for times which follow immediately after the macroscopic observation, unless we assume that the perturbation produced by this observation could be described by a series of measurements, as we indicated above: this would be interpreted in this case as an external effect of the surroundings on the system (criticism of the isolated system concept).

3. *Irreversibility of quantum measurement and macroscopic irreversibility*

This interpretation, however, does not take into account the nature of the macroscopic observation which enables us to ascertain only an objective state of the system without disturbing it (or with a very small perturbation). In the proof of the generalised $\bar{H}$-theorem, the decrease due to Klein's lemma will thus, in general, be negligible compared to the decrease due to the difference between the fine-grained density $\overline{\hat{\varrho}^{(\alpha)}}(t)$ and the coarse-grained density $\hat{P}^{(\alpha)}(t)$; this term could become important only in the case where the observation would lose its macroscopic nature. Moreover, $\overline{\hat{\varrho}^{(\alpha)}}(0)$ is diagonal, because of the initial equality of the fine-grained and coarse-grained densities and this condition is fulfilled also in classical theory. Further, the definition (VII.17) of H, and thus of the entropy of a system, does not involve the conditions of the macroscopic observation, which does not seem to us to be in accordance with Boltzmann's formula, while the definitions (VI.28) and (VI.40), which employ the coarse-grained density matrix

$$P_{ij}^{(\alpha)}(t) = \delta_{ij} \sum_{i=1}^{s_\nu(\alpha)} \frac{(\bar{\varrho}_d)_{ii}}{S_\nu^{(\alpha)}}, \tag{VII.19}$$

agree with Boltzmann's formula. It is necessary also to compare with one another the continuous change in the macroscopic entropy and the discrete change by "jumps" of the quantal entropy.

Finally, we note that if quantitative information about the irreversible evolution of a system is desired, we must resort to assumption (VII.16). However, this cannot be satisfied by a quantum measurement process; this is an assumption of a statistical nature, which originates from the definition adopted for ensembles of systems. (It corresponds to the application at each instant of the fundamental hypothesis of quantum statistical mechanics; this application must be justified, as we have seen that it can be contradictory to the laws of wave mechanics.) Hence, the analogy between the effects of macroscopic observations and of quantum measurements is particularly clear; in fact, contrary to this latter, macroscopic observation has no effect on the system and it does not put it in a "state". It is included in the reasoning only for constructing new statistical ensembles which constitute a mathematical tool and which enable us to establish at every instant a bridge between the macroscopic quantities and the microscopic description of the event; moreover, we have seen in § III.3 of Chapter V that macroscopic observation played exactly the same role in classical theory.

Summarising, we obtain in fact a description of macroscopic evolution by assuming successively that:

1. The quantities $P_\nu^{(\alpha)}(t)$, calculated according to (VII.13) at a time t close to the initial time $t = 0$ starting from $P_\nu^{(\alpha)}(0)$, represent completely the quantities actually observed macroscopically.

2. We must define a new statistical ensemble at $t + 0$, since the quantities $P_\nu^{(\alpha)}(t)$ then correspond to a real macroscopic state of the system; we assume that this ensemble can be determined by starting from the $P_\nu^{(\alpha)}(t)$ by the fundamental postulate of statistical mechanics, *without taking into account prior knowledge of* $P_\nu^{(\alpha)}(0)$. This last point is the cause of the difficulties in the theory of irreversible processes, as we have emphasised in Chapters V and VI and it leads us to try to describe the evolution of a macroscopic system by a Markovian stochastic process.

Here again we find the parallelism between the classical and quantum theories; apart from the differences in detail which we have noted throughout the book, the basic structure of the two theories is the same: the probabilities involved are classical probabilities which obey the classical rules of probability calculus; in the case of the irreversibility of the quantum measurement process (which is one of the fundamental differences between classical and wave mechanics), we have just seen that it does not appear to play an important role in the statistical interpretation of macroscopic irreversibility. This is connected with the remarks at the end of Chapter IV concerning the isolated system concept; we can consider the interaction of a system with its surroundings to be negligible, provided that the system has a large number of degrees of freedom (this is practically always so for systems envisaged in statistical mechanics); in this case, interactions with the surroundings are reduced generally to surface effects. Thus, contrary to the opinion of Landau and Lifshitz and other authors (see last footnote of Chapter IV) who assign an essential role to perturbations created by the external observation, it would appear reasonable to support the opinion expressed by Pauli (1949) in considering that "the difference between classical and quantum mechanics, which is so important in other respects, does not play an essential role in thermodynamic questions. Actually, the perturbation arising from macroscopic observations can be made small and a single macroscopic observation is, in principle, sufficient for proving whether or not the system has reached thermal equilibrium".

In conclusion, we have seen that classical thermodynamics and its reversible transformations find a satisfactory atomistic interpretation, thanks to the stationary ensembles of statistical mechanics—the use of which is justified by ergodic theory—more particularly in the form of the probability ergodic theorems. On the contrary, the study of the irreversible evolution of macroscopic systems encounters greater difficulties and leads us to make assumptions which are liable to contradict the laws of mechanics of macroscopic evolution; however, by not losing sight of the fact that observed macroscopic quantities are related to the definition of coarse-grained densities, we can hope that new methods for the approximate integration of Liouville's equation and a greater call on certain types of stochastic processes will enable us to give an account of macroscopic irreversible phenomena.

Appendix I

1. *Historical review of ergodic theory*

The ergodic theory has its origin in the problems encountered by the kinetic theory of gases. The development of kinetic theory occurred in the second part of the nineteenth century, mainly through the important work of Clausius, Maxwell, and Boltzmann; its basic tasks were to define, from the microscopic point of view, what it was necessary to understand by the equilibrium state of a system and to explain further the irreversible tendency of a non-equilibrium system to evolve towards equilibrium. We know that the method of the kinetic theory of gases leads us to interpret this equilibrium condition (to which the Maxwell–Boltzmann distribution corresponds) as the *most probable* state of the system and that it takes into account the irreversible evolution of a non-equilibrium system, thanks to the proof by Boltzmann in 1872 of the celebrated *H*-theorem. Subsequently, Loschmidt (1876) and Zermelo (1896) having shown that this theorem in its original form encounters insurmountable objections based on the reversible and quasi-periodic nature of the motion of Hamiltonian systems, Boltzmann was then led to highlighting the statistical aspects of the *H*-theorem and to show that this theorem was concerned only with the *most probable* behaviour of a system; this statistical aspect, moreover, had to be analysed in detail and examined thoroughly by P. and T. Ehrenfest in 1911.

However, in parallel with this work on the kinetic theory of gases, Boltzmann tried to open a new field of research by trying to establish that the equilibrium state of a system was equivalent to its average state. By this, he understood that the time-average over an infinitely long interval of time of some phase function must be equal to the value of this phase function at equilibrium. In addition, we note that this definition of the equilibrium state enables us to arrive at the identical conclusions to that of the kinetic theory of gases, by avoiding the paradoxes of the *H*-

theorem; actually, if the time-average of a phase function is equal to its value at equilibrium, it follows that the system will be, for most of the time, in an equilibrium state and that it will rapidly regain this state as soon as it deviates appreciably. It is in view of the calculation of these time-averages that Boltzmann was led as long ago as 1865 to introduce for the first time the hypothesis according to which the trajectory of a system passes through all points of the constant energy hypersurface. It was only in 1887 that the term ergodic theory (from the Greek ἔϱγον: work and ὁδός: path) was used for the first time by Boltzmann; the same theory was expressed by Maxwell (1879) under the name of "the assumption of the continuity of path". We shall now define this theory more precisely:

If $\mathscr{M}$ is the multiplicity of Γ-space (the intersection of the multiplicities corresponding to integrals of motion) over which the representative point of the system moves, it is said that the system is ergodic if the point P_t passes, during a sufficiently long interval of time, through *all* points of $\mathscr{M}$. In the case where the ergodic system only has a single uniform integral of motion, the multiplicity $\mathscr{M}$ is the constant energy hypersurface Σ and it can be shown easily that there is only a single trajectory over Σ, the different motions being distinguished only by the instant t_0 at which the point P_t passes through a point P_0 of the trajectory. It can be deduced from this property that the time the representative point P_t stays in a region $\mathscr{R}$ of Σ is proportional to the measure of $\mathscr{R}$ in Σ and that the time-average (taken over an infinitely large interval of time) of a function $f(P_t)$ is equal to its phase average over Σ.

Thus, the time-average of a phase function is replaced by its phase average over the whole hypersurface Σ, i.e. over a (microcanonical) *ensemble* of systems of the same energy. With this, the fundamental concepts of statistical mechanics were introduced implicitly for the first time, namely, concepts of ensembles of systems and averages over these ensembles, concepts which would be used and developed systematically by Gibbs some years later (1902). However, the progress of the analysis and, more particularly, of the theory of ensembles soon enabled one to show that the ergodic theory in the form given to it by Boltzmann must be abandoned. After the first objections by Lord Kelvin (1891) and Poincaré (1894), Rosenthal and Plancherel actually demonstrated the impossibility of the ergodic theory (1913) by invoking the theory of the measure of sets. The principle of the proof is as follows: since the trajectory of the representative point passes through all points of Σ, it is similar to a Peano curve and, consequently, it is a curve without tangent; this is in contradiction to the existence of the p_i defined by the Hamiltonian equations of motion.

In the face of the difficulties encountered by the Boltzmann–Maxwell ergodic theory and even before its impossibility had been effectively established, P. and T. Ehrenfest (1911) proposed the *quasi-ergodic theory*: this consists in assuming that the trajectories of the representative point P_t are not closed and that they pass as close as desired to every point of Σ; in other words, the trajectory of a quasi-ergodic system is everywhere dense on Σ (an example of such trajectories is provided by Lissajous figures, in the case where the ratio of the periods is not rational). The quasi-ergodic theory, however, was proved to be unsuitable for demonstrating the equality of the time-averages and phase-averages; this is why the many proofs of this equality which start from the quasi-ergodic theory, such as that of Rosenthal (1914), are generally vitiated. We point out, nevertheless, that Fermi demonstrated in 1923 the existence of a class of quasi-ergodic systems (kanonische Normalsysteme) but that he could not deduce the equality of the time- and phase-averages. In order to arrive at a satisfactory definition of the ergodic problem, it was again necessary to revert to exact mathematical definitions of sets which are everywhere dense and of zero-measure sets, and also to the fundamental concept of metric transitivity introduced by Birkhoff and von Neumann.

These latter tasks were destined to lead to the modern statements of the ergodic theorems, with almost certain convergence in Birkhoff's theorem (1931) or with convergence in quadratic mean in the case of von Neumann and Hopf's theorems (1932). The solution of the ergodic problem was thus found to be reduced to proving that the constant energy hypersurfaces Σ are, in the case of macroscopic physical systems, metrically transitive. In fact, there is no criterion which permits us to establish with certainty the property of metric transitivity, with the result that this property still appears as an assumption concerning the nature of the system. It must be emphasised, however, that Birkhoff's theorem opens a new field of research and that it is at the origin of numerous recent publications dealing with the study of the metric transitivity of hypersurfaces amongst which we mention in particular the important work of Oxtoby and Ulam (1941), who established the metric transitivity of a very general class of surfaces, polyhedrons of three and more dimensions, and show in this way that a topological objection to metric transitivity of systems cannot be raised; such surfaces, however, are very far removed from those encountered in physical systems, so that the ergodic problem still remains without a definitive solution. Before studying some aspects of the problem of the likelihood of the theory of metric transitivity, we shall review rapidly the proof of Birkhoff's theorem.

2. *Birkhoff's theorem*

We shall give here the general principles of the proof; for the details we can turn to either Birkhoff's original paper or Khinchin's account (for a more complete mathematical account see especially Blanc-Lapierre, Casal and Tortrat, 1959; Riesz, 1945). Birkhoff's theorem proves the existence of the limit

$$f^*(P_0) = \lim_{T\to\infty} \frac{1}{T} \int_{t_0}^{t_0+T} f(P_t)\, dt, \tag{1}$$

for almost all trajectories of an invariant ensemble Ω of finite measure and the independence of this limit of the initial instant t_0 (the point P_0 corresponds, obviously, to the position of P_t at t_0).

We start by proving the theorem for the case where the interval T varies by finite increments of duration τ; then we put $T = n\tau$ and we study the limit of the series

$$F_n(P_0) \equiv \frac{1}{n\tau} \int_{t_0}^{t_0+n\tau} f(P_t)\, dt \tag{2}$$

as $n \to \infty$.

Suppose that P_0 be an exceptional phase, i.e. such that $F_n(P_0)$ has no limit. In this case, the upper bound $\overline{F}(P_0)$ and the lower bound $\underline{F}(P_0)$ are different; therefore, there exists a pair of numbers α and β with $\alpha < \beta$, such that

$$\underline{F}(P_0) < \alpha, \quad \overline{F}(P_0) > \beta. \tag{3}$$

Let D be the ensemble of these exceptional phases: we must show that it is of zero measure. We use the property by virtue of which, if D is of positive measure, it is possible to find a pair of numbers (α, β) which satisfy relation (3) for all phases P_0 in a sub-ensemble D^* of positive measure (for the proof, see especially Khinchin, 1949); we show then that the property $\mu(D^*) > 0$ is contradictory to the inequality $\alpha < \beta$, which proves that the ensemble D is necessarily of zero measure.

In order to emphasise this contradiction, we consider a series of times $t_k = t_0 + k\tau$ and the series of corresponding phases $P_k = P_{t_k}$ and we put

$$f_k(P_0) = \frac{1}{\tau} \int_{t_k}^{t_{k+1}} f(P_t)\, dt. \tag{4}$$

In addition, by changing the time origin, we have

$$f_k(P_0) = f_0(P_k). \tag{5}$$

With these definitions the time-average $F_n(P_0)$ can be expressed as

$$F_n(P_0) = \frac{1}{n} \sum_{k=0}^{n-1} f_k(P_0). \tag{6}$$

We consider now a sub-ensemble $D_0^{(n)}$ of D^* which we shall define more exactly below. During motion, the phases P_0 of $D_0^{(n)}$ are transformed into P_k (after a time $k\tau$), and the ensemble $D_0^{(n)}$ is transformed into $D_k^{(n)}$. By integrating (6) over the set $D_0^{(n)}$, we have

$$n \int_{D_0^{(n)}} F_n(P_0)\, d\mu = \sum_{k=0}^{n-1} \int_{D_0^{(n)}} f_k(P_0)\, d\mu = \sum_{k=0}^{n-1} \int_{D_k^{(n)}} f_0(P)\, d\mu. \tag{7}$$

We now choose $D_0^{(n)}$ such that for every phase P_0 of $D_0^{(n)}$ we have

$$F_n(P_0) > \beta; \tag{8}$$

in this case, we derive from (7) the inequality

$$\sum_{k=0}^{n-1} \int_{D_k^{(n)}} f_0(P)\, d\mu > n\beta\mu\,(D_0^{(n)}). \tag{9}$$

Let us suppose, further, that the ensembles $D_k^{(n)}$ are disjoint and denote by $D^{(n)}$ their sum

$$D^{(n)} \equiv \sum_{k=0}^{n-1} D_k^{(n)}. \tag{10}$$

Since, according to Liouville's theorem, $\mu(D_k^{(n)}) = \mu(D_0^{(n)})$, inequality (9) can be rewritten as

$$\int_{D^{(n)}} f_0(P)\, d\mu > \beta\mu\,(D^{(n)}). \tag{11}$$

However, one can show that from such sums $D^{(n)}$ disjoint sets can be chosen for each value of n in such a way that they exhaust D^*. By summing inequalities such as (11) over n, we obtain

$$\int_{D^*} f_0(P)\, d\mu > \beta\mu\,(D^*), \tag{12}$$

as $n \to +\infty$.

Adopting the same argument with another definition of the ensembles $D_0^{(n)}$,we obtain

$$\int_{D^*} f_0(P)\, d\mu < \alpha\mu\,(D^*). \tag{13}$$

It can be seen that the two inequalities (12) and (13) contradict the inequality $\alpha < \beta$, if $\mu(D^*) > 0$, so that the ensemble D of the exceptional phases is of zero measure.

In order to complete the proof of Birkhoff's theorem, we must still compare the average for an arbitrary interval of time T with that taken over the interval $n\tau$ closest to T. We have successively

$$\lim_{n\to\infty} \left| \frac{1}{T} \int_{t_0}^{t_0+T} f(P_t)\, dt - \frac{1}{n\tau} \int_{t_0}^{t_0+T} f(P_t)\, dt \right| = 0 \tag{14}$$

and

$$\left| \frac{1}{n\tau} \int_{t_0}^{t_0+T} f(P_t)\, dt - \frac{1}{n\tau} \int_{t_0}^{t_0+n\tau} f(P_t)\, dt \right|$$

$$= \frac{1}{n\tau} \left| \int_{t_0+n\tau}^{t_0+T} f(P_t)\, dt \right| \leqq \frac{1}{n} |f_n(P_0)|. \tag{15}$$

We can verify easily that

$$\lim_{n\to\infty} \frac{1}{n} |f_n(P_0)| = 0 \tag{16}$$

almost everywhere. It is sufficient to show, using (5) and Liouville's theorem, that the set of phases P_0 for which $|f_n(P_0)| > n\varepsilon$ is of zero measure. We deduce then from (14), (15) and (16) the existence of a limit for the average $\frac{1}{T}\int_{t_0}^{t_0+T} f(P_t)\,dt$, except over a set of zero measure.

To conclude, it remains for us to show that this limit is independent of the initial instant t_0. We have

$$\lim \frac{1}{T}\int_{t_0}^{t_0+T} = \lim \frac{1}{T + t_1 - t_0}\int_{t_0}^{t_1+T} = \lim \frac{1}{T}\int_{t_0}^{t_1+T} \tag{17}$$

since the difference between the two latter expressions approaches zero with $(t_1 - t_0)/T$. Finally, we see immediately that the difference

$$\frac{1}{T}\int_{t_0}^{t_1+T} - \frac{1}{T}\int_{t_1}^{t_1+T} = \frac{1}{T}\int_{t_0}^{t_1} \tag{18}$$

approaches zero with $1/T$, which completes our proof. It follows that the limit $f^*(P_0)$ depends only on the trajectory considered and not on the position of P_0 along this trajectory: $f^*(P_0)$ is therefore an integral of motion. Since, further, every integral of motion is equal to its time-average, we can conclude that *the set of time-averages is identical with the set of integrals of motion.*

3. *Notes on the metric transitivity of hypersurfaces*

We recall first of all the definition of metric transitivity: if Ω is an invariant set of finite measure in Γ-space, it will be metrically transitive (and the group of automorphisms which describes the motion will be metrically transitive) if Ω cannot be divided into two invariant sub-sets Ω_1 and Ω_2 of non-zero measure. As we have already pointed out in Chapter I, the property of metric transitivity implies that the limits $f^*(P)$ of the time-averages are constant almost everywhere over Ω (unless we could decompose Ω into two invariant sub-sets of non-zero measure, since the $f^*(P)$ are integrals of motion).

For applications to statistical mechanics, the invariant set Ω is generally the hypersurface Σ (we shall return to this point a little further on) and it can then be proved easily that the time-average of any summable function $f(P)$ is equal for almost all trajectories of Σ to the phase-average of $f(P)$. For this, it is sufficient to use equation (I.24′):

$$\int_{\Sigma} f^*(P) \frac{d\Sigma}{\text{grad } E} = \int_{\Sigma} f(P) \frac{d\Sigma}{\text{grad } E}$$

which can be written as

$$f^*(P) = \frac{1}{\mu(\Sigma)} \int_{\Sigma} f(P) \frac{d\Sigma}{\text{grad } E} \tag{19}$$

since $f^*(P)$ is constant almost everywhere. Thus, metric transitivity is a sufficient condition for ensuring equality of time- and phase-averages.

We show now that it is also necessary. In fact, let us suppose that the equality in question is satisfied for almost all trajectories and that nevertheless the hypersurface Σ can be decomposed into two invariant sub-sets Σ_1 and Σ_2 of non-zero measure; we consider, as a summable function, the characteristic function of the sub-ensemble Σ_1 which takes the value 1 on Σ_1 and 0 on Σ_2: its time-average is 0 or 1 whilst its phase-average is a number between 0 and 1 (which is weighted by the relative measure of the two subsets), which is contrary to our hypothesis and which proves that the property of metric transitivity is necessary to ensure equality of the two types of averages. We point out, in addition, that this argument shows that metric transitivity is a consequence of *the ergodic hypothesis*, according to which all time-averages $f^*(P)$ are constant almost everywhere; in this way we have established the equivalence between the property of metric transitivity and the ergodic hypothesis. Metric transitivity (or the ergodic hypothesis) is thus the essential and sufficient condition for ensuring the validity of the fundamental relation (19); this relation allows us to calculate the time-averages $f^*(P)$ without having to determine the mechanical trajectory of the system.

Having done this, it now remains to examine two important questions:

(a) Why do we generally limit ourselves by considering only the energy as an integral of motion?

(b) Is the ergodic theory, in its modern form, probable, i.e. is it not incompatible with the known properties of Hamiltonian systems? We shall see that the answer to these two questions depends essentially on the study of the physical meaning of the integrals of motion and on an extension of the concept of metric transitivity.

(a) *Controllable integrals.* Our Hamiltonian system with N degrees of freedom has $2N - 1$ independent integrals of motion which allow us to determine the trajectory originating from a given point P_0; there remains a first-order differential equation for determining the motion of the point P_t along this trajectory. Since we can equate the set of time-averages and the set of integrals of motion (as we have shown in the previous section), it can be seen that it is necessary to assume that a large number of integrals of motion have no physical significance; otherwise, the macroscopic state of our system would depend, in fact, on $2N - 1$ independent macroscopic quantities (corresponding to different time-averages), which is obviously contrary to experience. Therefore, we must define precisely what conditions must be fulfilled in order that an integral of motion should have a physical meaning.

This condition can be obtained by requiring that a phase function representing a physical quantity assumes the same well-defined value for all phases corresponding to the same physical state of the system. We note, in fact, that the same physical state is very often represented by more than one point of phase space. This is the case, for example, with angular variables, where we must consider as identical, states for which the values of the variable differ by an integral multiple of 2π. Another important example in physics is that of the difference between specific and generic phases for systems composed of identical particles (we mentioned this difference in Chapter I when defining grand-canonical ensembles): if the physical state of a system is described by the generic phase, all the specific phases which constitute this generic phase represent one and the same state of the system. Thus, a set of points of phase space often corresponds to the same physical state; however, the phase functions which describe the physical quantities must take the same value for all phases which represent the same physical state. The phase functions which possess this property are called *uniform* (or *normal* according to Khinchin): they are periodic with respect to the angular variables and, if the state of the system is defined by its generic phase, they are symmetrical with respect to the groups of variables describing the various particles. All the physical quantities of a system must be represented, therefore, by uniform phase functions. Also, we generally impose the condition that the phase function must be continuous, which allows us to allocate to it neighbouring values for neighbouring microscopic states.

In the majority of cases, these properties are only fulfilled by a limited number, k, of integrals of motion $I_1, \ldots, I_k$, which represent independent physical quantities whose value can be determined by macroscopic observation: these are the controllable integrals of the system. To them there corresponds an invariant set $\mathcal{M}_k$ with $2N - k$ dimensions, on which we can define an invariant measure and develop the ergodic theory. The ergodic hypothesis can then be stated as: all the limits $f^*(P)$ are constant almost everywhere on the set $\mathcal{M}_k$; they are therefore functions of the k controllable integrals I_k which determine the macroscopic state of the system. It is clear that the other (non-uniform) integrals of motion can then take any values since they are devoid of physical meaning: these are Khinchin's free integrals (Khinchin, 1949, pp. 50–51).

If the energy is the only uniform integral of motion, the sub-set $\mathcal{M}_k$ is the same as the hypersurface Σ; this is what we generally assume in statistical mechanics. We can show occasionally that this is so because of the mechanical properties of the system: this would be the case, for example, with an isolated system for which there would be no conservation of momentum and angular momentum because of wall effects; otherwise, we must accept it as an assumption which is justified by its consequences. For such an ergodic system the time-averages $f^*(P)$ are constant almost everywhere on Σ and are, therefore, functions of the energy E.

It is important now to recall that—as we have seen in Chapter I (Section III)—the phase functions having a physical significance usually have a structure such that they are constant over the larger part of Σ, except over a set of very small measure. It follows that for the majority of trajectories over Σ, the time-averages of these functions are close to one another and almost equal to their phase average over Σ. Consequently, it is not necessary to fix the value of these integrals and only those integrals whose value, differing considerably from the phase average, would have been determined by macroscopic observation will be retained as controllable integrals. All other integrals of motion can be considered as free integrals: these include non-uniform integrals devoid of physical meaning and uniform (or normal) integrals whose value has not been fixed by macroscopic observation. These remarks allow us further to reduce the number of controllable integrals and in this way to justify in the majority of cases the special role of the energy integral of motion (which amounts to saying that the macroscopic state of a system which occupies a volume V is determined by a single macroscopic parameter which depends on the total energy of the system). It remains now for us to use the foregoing considerations in order to answer question (b) relating to the likelihood of the ergodic hypothesis.

(b) *Metric transitivity in the physical sense.* We shall see first of all, by elementary arguments, that the concept of metric transitivity must be modified in order to make it likely. We consider, in fact, an integral of motion of a Hamiltonian system which is not the energy and which does not contain the time explicitly. It cannot be constant over the entire hypersurface Σ, otherwise its value would be determined completely by a knowledge of the energy, which is contrary to our hypothesis (and, since the function is continuous, it cannot be constant almost everywhere). It is then always possible to find a value α such that the relations $f(P) \leqq \alpha$ and $f(P) > \alpha$ define two sub-sets of non-zero measure (see Khinchin, 1949, p. 30) which is contrary to our hypothesis [$f(P)$ is an integral of motion and the two sub-sets are invariant]. Thus, we can see that if it were not modified considerably, the hypothesis of metric transitivity of the hypersurface Σ would encounter an impossibility, just like Boltzmann's ergodic hypothesis.

It is the limitation on the nature of the functions describing the physical quantities which will allow us to modify the formulation of the metric transitivity hypothesis, by assuming that only uniform (or normal) sum functions satisfy the ergodic hypothesis. For this purpose, we shall call any subdivision of the constant energy hypersurface into two sets of positive measure a normal one,when all the equivalent physical phases belong to the same set. A hypersurface will be metrically transitive in the physical sense if there does not exist any normal subdivision. With this new definition, the condition which is necessary and sufficient for the time-average of a summable and uniform phase function to be equal to its phase average is that the hypersurface Σ be metrically transitive *in the physical sense.*

Under these conditions, the ergodic problem is reduced to showing that constant energy hypersurfaces possess the property of metric transitivity in the physical sense. This modification of the statement of the problem will not affect our previous argument for the uniform integrals I_k: these define, in fact, a normal subdivision of the hypersurface Σ, since they take the same value for all physically equivalent points. This proves simply, as we have seen already in the preceding section, that in this case it is necessary to use not the hypersurface Σ but the sets $\mathscr{M}_k$. This is not the case for non-uniform integrals, since we cannot arbitrarily divide the hypersurface into two sets corresponding to the separation of all values of f into two parts, since this subdivision would not be normal. In order to have a normal subdivision, it would be necessary that the values taken by $f(P)$ at physically equivalent points belong to the same section of $f(P)$ values. As we can see from examples, this condition may be incompatible with the existence of two invariant sub-sets of positive measure. The foregoing argument is therefore no longer valid for non-uniform integrals and transitivity in the physical sense then becomes possible. However, since the Hamiltonian must satisfy certain conditions which cannot be stated in general terms, it follows that metric transitivity in the physical sense remains for the present a hypothesis concerning the nature of the system.

Be that as it may, the foregoing remarks show that it is necessary to take account of all uniform integrals whose value must be fixed (controllable integrals): these define a set $\mathscr{M}_k$ which we can then assume to be metrically transitive in the physical sense; this set reduces to the hypersurface Σ if the only uniform integral is the energy of the system. Nevertheless, in practice, all the uniform integrals different from the energy are constant over a very large part of Σ: the regions over which they differ significantly from this constant value are of very small measure and can be neglected, and the phase averages can be taken over the whole hypersurface Σ, which is the same as assuming that the only controllable integral is the energy of the system.

In the language of probability calculus we can say also that the localisation of a representative point of the system on the set Σ (or $\mathscr{M}_k$) is a random event, so that there is a very small probability of finding this point in an ensemble of very small measure. The introduction of this point of view is also fully justified if we note that there is a very large number of possible trajectories with $2N - k - 1$ parameters on the set $\mathscr{M}_k$ associated with k controllable integrals, which cannot be distinguished macroscopically since they give the same value $f^*(P)$ to a macroscopic observable. If an experiment is carried out on the system (for example, the recording of a physical quantitity), one of these trajectories is chosen but this choice does not affect the value of $f^*(P)$: this is the phenomenon of *macroscopic reproducibility*. The choice between all the possible trajectories on $\mathscr{M}_k$ can thus be likened to a test in the sense of probability calculus: consequently, it is legitimate to consider $P(t_0)$ as a random point (with t_0 fixed) and $P(t)$ as a random vector function of t, the probability on $\mathscr{M}_k$ being defined by an invariant measure.

In conclusion, we shall illustrate the foregoing discussions by an example. We consider a system with two degrees of freedom described by two uncoupled rotations round fixed axes; let φ_1 and φ_2 be the corresponding azimuths and suppose for simplicity that the two moments of inertia are equal to unity. If p_1 and p_2 are the canonically conjugate moments of the variables φ_1 and φ_2, the Hamiltonian of the system can be written as

$$H = \frac{1}{2}(p_1^2 + p_2^2) \tag{20}$$

and the Hamiltonian equations of motion give us

$$\frac{d\varphi_1}{dt} = p_1, \quad \frac{d\varphi_2}{dt} = p_2, \quad \frac{dp_1}{dt} = 0, \quad \frac{dp_2}{dt} = 0. \tag{21}$$

It can be seen from (21) that the conjugate moments p_1 and p_2 are uniform integrals of motion and we shall assign to them the values ω_1 and ω_2, which give the value

$$H = \frac{1}{2}(\omega_1^2 + \omega_2^2) \tag{22}$$

to the energy of the system.

The set containing the trajectory is thus reduced to the (φ_1, φ_2) plane and the trajectory in this plane is a straight line defined by the equations

$$\varphi_1 = \omega_1 t + \varphi_1^0, \quad \varphi_2 = \omega_2 t + \varphi_2^0. \tag{23}$$

This plane is divided by the two sets of lines $\varphi_1 = 2m\pi$ and $\varphi_2 = 2n\pi$ (m and n integral) into a system of squares which contain all phases physically equivalent to one another. There is an *a priori* difficulty for using the ergodic theory, since the set on the (φ_1, φ_2) plane is infinite: in fact, we can restrict our considerations to a single square of the system. For this purpose we define the trajectory, which in this square is equivalent to the trajectory (23), by transferring into the square considered all the segments of the trajectory contained in the other squares; the equivalent trajectory obtained in this way is an ensemble of parallel segments; this ensemble is finite or infinite according as the ratio ω_1/ω_2 is rational or irrational.

We note now that we have a fourth time-independent integral given by

$$M = \varphi_2 p_1 - \varphi_1 p_2. \tag{24}$$

It is easy to verify that it is not uniform; we have, in fact, for the initial phases φ_1^0 and φ_2^0,

$$M_0 = \varphi_2^0 \omega_1 - \varphi_1^0 \omega_2, \tag{25}$$

and for the phase $\varphi_1^0 + 2m\pi$, $\varphi_2^0 + 2n\pi$ (which is physically equivalent to φ_1^0, φ_2^0),

$$M = M_0 + 2\pi(n\omega_1 - m\omega_2), \tag{26}$$

which proves the non-uniformity of M.

Finally, starting from the preceding remarks, we can show (Khinchin, 1949, pp. 60–61) that each square is metrically transitive in the physical sense, provided that the ratio ω_1/ω_2 is irrational. Since all squares are physically equivalent, this result is valid for the whole (φ_1, φ_2) plane: thus, we have an example of an infinite set, metrically transitive in the physical sense, and defined by assigning specified values to the uniform integrals; in addition, we note that general examples can be constructed of the same type as the example above (Hopf, 1937).

4. *Structure functions in classical statistical mechanics*

(a) *Definitions.* We have just seen the role played in classical mechanics by constant energy hypersurfaces corresponding to the integral of motion $H = E(q_1, \ldots, q_N; p_1, \ldots, p_N)$: we shall now briefly discuss the properties of these hypersurfaces and on their use in the actual calculation of phase averages.

First of all, it is generally assumed that physical systems considered in the applications are such that the region of Γ-space defined by the inequality $E < x$ (with $x > 0$) is a simply-connected domain, bounded by a closed hypersurface Σ_x (corresponding to $H = x$) which is sufficiently regular to permit the use of analytical methods. Consequently, the surface Σ_{x_1}, which corresponds to $x_1 < x_2$, is located entirely inside Σ_{x_2} and the ensemble of hypersurfaces thus appears as a family of surfaces all contained one within the other.

Bearing this in mind, the first step is to define an invariant measure on the hypersurfaces Σ. In fact, if we have defined the probability for $P_0 \in A_0$ at t_0, where A_0 is some set on Σ, this is naturally equal to the probability of finding $T_t P_0 \in A_t$, where A_t follows from A_0 by the transformation T_t; it follows that the probability with respect to Σ must be an invariant measure under T_t. In order to construct such a measure we depend on the following theorem:†

Let $\mathscr{V}_x$ be the region of Γ-space corresponding to the inequality $E < x$, and let $\mathscr{V}(x)$ be its volume; it is a monotonic function which increases from 0 to infinity with x. If $f(P)$ is a summable function over $\mathscr{V}_x$, we have the relation

$$\frac{d}{dx} \int_{\mathscr{V}_x} f(P)\, d\Gamma = \int_{\Sigma_x} f(P) \frac{d\Sigma}{\operatorname{grad} E}, \tag{27}$$

with

$$\operatorname{grad} E = \left\{ \sum_{i=1}^{N} \left[\left(\frac{\partial E}{\partial q_i} \right)^2 + \left(\frac{\partial E}{\partial p_i} \right)^2 \right] \right\}^{\frac{1}{2}}. \tag{28}$$

If we consider a set A on Σ_x, it will be transformed by the motion T_t into another set on Σ_x, because of the invariance of Σ_x; if, however, we were to define its measure by $\mu(A) = \int_A d\Sigma$, this

† For the proof, see Khinchin, 1949, p. 34.

would not be generally conserved. In this case we consider at each point of A the normal to Σ_x and its intersection with $\Sigma_{x+\Delta x}$; the ensemble of these segments constitutes an invariant region $\mathcal{N}$ in Γ-space whose volume $\mu(\mathcal{N})$ is also invariant and defined by

$$\mu(\mathcal{N}) = \int\limits_{x \leqq E < x+\Delta x} \varphi(P)\, d\Gamma, \tag{29}$$

where $\varphi(P)$ is the characteristic function of the set $\mathcal{N}$. The ratio of this volume to Δx and the limit of this ratio are likewise invariant; according to the previous theorem, this limit can be written as

$$\int\limits_{\Sigma_x} \varphi(P) \frac{d\Sigma}{\operatorname{grad} E} = \int\limits_{A} \frac{d\Sigma}{\operatorname{grad} E}. \tag{30}$$

Thus, $d\Sigma/\operatorname{grad} E$ is an invariant definition of the measure on Σ_x. By denoting the measure of the whole hypersurface Σ_x by $\Omega(x)$ we have according to (27)

$$\Omega(x) = \int\limits_{\Sigma_x} \frac{d\Sigma}{\operatorname{grad} E} = \mathcal{V}'(x), \tag{31}$$

and the phase average of a summable function on Σ_x takes the form

$$\bar{f} = \frac{1}{\Omega(x)} \int\limits_{\Sigma_x} f(P) \frac{d\Sigma}{\operatorname{grad} E} = \frac{1}{\Omega(x)} \frac{d}{dx} \int\limits_{\mathcal{V}_x} f(P)\, d\Gamma. \tag{32}$$

The function $\Omega(x)$ is a monotonic function which increases from 0 to infinity with x. Since it completely determines the fundamental properties of a mechanical system, it is called the *structure function* of the system being considered. We note that if the value of a phase function is completely defined by the energy of the system at the corresponding point of Γ-space, the integral of this function over the region $\Delta\mathcal{V}$ bounded by the two hypersurfaces Σ_{x_1} and Σ_{x_2} can be written simply as:

$$\int\limits_{\Delta\mathcal{V}} f(P)\, d\Gamma = \int\limits_{x_1}^{x_2} f(x)\, \Omega(x)\, dx. \tag{33}$$

(b) *Systems with weakly coupled components. The law for combining structure functions.* We shall assume now that the energy E of our system can be represented by the sum of two terms E_1 and E_2, where E_1 depends on one set of dynamic coordinates and E_2 on the other coordinates; in this case, we can say that the system is divided into two *components* which are energetically uncoupled. We note in passing that this definition, which is useful from the practical point of view, leads to a paradoxical situation from the point of view of principles, since these require that there should be a continuous exchange of energy between the various components of a given system (for example, between the constituent particles): in particular, a system which can be split into two parts, in the sense of the foregoing definition, cannot be ergodic since the energy of each component remains constant during motion. This paradox can be avoided by assuming that the components of the system are only approximately isolated and, consequently, *weakly coupled* but with an energy of interaction which is negligible compared with E_1 and E_2, except on a set of very small measure. (For the mathematical implications of the definition of systems with weakly coupled components, see especially Blanc-Lapierre, Casal and Tortrat, 1959.) On the one hand, this interaction ensures the ergodicity of the system and, on the other hand, it is sufficiently small that its contribution can be neglected in calculating phase averages.

Having made these reservations, let $\mathscr{V}(x)$ and $\Omega(x)$ be the functions defined previously for the total system and let $\mathscr{V}_1(x)$, $\Omega_1(x)$, $\mathscr{V}_2(x)$ and $\Omega_2(x)$ be the corresponding functions for the two components considered. First of all, we have

$$\mathscr{V}(x) = \int\limits_{\mathscr{V}_x} d\Gamma = \int\limits_{(\mathscr{V}_x)_1} d\Gamma_1 \int\limits_{(\mathscr{V}_{x-E_1})_2} d\Gamma_2 = \int\limits_{(\mathscr{V}_x)_1} \mathscr{V}_2(x - E_1)\, d\Gamma_1, \tag{34}$$

where $(\mathscr{V}_x)_1$ denotes the set of points of Γ_1 for which $E_1 < x$ and where $(\mathscr{V}_{x-E_1})_2$ is defined in the same way. Since the phase function $\mathscr{V}_2(x - E_1)$ depends only on the energy E_1 of the first component, we have, according to (33),

$$\mathscr{V}(x) = \int\limits_0^x \mathscr{V}_2(x - E_1)\, \Omega_1(E_1)\, dE_1 = \int\limits_0^\infty \mathscr{V}_2(x - y)\, \Omega_1(y)\, dy, \tag{35}$$

since $\mathscr{V}_2(x - E_1) = 0$ for $E_1 > x$. Differentiating with respect to x, we have finally

$$\Omega(x) = \int_0^\infty \Omega_1(y)\, \Omega_2(x - y)\, dy; \tag{36}$$

This is the law for combining structure functions. It can immediately be extended to the case of a system consisting of N components; in this case we have

$$\Omega(x) = \int \left\{ \prod_{i=1}^{N-1} \Omega_i(x_i)\, dx_i \right\} \Omega_n \left(x - \sum_{i=1}^{N-1} x_i \right), \tag{37}$$

where the integration is extended over the entire $(N - 1)$-dimensional space. These formulae are important for calculating average values for a system of weakly coupled components; we shall conclude this brief review by pointing out the most important consequences of the foregoing formulae.

(c) *Distribution law for one component.* If we are dealing with a two-component system and if A_1 is some set in the phase-space of component 1, the probability that the point P_1 (representing the state of component 1) belongs to A_1 is given by:

$$\text{Prob}\,(P_1 \in A_1) = \frac{1}{\Omega(a)} \int_{A_1} \Omega_2(a - E_1)\, d\Gamma_1, \tag{38}$$

where Ω_2 is the structure function of component 2 and where a is the total energy of the system. The distribution law for component 1 in Γ_1 is thus given by

$$\frac{\Omega_2(a - E_1)}{\Omega(a)} \tag{39}$$

and the average value of a function $f(1)$, depending only on the coordinates of component 1, can be written as

$$\overline{f(1)} = \frac{1}{\Omega(a)} \int_{\Gamma_1} f(1)\, \Omega_2(a - E_1)\, d\Gamma_1 . \tag{40}$$

In particular, for the average energy $\overline{E}_1$, we have

$$\bar{E}_1 = \frac{1}{\Omega(a)} \int\limits_{\Gamma_1} E_1 \Omega_2(a - E_1)\, d\Gamma_1 = \frac{1}{\Omega(a)} \int\limits_0^\infty x\Omega_1(x)\, \Omega_2(a - x)\, dx, \tag{41}$$

which shows that the energy distribution law for component 1 is given by the density $\Omega_1(x)\,\Omega_2(a-x)/\Omega(a)$. If formula (39) is applied to the case of a monatomic gas composed of N identical atoms enclosed in a volume V and if one chooses one atom as a component, we can find Boltzmann's law in the limit as $N \to \infty$; in fact, we have in this case

$$\Omega_{(N)}(x) = V^N \frac{(2\pi m)^{3N/2} \dfrac{3N}{2}}{\Gamma\left(\dfrac{3N}{2} + 1\right)} x^{3N/2-1} \tag{42}$$

and the distribution law in the 3-dimensional physical space of an atom with energy x is then given by $\Omega_{(N-1)}(a-x)/\Omega_{(N)}(a)$, where a is the total energy of the gas.† Putting $a = \frac{3}{2} NkT$ and using Stirling's asymptotic formula for the factorials, then going to the limit as $N \to \infty$, we have

$$\frac{\Omega_{(N-1)}(a-x)}{\Omega_{(N)}(a)} = \frac{1}{V} \frac{1}{(2\pi mkT)^{3/2}} \left(1 - \frac{x}{\frac{3}{2} NkT}\right)^{3N/2} \times$$

$$\times \left(1 - \frac{x}{\dfrac{3}{2} NkT}\right)^{-5/2} \left(\frac{N-1}{N}\right)^{3/2} \simeq \frac{1}{V} \frac{1}{(2\pi mkT)^{3/2}} e^{-x/kT} \tag{43}$$

This is Boltzmann's formula; the energy distribution is obtained by multiplying (43) by $\Omega_1(x) = 2\pi V(2m)^{3/2} x^{1/2}$. These considerations show that the Maxwell–Boltzmann distribution is linked, in fact, with the asymptotic geometrical properties of the hypersurfaces of n-dimensional space. These properties, and especially those of the unit hypersphere (cf. Appendix II), form the basis of the probability

† We use the notation $\Omega_{(N)}$ here as a reminder that the system is composed of N particles; it must not be confused with the notation $\Omega_1, \Omega_2, \ldots$, which refers to the components of the system.

ergodic theory and of the proof of the H-theorem (cf. Chapter I, section III; Chapter IV, section IV; and Chapter V, section III).

(d) *Application of the central limit theorem.* With a view to applying the previous results to the calculation of the average values of sum functions and to deriving the canonical distribution, it is convenient to use the formal analogy between (37) and the law of compounding of probabilities and to try to use the *central limit theorem* of the calculus of probabilities. However, the measures $\Omega_i(x)$ cannot be identified directly with probability distributions, since these measures—whilst they are finite—are not generally bounded and increase like a positive power of the energy [see, for example, (42)]; we are thus led to associating with $\Omega_i(x)$ the conjugate distributions $u_i^{(\alpha)}(x)$, defined by

$$u_i^{(\alpha)}(x) = \frac{1}{\Phi_i(\alpha)} e^{-\alpha x} \Omega_i(x) \quad (\alpha > 0), \tag{44}$$

where the *generating function* $\Phi_i(\alpha)$ satisfies the relation:

$$\Phi_1(\alpha) = \int e^{-\alpha x} \Omega_i(x)\, dx; \tag{45}$$

this is the Laplace transform of the structure function $\Omega_i(x)$.† Thus, the normalisation of the distributions (44), which satisfy a law of compounding similar to (37), is assured. In this case we can apply the central limit theorem of the calculus of probabilities and from it deduce numerous applications such as the distribution law for a component of a system, the calculation of the average value and of the dispersion of the sum functions (results that we have used in Chapter I, Section III) and the derivation of the canonical distribution, as we have seen in Chapter IV (Section V); in particular, we obtain directly the value of the average occupation numbers $\bar{n}_i$.

In conclusion, we mention certain results (For the proofs, see Khinchin, 1949; Blanc-Lapierre, Tortrat and Casal, 1959.) For a system containing N components, the central limit theorem leads

† It can also be identified with the partition function of statistical mechanics, starting from which all the thermodynamic properties of a system can be derived.

to the following asymptotic expression (as $N \to \infty$):

$$U^{(\alpha)}(x) \simeq \frac{1}{(2\pi B)^{\frac{1}{2}}} e^{-(x-A)^2/2B}, \Omega_{(N)}(x) = \Phi(\alpha)\, e^{\alpha x} U^{(\alpha)}(x), \quad (46)$$

for the conjugate distribution $U^{(\alpha)}(x)$ associated with the total structure function $\Omega_{(N)}(x)$ with

$$A = \sum_{k=1}^{N} a_k, \quad B = \sum_{k=1}^{N} b_k, \quad \Phi(\alpha) = \prod_{k=1}^{N} \Phi_k(\alpha);$$

a_k and b_k are the mathematical expectation and the dispersion of each of the components of the system, and a value $\alpha = \theta$ must be chosen for the parameter α, such that

$$A = -\left(\frac{d \ln \Phi(\alpha)}{d\alpha}\right)_{\alpha=\theta}, \quad B = \left(\frac{d^2 \ln \Phi(\alpha)}{d\alpha^2}\right)_{\alpha=\theta}. \quad (47)$$

In particular, we find for the law of distribution for a component, according to (39) and (46)

$$\frac{\Omega_2(a - E_1)}{\Omega(a)} \simeq \frac{e^{-\theta E_i}}{\Phi_1(\theta)} \quad (48)$$

and for the average value of a sum function $\left[f(P) = \sum_{i=1}^{N} f_i(P_i)\right]$

$$\bar{f} = \sum_{i=1}^{n} \bar{f}_i \simeq \sum_{i=1}^{N} \int_{\Gamma_i} f_i(P_i) \frac{e^{-\theta E_i}}{\Phi_i(\theta)} d\Gamma_i. \quad (49)$$

We refer the reader to Chapter VIII of Khinchin's book (1949) for other applications and, especially, for the calculation of the dispersion of sum functions. Finally, we point out that the foregoing methods can be applied similarly to quantum statistical mechanics, by means of certain adaptations which are necessary for taking account of the indistinguishability of particles and of the exclusion principle [see, for example, Chapter IV (Section V) and also Chapter V of Blanc-Lapierre, Tortrat and Casal, 1959].

APPENDIX II

Probability Laws in Real *n*-Dimensional Euclidean Space

We shall study in this Appendix the definition of the laws of probability which are uniform on the unit hyperspheres of real Euclidean space with n and $2n$ dimensions, and then we shall calculate the average values $\overline{r_i^4}^m$ and $\overline{r_i^2 r_j^2}^m$ used in Chapter IV. Next, we shall deal with problems related to the subdivision of Hılbert space into cells and to the calculation of the average values

$$\langle |C_{ki}^{(\alpha)}|^4 \rangle, \quad \langle |C_{ki}^{(\alpha)}|^2 \, |C_{kj}^{(\alpha)}|^2 \rangle, \ldots$$

where the symbol $\langle\rangle$ denotes the averages taken over the collection of all possible subdivisions; we shall deduce from it the laws of probability for the quantities $\mu_{ij}^{(\alpha)}$ and $\mu_{ii}^{(\alpha)}$ encountered in Chapter III.

1. *The unit hypersphere in n-dimensional space*

This hypersphere is defined by the equation

$$x_1^2 + x_2^2 + \cdots + x_n^2 = 1 \tag{1}$$

and we shall try to define a uniform probability density on this hypersphere. For this purpose, we take parametric coordinates over the hypersphere (1) by putting

$$\left.\begin{aligned}
x_1 &= \cos\theta_1, \\
x_2 &= \sin\theta_1 \cos\theta_2, \\
x_3 &= \sin\theta_1 \sin\theta_2 \cos\theta_3, \\
&\ldots\ldots\ldots\ldots\ldots\ldots\ldots\ldots, \\
x_i &= \sin\theta_1 \ldots \sin\theta_{i-1} \cos\theta_i, \\
&\ldots\ldots\ldots\ldots\ldots\ldots\ldots\ldots\ldots, \\
x_n &= \sin\theta_1 \ldots \sin\theta_{n-1},
\end{aligned}\right\} \tag{2}$$

whence we derive

$$\frac{D(x_1, x_2, \ldots, x_{n-1})}{D(\theta_1, \theta_2, \ldots, \theta_{n-1})} = (-1)^{n-1} \sin^{n-1}\theta_1 \sin^{n-2}\theta_2 \ldots \sin\theta_{n-1}. \tag{3}$$

The surface element on the hypersphere (1) can then be written as

$$d\sigma_n = \frac{1}{x_n} dx_1\, dx_2 \ldots dx_{n-1} = \sin^{n-2}\theta_1 \sin^{n-3}\theta_2 \ldots \sin\theta_{n-2}$$
$$d\theta_1\, d\theta_2 \ldots d\theta_{n-1}. \tag{4}$$

In order to cover the whole hypersphere, the parameters $\theta_1, \theta_2, \ldots, \theta_{n-2}$ vary from 0 to π and θ_{n-1} from 0 to 2π; if we denote the total area of (1) by σ_n, we have

$$\sigma_n = \int_0^\pi \sin^{n-2}\theta_1\, d\theta_1 \int_0^\pi \sin^{n-3}\theta_2\, d\theta_2 \ldots \int_0^\pi \sin\theta_{n-2}\, d\theta_{n-2} \int_0^{2\pi} d\theta_{n-1}$$
$$= 2\pi \prod_{k=1}^{n-2} \left(\int_0^\pi \sin^k\theta\, d\theta \right). \tag{5}$$

We know that the integrals $J_k = \int_0^\pi \sin^k\theta\, d\theta$ take different values according as k is even or odd. We have (see, for example, Borel, 1914)

$$\left.\begin{aligned} J_{2p} &= \pi \frac{1.3.5 \ldots (2p-1)}{2.4.6 \ldots 2p} \quad \text{for } k = 2p, \\ J_{2p+1} &= 2 \frac{2.4.6 \ldots 2p}{1.3.5 \ldots (2p+1)} \quad \text{for } k = 2p+1. \end{aligned}\right\} \tag{6}$$

Thus, we find for the area σ_n:

$$\sigma_n = \begin{cases} \dfrac{(2\pi)^{\frac{n}{2}}}{2.4.6 \ldots (n-2)} & \text{if } n = 2p, \\ \dfrac{2(2\pi)^{\frac{n-1}{2}}}{1.3.5 \ldots (n-2)} & \text{if } n = 2p+1, \end{cases} \tag{7}$$

and we take for the uniform probability density over the hypersphere (1) the following expression

$$dP_n = \frac{d\sigma_n}{\sigma_n} = \frac{1}{\sigma_n}\sin^{n-2}\theta_1 \sin^{n-3}\theta_2 \ldots \sin\theta_{n-2}\, d\theta_1 \ldots d\theta_{n-1}. \tag{8}$$

We can easily use it to derive the probability that x_1 be contained between x_1 and $x_1 + dx_1$; it is sufficient to calculate the probability $P(\theta_1)\, d\theta_1$ such that θ_1 be contained between θ_1 and $\theta_1 + d\theta_1$. According to (8) it becomes

$$P(\theta_1)\, d\theta_1 = \frac{1}{\sigma_n}\sin^{n-2}\theta_1\, d\theta_1 \int_0^\pi \sin^{n-3}\theta_2\, d\theta_2 \ldots$$
$$\int_0^\pi \sin\theta_{n-2}\, d\theta_{n-2} \int_0^{2\pi} d\theta_{n-1} = \frac{1}{J_{n-2}}\sin^{n-2}\theta_1 d\theta_1. \tag{9}$$

Since $x_1 = \cos\theta_1$, $dx_1 = -\sin\theta_1\, d\theta_1$ and $\sin^2\theta_1 = 1 - x_1^2$, equation (9) can be rewritten to become

$$P(x_1)\, dx_1 = \frac{1}{J_{n-2}}(1 - x_1^2)^{\frac{n-3}{2}}\, dx_1. \tag{10}$$

An interesting application of the foregoing results consists in showing that the law of probability for the component u of the velocity of a molecule in a gas tends to the Maxwell–Boltzmann distribution (see Borel, 1925, pp. 46ff.). It is sufficient to consider a monatomic gas consisting of N atoms of mass m, whose interactions can be neglected; if we denote by u_i the velocity components of the molecules, the conservation of the total kinetic energy of the gas can be written as

$$\tfrac{1}{2} m \sum_{i=1}^{n} u_i^2 = E, \quad \text{with} \quad n = 3N, \tag{11}$$

which defines a hypersphere of radius $(2E/m)^{1/2}$ (in our problem this is the constant energy hypersurface). If we put $E = \frac{n\theta}{2}$ (where $\theta/2 = kT/2$ is the average kinetic energy of a gas molecule), we have, by replacing x_1^2 in (10) by $mu_1^2/n\theta$,

$$P(u_1)\, du_1 = \frac{1}{J_{n-2}}\left(1 - \frac{mu_1^2}{n\theta}\right)^{\frac{n}{2}}\left(1 - \frac{mu_1^2}{n\theta}\right)^{-\frac{3}{2}}\sqrt{\frac{m}{n\theta}}\, du_1. \tag{12}$$

Since n is very large we take the limit of (12) as $n \to \infty$; the asymptotic form of J_{n-2} is

$$J_{n-2} \sim \sqrt{\frac{4\pi}{2n}} = \sqrt{\frac{2\pi}{n}},$$

and we have finally

$$P(u_1)\, du_1 = \sqrt{\frac{m}{2\pi\theta}}\, e^{-\frac{mu_1{}^2}{2\theta}}\, du_1. \tag{13}$$

This is the Maxwell–Boltzmann law for a velocity component of a molecule; the calculation that we have just carried out is, in fact, a special case of the general method sketched in Appendix I (§ 4) and shows that the law (13) depends, as we have already mentioned, on the asymptotic geometrical properties of the unit hypersphere in n-dimensional space.

2. *The unit hypersphere in 2n-dimensional space*

Obviously, the results of the preceding section are applicable to this case, since they are valid whatever be the number of dimensions of the space. However, it will be more convenient for us to write the equation for the hypersphere considered in the form

$$x_1^2 + y_1^2 + x_2^2 + y_2^2 + \cdots + x_n^2 + y_n^2 = 1 = \sum_{i=1}^{n} r_i^2 \quad (r_i^2 = x_i^2 + y_i^2), \tag{14}$$

as we have in mind to calculate the average values of expressions such as r_i^4 and $r_i^2 r_j^2$ encountered in Chapter IV. We shall carry out the change of variables

$$x_i = r_i \cos\alpha_i, \quad y_i = r_i \sin\alpha_i, \quad (i = 1, 2, \ldots, n), \tag{15}$$

with

$$r_n^2 = 1 - \sum_{i=1}^{n-1} r_i^2;$$

its advantage is to introduce explicitly the angles α_i, which are the arguments of the complex coefficients $x_i + iy_i = r_i\, e^{i\alpha_i}$ of the expansion of the wave function in terms of eigenfunctions. Since the surface element on the hypersphere (14) can be written as

$$d\sigma_{2n} = \frac{dx_1\, dy_1 \ldots dx_{n-1}\, dy_{n-1}\, dx_n}{y_n}, \tag{16}$$

we must calculate the functional determinant

$$\frac{D(x_1 y_1, \ldots, x_{n-1} y_{n-1}, x_n)}{D(r_1, \alpha_1, r_2, \alpha_2, \ldots, r_{n-1}, \alpha_{n-1}, \alpha_n)};$$

we obtain easily

$$\frac{D(x_1 y_1, \ldots, x_{n-1} y_{n-1}, x_n)}{D(r_1 \alpha_1 \ldots r_{n-1} \alpha_{n-1}, \alpha_n)} = -r_1 \ldots r_{n-1} r_n \sin \alpha_n, \tag{17}$$

whence, for the surface element $d\sigma_{2n}$,

$$\begin{aligned} d\sigma_{2n} &= \frac{D(x_1 y_1, \ldots, x_{n-1} y_{n-1}, x_n)}{D(r_1 \alpha_1 \ldots r_{n-1} \alpha_{n-1}, \alpha_n)} \frac{1}{y_n} dr_1 \ldots dr_{n-1}\, d\alpha_1 \ldots d\alpha_n \\ &= \prod_{i=1}^{n-1} r_i\, dr_i \prod_{i=1}^{n} d\alpha_i . \end{aligned} \tag{18}$$

In order to calculate the area σ_{2n} of the hypersphere (14), we must integrate over all possible values of the α_i and r_i; the angular variables α_i are independent and vary from 0 to 2π, whilst the r_i traverse the region defined by the inequality

$$r_1^2 + \ldots + r_{n-1}^2 \leqq 1, \quad \text{with} \quad r_n^2 = 1 - (r_1^2 + \ldots + r_{n-1}^2).$$

It is therefore completely in order to effect once again the change of variables (2) by replacing the x_i by the r_i. According to the results of the previous section, we have

$$\begin{aligned} dr_1 \ldots dr_{n-1} &= \frac{D(r_1 \ldots r_{n-1})}{D(\theta_1 \ldots \theta_{n-1})} d\theta_1 \ldots d\theta_{n-1} \\ &= (-1)^{n-1} \sin^{n-1} \theta_1 \sin^{n-2} \theta_2 \ldots \sin \theta_{n-1}\, d\theta_1 \ldots d\theta_{n-1}, \end{aligned} \tag{19}$$

whence

$$\begin{aligned} \prod_{i=1}^{n-1} r_i\, dr_i &= \sin^{2n-3}\theta_1 \cos \theta_1 \sin^{2n-5}\theta_2 \cos \theta_2 \ldots \sin^3 \theta_{n-2} \cos \theta_{n-2} \\ &\quad \times \sin \theta_{n-1} \cos \theta_{n-1}\, d\theta_1\, d\theta_2 \ldots d\theta_{n-2}\, d\theta_{n-1}. \end{aligned} \tag{20}$$

Since the variables r_i here are always positive, we can vary the parameters θ from 0 to $\pi/2$, and for the area σ_{2n} we have

$$\sigma_{2n} = (2\pi)^n \int_0^{\pi/2} \sin^{2n-3}\theta_1 \cos\theta_1\, d\theta_1 \int_0^{\pi/2} \sin^{2n-5}\theta_2 \cos\theta_2\, d\theta_2 \ldots \times$$

$$\times \int_0^{\pi/2} \sin\theta_{n-1} \cos\theta_{n-1}\, d\theta_{n-1} = \frac{(2\pi)^n}{(2n-2)(2n-4)\ldots 4.2}$$

$$= \frac{2\pi^n}{(n-1)!}; \tag{21}$$

obviously, the first formula (7) is found again by putting $n = p$. The uniform probability density over the hypersphere (14) is, therefore, given by

$$dP_{2n} = \frac{d\sigma_{2n}}{\sigma_{2n}} = \frac{(n-1)!}{2\pi^n} \prod_{i=1}^{n-1} r_i\, dr_i \prod_{i=1}^{n} d\alpha_i; \tag{22}$$

if we are interested solely in the distribution $dP_n(\ldots r_i \ldots)$ of the r_i, we must integrate over all the α, whence we obtain

$$dP_n(\ldots r_i \ldots) = \frac{(2\pi)^n}{\sigma_{2n}} \prod_{i=1}^{n-1} r_i\, dr_i = 2^{n-1}(n-1)! \prod_{i=1}^{n-1} r_i\, dr_i. \tag{23}$$

In particular, we have for the probability law $P(\theta_1)\, d\theta_1$:

$$P(\theta_1)\, d\theta_1 = \frac{\sin^{2n-3}\theta_1 \cos\theta_1\, d\theta_1}{\int_0^{\pi/2} \sin^{2n-3}\theta_1 \cos\theta_1\, d\theta_1}$$

$$= 2(n-1) \sin^{2n-3}\theta_1 \cos\theta_1\, d\theta_1, \tag{24}$$

whence we derive the marginal probability law $P(r_1)\, dr_1$:

$$P(r_1)\, dr_1 = 2(n-1)\, r_1(1 - r_1^2)^{n-2}\, dr_1, \tag{25}$$

which can be rewritten as

$$P(u)\, du = (n-1)(1-u)^{n-2}\, du \tag{26}$$

by putting $u = r_i^2$.

It is now easy to calculate the average values $\overline{r_i^{2m}}$, $\overline{r_i^{4m}}$ and $\overline{r_i^2 r_j^{2m}}$ of Chapter IV, starting from (23) and (20). First of all, by means of the preceding results, it can easily be established that

$$\overline{r_i^2}^m = \frac{1}{\sigma_{2n}} \int r_i^2 \, d\sigma_{2n} = \frac{1}{n}. \tag{27}$$

Thus, we verify that the probability distribution (22) corresponds really to the microcanonical distribution of the energy states in a given energy range. We show now how to calculate the expressions $\overline{r_i^4}^m$ and $\overline{r_i^2 r_j^2}^m$; we have successively:

$$\begin{aligned}
\overline{r_i^4}^m &= \frac{1}{\sigma_{2n}} \int r_i^4 \, d\sigma_{2n} \\
&= \frac{(2\pi)^n}{\sigma_{2n}} \int_0^{\pi/2} \sin^{2n+1}\theta_1 \cos\theta_1 \, d\theta_1 \int_0^{\pi/2} \sin^{2n-1}\theta_2 \cos\theta_2 \, d\theta_2 \ldots \times \\
&\times \int_0^{\pi/2} \sin^{2(n-i)+5}\theta_{i-1} \cos\theta_{i-1} \, d\theta_{i-1} \int_0^{\pi/2} \sin^{2(n-i)-1}\theta_i \cos^5\theta_i \, d\theta_i \times \\
&\times \int_0^{\pi/2} \sin^{2(n-i)-3}\theta_{i+1} \cos\theta_{i+1} \, d\theta_{i+1} \ldots \int_0^{\pi/2} \sin\theta_{n-1} \cos\theta_{n-1} \, d\theta_{n-1}
\end{aligned} \tag{28}$$

and

$$\begin{aligned}
\overline{r_i^2 r_j^2}^m &= \frac{1}{\sigma_{2n}} \int r_i^2 r_j^2 \, d\sigma_{2n} \\
&= \frac{(2\pi)^n}{\sigma_{2n}} \int_0^{\pi/2} \sin^{2n+1}\theta_1 \cos\theta_1 \, d\theta_1 \ldots \int_0^{\pi/2} \sin^{2(n-j)+5}\theta_{j-1} \cos\theta_{j-1} \, d\theta_{j-1} \\
&\times \int_0^{\pi/2} \sin^{2(n-j)+1}\theta_j \cos^3\theta_j \, d\theta_j \int_0^{\pi/2} \sin^{2(n-j)-1}\theta_{j+1} \cos\theta_{j+1} \, d\theta_{j+1} \ldots \\
&\times \int_0^{\pi/2} \sin^{2(n-i)+3}\theta_{i-1} \cos\theta_{i-1} \, d\theta_{i-1} \int_0^{\pi/2} \sin^{2(n-i)-1}\theta_i \cos^3\theta_i \, d\theta_i \times \\
&\times \int_0^{\pi/2} \sin^{2(n-i)-3}\theta_{i+1} \cos\theta_{i+1} \, d\theta_{i+1} \ldots \int_0^{\pi/2} \sin\theta_{n-1} \cos\theta_{n-1} \, d\theta_{n-1}.
\end{aligned} \tag{29}$$

Since we have

$$\int_0^{\pi/2} \sin^{k+1}\theta \cos\theta \, d\theta = \frac{1}{k+2},$$

we can see that it remains for us to evaluate two integrals of the form $\int_0^{\pi/2} \sin^k\theta \cos^3\theta \, d\theta$ and $\int_0^{\pi/2} \sin^k\theta \cos^5\theta \, d\theta$. Integrating by parts, we obtain easily

$$\int_0^{\pi/2} \sin^k\theta \cos^3\theta \, d\theta = \frac{2}{k+1}\int_0^{\pi/2} \sin^{k+2}\theta \cos\theta \, d\theta = \frac{2}{(k+1)(k+3)}$$

and

$$\int_0^{\pi/2} \sin^k\theta \cos^5\theta \, d\theta = \frac{4}{k+1}\int_0^{\pi/2} \sin^{k+2}\theta \cos^3\theta \, d\theta$$

$$= \frac{8}{(k+1)(k+3)(k+5)}.$$

Expressions (28) and (29) can then be calculated completely, starting from these integrals, and we obtain

$$\left.\begin{aligned} \overline{r_i^4}^m &= \frac{(2\pi)^n}{\sigma_{2n}} \frac{2^3}{2^{n+1}(n+1)!} = \frac{2}{n(n+1)}, \\ \overline{r_i^2 r_j^2}^m &= \frac{(2\pi)^n}{\sigma_{2n}} \frac{2^2}{2^{n+1}(n+1)!} = \frac{1}{n(n+1)}; \end{aligned}\right\} \tag{30}$$

whence

$$\left.\begin{aligned} \overline{r_i^4}^m - \left(\overline{r_i^2}^m\right)^2 &= \frac{2}{n(n+1)} - \frac{1}{n^2} = \frac{(n-1)}{n^2(n+1)}, \\ \overline{r_i^2 r_j^2}^m - \left(\overline{r_i^2}^m\right)\left(\overline{r_j^2}^m\right) &= \frac{1}{n(n+1)} - \frac{1}{n^2} = -\frac{1}{n^2(n+1)}. \end{aligned}\right\} \tag{31}$$

3. *Probability laws for the quantities $D_{ii}^{(\nu)}$ and $\mu_{ij}^{(\alpha)}$*

We have seen in Chapter II that the proof of the ergodic theorems of von Neumann and Pauli–Fierz depends ultimately on the fact that a non-thermodynamic macroscopic observer would be extremely improbable; in other words, that the probability of a subdivision of Hilbert space into macroscopic cells assigning to $\mu_{ij}^{(\alpha)}$ a value which is larger than a certain given quantity would be very small. We shall give here the detailed proof of this by following the method given by Pauli-Fierz (1937);† we shall study in the first place the distribution in probability of $\mu_{ij}^{(\alpha)}$ for the case where $i \neq j$ and afterwards we shall proceed to study the distribution of the $\mu_{ii}^{(\alpha)}$.

(a) *Probability law for $D_{ii}^{(\nu)}$ and average values.* First of all, we must evaluate in a unitary space of $S^{(\alpha)}$ dimensions the probability that $D_{ii}^{(\nu)}$, defined by

$$D_{ij}^{(\nu)} = \sum_{k=1}^{s_\nu(\alpha)} C_{ki}^{(\alpha)*} C_{kj}^{(\alpha)}, \tag{32}$$

has its value contained between u and $u + du$, assuming that all possible subdivisions in macroscopic cells are equally probable. We deduce from them the average values

$$\langle |C_{ki}^{(\alpha)}|^4 \rangle, \quad \langle |C_{ki}^{(\alpha)}|^2 \, |C_{kj}^{(\alpha)}|^2 \rangle, \ldots,$$

over all the possible bases $\Omega_k^{(\alpha)}$, that is to say, over all possible macroscopic observers.

Because of the orthonormality relations which they satisfy (see Chapter II, § 2), the $C_{ki}^{(\alpha)}$ can be considered as components of unit vectors $\boldsymbol{C}_i$ on a complex unitary space with $S^{(\alpha)}$ dimensions and, according to (32), $D_{ij}^{(\nu)}$ is the scalar product of the projections of the vectors $\boldsymbol{C}_i$ and $\boldsymbol{C}_j$ on a sub-space with $s_\nu^{(\alpha)}$ dimensions. The averages must be taken over all possible subdivisions of the space, i.e. over all systems of coordinates $C_{ki}^{(\alpha)}$ satisfying the relation

$$\sum_{i=1}^{S(\alpha)} |C_{ki}^{(\alpha)}|^2 = 1 .$$

Since the unitary nature of the space is not involved here, this problem can be dealt with also in real $2S^{(\alpha)}$-dimensional space, where we shall be concerned with a uniform probability distribution over the unit hypersphere.

We consider now, in this space, a unit vector Ω decomposed according to $\Omega = \varrho + \sigma$ into two orthogonal components ϱ and σ respectively with $2s_\nu^{(\alpha)}$ and $2(S^{(\alpha)} - s_\nu^{(\alpha)})$ dimensions and we can

† We point out that, if the conclusions of this paper are accurate, certain intermediate calculations must be viewed with caution.

find the probability density that the square of the length $|\varrho|^2$ of the component ϱ be contained between u and $u + du$. It is given by the area of a $2s_\nu^{(\alpha)}$-dimensional portion of the surface on the $2S^{(\alpha)}$-dimensional unit sphere; we have thus for this probability

$$W(u)\,du = \frac{\Gamma(S^{(\alpha)})}{\Gamma(s_\nu^{(\alpha)})\,\Gamma(S^{(\alpha)} - s_\nu^{(\alpha)})}\,u^{s_\nu(\alpha)-1}(1-u)^{S(\alpha)-s_\nu(\alpha)-1}\,du,$$

$$0 \leqq u \leqq 1. \tag{33}$$

This function assumes a maximum for

$$u_0 = \frac{s_\nu^{(\alpha)} - 1}{S^{(\alpha)} - 2} = \frac{m_\nu}{M^{(\alpha)}}, \tag{34}$$

by putting

$$s_\nu^{(\alpha)} - 1 = m_\nu \quad \text{and} \quad S^{(\alpha)} - 2 = M^{(\alpha)}.$$

We note that, in the case where $s_\nu^{(\alpha)} = 1$, formula (26) of the preceding section is found again and that (33) represents, similarly, the probability that $D_{ii}^{(\nu)}$ is contained between u and $u + du$, since by definition

$$D_{ii}^{(\nu)} = \sum_{k=1}^{s_\nu(\alpha)} |C_{ki}^{(\alpha)}|^2$$

represents the square of the length of the projection of the unit vector $\mathbf{C}_i$ on a sub-space of $2s_\nu^{(\alpha)}$ dimensions.

Having done this, it is easy to calculate from (33) various average values. First of all, multiplying (33) by $u^n\,du$ and integrating from 0 to 1, we have

$$\langle \varrho^{2n} \rangle = \frac{(s_\nu^{(\alpha)} + n - 1)\,(s_\nu^{(\alpha)} + n - 2)\,\ldots\,s_\nu^{(\alpha)}}{(S^{(\alpha)} + n - 1)\,(S^{(\alpha)} + n - 2)\,\ldots\,S^{(\alpha)}}. \tag{35}$$

In the particular case $n = s_\nu^{(\alpha)} = 1$, we have

$$\langle \varrho^2 \rangle = \langle |C_{ki}^{(\alpha)}|^2 \rangle = \frac{1}{S^{(\alpha)}}, \tag{36}$$

and similarly, for $s_\nu^{(\alpha)} = 1$ and $n = 2$, we have

$$\langle \varrho^4 \rangle = \langle |C_{ki}^{(\alpha)}|^4 \rangle = \frac{2}{S^{(\alpha)}(S^\alpha + 1)}. \tag{37}$$

From this we can deduce immediately other average values by using especially the orthogonality relations satisfied by the $C_{ki}^{(\alpha)}$. In this case, according to (32), we have

$$(D_{ii}^{(\nu)})^2 = \left(\sum_{k=1}^{s_\nu(\alpha)} |C_{ki}^{(\alpha)}|^2\right)^2 = \sum_{k=1}^{s_\nu(\alpha)} |C_{ki}^{(\alpha)}|^4 + \sum_{k \neq k'} |C_{ki}^{(\alpha)}|^2 \, |C_{k'i}^{(\alpha)}|^2, \tag{38}$$

and, putting $n = 2$ in (35), we have

$$\langle (D_{ii}^{(\nu)})^2 \rangle = \langle \varrho^4 \rangle = \frac{s_\nu^{(\alpha)}(s_\nu^{(\alpha)} + 1)}{S^{(\alpha)}(S^{(\alpha)} + 1)}, \tag{39}$$

whence, by using (37),

$$\frac{s_\nu^{(\alpha)}(s_\nu^{(\alpha)} + 1)}{S^{(\alpha)}(S^{(\alpha)} + 1)} = \frac{2s_\nu^{(\alpha)}}{S^{(\alpha)}(S^{(\alpha)} + 1)} + s_\nu^{(\alpha)}(s_\nu^{(\alpha)} - 1) \, \langle |C_{ki}^{(\alpha)}|^2 \, |C_{k'i}^{(\alpha)}|^2 \rangle,$$

and, consequently,

$$\langle |C_{ki}^{(\alpha)}|^2 \, |C_{k'i}^{(\alpha)}|^2 \rangle = \frac{1}{S^{(\alpha)}(S^{(\alpha)} + 1)} \qquad (k \neq k'). \tag{40}$$

Similarly, by squaring the relationship $\sum_i |C_{ki}^{(\alpha)}|^2 = 1$ [see Chapter II, formula (II.61)], it becomes

$$S^{(\alpha)} \langle |C_{ki}^{(\alpha)}|^4 \rangle + S^{(\alpha)}(S^{(\alpha)} - 1) \, \langle |C_{ki}^{(\alpha)}|^2 \, |C_{kj}^{(\alpha)}|^2 \rangle = 1,$$

whence

$$\langle |C_{k'i}^{(\alpha)}|^2 \, |C_{kj}^{(\alpha)}|^2 \rangle = \frac{1}{S^{(\alpha)}(S^{(\alpha)} + 1)} \qquad (i \neq j). \tag{41}$$

In addition, relations (II.62) and (II.63) give successively

$$S^{(\alpha)} \langle |C_{ki}^{(\alpha)}|^2 \, |C_{k'i}^{(\alpha)}|^2 \rangle + S^{(\alpha)}(S^{(\alpha)} - 1) \, \langle |C_{ki}^{(\alpha)}|^2 \, |C_{k'j}^{(\alpha)}|^2 \rangle = 1,$$

or

$$\langle |C_{ki}^{(\alpha)}|^2 \, |C_{k'j}^{(\alpha)}|^2 \rangle = \frac{1}{S^{(\alpha)2} - 1} \qquad (i \neq j, \; k \neq k'), \tag{42}$$

then

$$S^{(\alpha)}(S^{(\alpha)} - 1) \, \langle C_{ki}^{(\alpha)*} C_{k'i}^{(\alpha)} C_{kj}^{(\alpha)} C_{k'j}^{(\alpha)*} \rangle = -S^{(\alpha)} \langle |C_{ki}^{(\alpha)}|^2 \, |C_{k'i}^{(\alpha)}|^2 \rangle,$$

whence

$$\langle C_{ki}^{(\alpha)*} C_{k'i}^{(\alpha)} C_{kj}^{(\alpha)} C_{k'j}^{(\alpha)*} \rangle = -\frac{1}{S^{(\alpha)}(S^{(\alpha)2} - 1)} \qquad (i \neq j, \; k \neq k'). \tag{43}$$

Finally, symmetry considerations show that we have the relations:

$$\langle C_{kj}^{(\alpha)*} C_{ki}^{(\alpha)} \rangle = \langle C_{ki}^{(\alpha)*} C_{k'i}^{(\alpha)} \rangle = 0 \quad (i \neq j,\ k \neq k'),$$

$$\langle |C_{ki}^{(\alpha)}|^2 \, C_{k'j}^{(\alpha)*} C_{k'l}^{(\alpha)} \rangle = 0 \quad (i \neq j \neq l). \tag{43'}$$

Thus we know all the average values that are used in Chapters III and IV. In addition, we can deduce the average value of the $\mu_{ij}^{(\alpha)}$ over all possible subdivisions of Hilbert space; we find easily that

$$\langle \mu_{ij}^{(\alpha)} \rangle = \left\langle \sum_{\nu=1}^{N(\alpha)} \frac{S^{(\alpha)}}{s_\nu^{(\alpha)}} |D_{ij}^{(\nu)}|^2 \right\rangle = \frac{N^{(\alpha)} - 1}{S^{(\alpha)} - \dfrac{1}{S^{(\alpha)}}}, \tag{44}$$

$$\langle \mu_{ii}^{(\alpha)} \rangle = \left\langle \sum_{\nu=1}^{N(\alpha)} \frac{S^{(\alpha)}}{s_\nu^{(\alpha)}} \left(D_{ii}^{(\nu)} - \frac{s_\nu^{(\alpha)}}{S^{(\alpha)}} \right)^2 \right\rangle = \frac{N^{(\alpha)} - 1}{S^{(\alpha)} + 1}; \tag{44'}$$

it can be seen that these average values are of the order of $N^{(\alpha)}/S^{(\alpha)}$, therefore inversely proportional to the average number of states per cell. We are now able to take up the study of the laws of probability for the quantities $\mu_{ij}^{(\alpha)}$ and $\mu_{ii}^{(\alpha)}$.

(b) *Probability law for $\mu_{ij}^{(\alpha)}$ with $i \neq j$.* In order to give an estimate for this probability law, we shall evaluate the average values of the powers of $\mu_{ij}^{(\alpha)}$, i.e. of the sums of expressions of the type

$$\left(\sum_{k=1}^{s_\nu(\alpha)} x_k y_k \right)^{2n}, \tag{45}$$

where x_k and y_k are the components in the $S^{(\alpha)}$-dimensional complex unitary space of two orthogonal unit vectors $\boldsymbol{x}$ and $\boldsymbol{y}$. As before, we can also deal with this problem in a real orthogonal space of $2S^{(\alpha)}$ dimensions. We denote by r_ν^2 the length of the projection of $\boldsymbol{x}$ on the subspace with $s_\nu^{(\alpha)}$ dimensions $\left(r_\nu^2 = \sum_{k=1}^{s_\nu(\alpha)} x_k^2 \right)$; if we take a system of coordinates perpendicular to $\boldsymbol{x}$, we have in such a system

$$\left(\sum_{k=1}^{s_\nu(\alpha)} x_k y_k \right)^{2n} = r_\nu^{2n}(1 - r_\nu^2)^n z^{2n}, \tag{46}$$

where z is a component of a unit vector perpendicular to $\boldsymbol{x}$ in $(S^{(\alpha)}-1)$-dimensional space. We calculate the average value of (45) at first keeping r_ν fixed and varying z, then by taking the average with respect to r_ν. In this case, by (35), we have

$$\langle z^{2n}\rangle = \frac{1}{(S^{(\alpha)}-1)\,S^{(\alpha)}(S^{(\alpha)}+1)\dots(S^{(\alpha)}+n-2)}, \tag{47}$$

and thus

$$\langle r_\nu^{2n}\rangle = \frac{s_\nu^{(\alpha)}(s_\nu^{(\alpha)}+1)\dots(s_\nu^{(\alpha)}+2n-1)}{S^{(\alpha)}(S^{(\alpha)}+1)\dots(S^{(\alpha)}+2n-1)}. \tag{47'}$$

Since $(1-r_\nu^2)^n \leqq 1$, we have

$$\left(\sum x_k y_k\right)^{2n} \leqq r_\nu^{2n} z^{2n},$$

or

$$\left\langle \left(\sum_{k=1}^{s_\nu(\alpha)} x_k y_k\right)^{2n} \right\rangle$$

$$\leqq \frac{s_\nu^{(\alpha)}(s_\nu^{(\alpha)}+1)\dots(s_\nu^{(\alpha)}+2n-1)}{S^{(\alpha)}(S^{(\alpha)}+1)\dots(S^{(\alpha)}+2n-1)\,(S^{(\alpha)}-1)\,S^{(\alpha)}\dots(S^{(\alpha)}+n-2)}. \tag{48}$$

Let us evaluate now the average value of the powers of $\mu_{ij}^{(\alpha)}$; we have

$$(\mu_{ij}^{(\alpha)})^n = \left(\sum_{\nu=1}^{N(\alpha)} \frac{S^{(\alpha)}}{s_\nu^{(\alpha)}}\,|D_{ij}^{(\nu)}|^2\right)^n, \tag{49}$$

whence we derive

$$(\mu_{ij}^{(\alpha)})^n \leqq N^{(\alpha)n-1} \sum_{\nu=1}^{N(\alpha)} \left(\frac{S^{(\alpha)}}{s_\nu^{(\alpha)}}\right)^n |D_{ij}^{(\nu)}|^{2n}, \tag{50}$$

since the sum (49) contains only non-negative terms. By taking the average values, we have further,

$$\langle(\mu_{ij}^{(\alpha)})^n\rangle \leqq N^{(\alpha)n-1} \sum_{\nu=1}^{N(\alpha)} \left(\frac{S^{(\alpha)}}{s_\nu^{(\alpha)}}\right)^n \langle|D_{ij}^{(\nu)}|^{2n}\rangle. \tag{51}$$

However, we calculated in (48) the average values of the quantities $|D_{ij}^{(\nu)}|^{2n}$; in fact, we have in real space with $2S^{(\alpha)}$ dimensions,

$$|D_{ij}^{(\nu)}|^{2n} = \left(\sum_{k=1}^{2s_\nu^{(\alpha)}} x_k y_k \right)^{2n}.$$

We can therefore write

$$\langle (\mu_{ij}^{(\alpha)})^n \rangle \leq N^{(\alpha)n-1} \sum_{\nu=1}^{N(\alpha)} \left(\frac{S^{(\alpha)}}{s_\nu^{(\alpha)}} \right)^n \times$$

$$\times \frac{s_\nu^{(\alpha)}(s_\nu^{(\alpha)}+1)\ldots(s_\nu^{(\alpha)}+2n-1)}{S^{(\alpha)}(S^{(\alpha)}+1)\ldots(S^{(\alpha)}+2n-1)(S^{(\alpha)}-1)\ldots(S^{(\alpha)}+n-2)}, \tag{52}$$

or

$$\langle (\mu_{ij}^{(\alpha)})^n \rangle < \frac{N^{(\alpha)n-1}}{(S^{(\alpha)}-1)^n (S^{(\alpha)})^n} \sum_{\nu=1}^{N(\alpha)} \frac{(s_\nu^{(\alpha)}+2n)^{2n}}{(s_\nu^{(\alpha)})^n},$$

$$= \frac{N^{(\alpha)n-1}}{(S^{(\alpha)}-1)^n (S^{(\alpha)})^n} \sum_{\nu=1}^{N(\alpha)} (s_\nu^{(\alpha)})^n \left(1 + \frac{2n}{s_\nu^{(\alpha)}}\right)^{2n}, \tag{53}$$

and, by noting that

$$\sum_\nu \frac{s_\nu^{(\alpha)}}{S^{(\alpha)}} = 1 \quad \text{and} \quad \sum_\nu \left(\frac{s_\nu^{(\alpha)}}{S^{(\alpha)}} \right)^n \leq 1 \quad \text{if} \quad n > 1,$$

we have finally

$$\langle (\mu_{ij}^{(\alpha)})^n \rangle \leq \frac{N^{(\alpha)n-1}}{(S^{(\alpha)}-1)^n} (1+2n)^{2n}. \tag{54}$$

In order to deduce from this result the probability law for the $\mu_{ij}^{(\alpha)}$, we use the following relation: if $f(x)$ is a monotonically increasing function which is nowhere negative, we have

$$\int_0^\infty f(x)\, W(x)\, dx = \bar{f} > \int_\xi^\infty f(x)\, W(x)\, dx > f(\xi) \int_\xi^\infty W(x)\, dx,$$

whence we derive an upper limit for the probability that $x > \xi$; we have

$$\int_\xi^\infty W(x)\, dx < \frac{\bar{f}}{f(\xi)}.$$

Let us apply this relation by choosing as the function $f(x) = e^{\alpha\sqrt{x}}$; we have, using (54),

$$\langle e^{\alpha\sqrt{\mu_{ij}(\alpha)}}\rangle = \sum_{n=0}^{\infty}\frac{1}{n!}\alpha^n\langle(\mu_{ij}^{(\alpha)})^{n/2}\rangle < 2\sum_{n=0}^{\infty}\frac{1}{2n!}\alpha^{2n}\langle(\mu_{ij}^{(\alpha)})^n\rangle$$

$$< 2 + \frac{2}{N^{(\alpha)}}\sum_{n=1}^{\infty}\frac{1}{2n!}\alpha^{2n}\left(\frac{N^{(\alpha)}}{S^{(\alpha)}-1}\right)^n(1+2n)^{2n}. \qquad (55)$$

We choose α such that the series is convergent; since the ratio of two consecutive terms is given by

$$\frac{\alpha^2 N^{(\alpha)}}{S^{(\alpha)}-1}\,\frac{(2n+3)^{2n+2}}{(2n+2)(2n+1)^{2n+1}} < \frac{\alpha^2 N^{(\alpha)}}{S^{(\alpha)}-1}\left(1+\frac{2}{2n+1}\right)^{2n+2}$$

$$< \frac{\alpha^2 N^{(\alpha)}}{S^{(\alpha)}-1}\,e^2,$$

we shall choose $\frac{\alpha^2 N^{(\alpha)}}{S^{(\alpha)}-1}e^2 < 1$ in order that $\langle e^{\alpha\sqrt{x}}\rangle$ is finite. Putting

$$\alpha = \frac{1}{e\sqrt{2}}\sqrt{\frac{S^{(\alpha)}-1}{N^{(\alpha)}}},$$

we have

$$\langle e^{\alpha\sqrt{\mu_{ij}(\alpha)}}\rangle < 2 + \frac{9}{N^{(\alpha)}e^2}$$

and

$\text{Prob}\{\mu_{ij}^{(\alpha)} > x\} =$

$$\int_x^{\infty} W(\xi)\,d\xi \leq \left(2+\frac{9}{N^{(\alpha)}e^2}\right)\exp\left[-\frac{1}{e\sqrt{2}}\sqrt{\frac{S^{(\alpha)}-1}{N^{(\alpha)}}\,x}\right]. \quad (56)$$

Knowing the probability law for $\mu_{ij}^{(\alpha)}$, it is easy to derive the probability that the quantity (encountered in Chapter III, § IV.1)

$$M = \frac{1}{W^{(\alpha)}}\sum_{i\neq j}^{(\alpha)} r_i^2 r_j^2 \mu_{ij}^{(\alpha)}$$

is greater than $W^{(\alpha)}x$.† We have

$$\text{Prob}\{M > W^{(\alpha)}x\}$$
$$\leqq \left(2 + \frac{9}{N^{(\alpha)}e^2}\right) \exp\left\{2 \ln S^{(\alpha)} - \frac{1}{e\sqrt{2}}\sqrt{\frac{S^{(\alpha)} - 1}{N^{(\alpha)}}x}\right\}, \quad (57)$$

which solves the first part of our problem.

(c) *Probability law for the* $\mu_{ii}^{(\alpha)}$. First of all, we begin by studying the probability that

$$\frac{S^{(\alpha)}}{s_\nu^{(\alpha)}}\left(D_{ii}^{(\nu)} - \frac{s_\nu^{(\alpha)}}{S^{(\alpha)}}\right)^2 > a$$

and we derive from it the probability law for the $\mu_{ii}^{(\alpha)}$ by applying the theorem mentioned earlier. We shall put henceforth, in order to simplify the writing, $s_\nu^{(\alpha)}/S^{(\alpha)} = g_\nu^{(\alpha)}$, and we shall use the probability law for the $D_{ii}^{(\nu)}$ given by $W(u)\,du$ of (33).

Let us consider now the inequality

$$\frac{1}{g_\nu^{(\alpha)}}(D_{ii}^{(\nu)} - g_\nu^{(\alpha)})^2 > a; \quad (58)$$

it has two solutions:

$$D_{ii}^{(\nu)} > \sqrt{ag_\nu^{(\alpha)}} + g_\nu^{(\alpha)} \quad \text{when} \quad \sqrt{ag_\nu^{(\alpha)}} + g_\nu^{(\alpha)} < 1,$$

† For this purpose we use the following theorem: *If we have N positive random variables* $x_1, \ldots, x_N$, and if

$$\text{Prob}\{x_k > a\} = F_k(a),$$

the probability that the sum $x_1 + x_2 + \ldots x_N$ *is greater than a, is less than* $W(a) = F_1(ap_1) + \cdots + F_N(ap_N)$, *where the* p_k *are positive weights such that* $\sum_{k=1}^{N} p_k = 1$.

This is satisfied immediately for $N = 2$ and we prove it by induction for any value of N. Similarly, we establish that the probability that

$$p_1x_1 + \cdots + p_Nx_N > a,$$

is less than

$$W'(a) = F_1(a) + \cdots + F_N(a)$$

where the p_k and $F_k(a)$ have the same meaning as before.

and

$$D_{ii}^{(\nu)} < g_\nu^{(\alpha)} - \sqrt{a g_\nu^{(\alpha)}} \quad \text{when} \quad a < g_\nu^{(\alpha)};$$

we must therefore evaluate the two integrals

$$\int_{g_\nu^{(\alpha)} + \sqrt{a g_\nu^{(\alpha)}}}^{1} W(u)\,du \quad \text{and} \quad \int_{0}^{g_\nu^{(\alpha)} - \sqrt{a g_\nu^{(\alpha)}}} W(u)\,du.$$

If we try to compare the integration limits with the value of u_0 which makes (33) maximum, we have, by evaluating $|u_0 - g_\nu^{(\alpha)}|$,

$$|u_0 - g_\nu^{(\alpha)}| = \left| \frac{2s_\nu^{(\alpha)} - S^{(\alpha)}}{S^{(\alpha)}(S^{(\alpha)} - 2)} \right| \leq \frac{1}{S^{(\alpha)}}$$

so that, if $a > 1/S^{(\alpha)}$, we can state that the maximum of $W(u)$ is located outside the two integration paths; it follows that the integrals are monotonic and that they can be overestimated by replacing $W(u)$ by the value it takes at the lower or the upper limit.

We must look, therefore, for a suitable evaluation of $W(u)$ for the values considered. We shall begin by finding first of all a bound for $W(u_0)$. With the notation introduced in (34) and writing the Γ-functions out explicitly, we have

$$\begin{aligned} W(u_0) &= (M^{(\alpha)} + 1)\binom{M^{(\alpha)}}{m_\nu}\left(\frac{m_\nu}{M^{(\alpha)}}\right)^{m_\nu}\left(\frac{M^{(\alpha)} - m_\nu}{M^{(\alpha)}}\right)^{M^{(\alpha)} - m_\nu} \\ &= f(m_\nu). \end{aligned}$$

Since $f(m_\nu) = f(M^{(\alpha)} - m_\nu)$, it is sufficient to consider the case in which $2m_\nu < M^{(\alpha)}$. We shall prove that $f(0) = M^{(\alpha)} + 1 \geqq f(m_\nu)$; this results from the monotonic decrease of $f(m_\nu)$, from $m_\nu = 0$ to $m_\nu = M^{(\alpha)}/2$. In fact, we have

$$\frac{f(m_\nu)}{f(m_\nu + 1)} = \left(\frac{m_\nu}{m_\nu + 1}\right)^{m_\nu}\left(\frac{M^{(\alpha)} - m_\nu}{M^{(\alpha)} - m_\nu - 1}\right)^{M^{(\alpha)} - m_\nu - 1}$$

and

$$\begin{aligned} \ln \frac{f(m_\nu)}{f(m_\nu + 1)} &= m_\nu(\ln m_\nu - \ln (m_\nu + 1)) \\ &\quad + (M^{(\alpha)} - m_\nu - 1)\,[\ln (M^{(\alpha)} - m_\nu) - \ln (M^{(\alpha)} - m_\nu - 1)]. \end{aligned}$$

If we use the function

$$L(x, y) = x(\ln x - \ln y) - x + y,$$

we have finally

$$\ln \frac{f(m_\nu)}{f(m_\nu + 1)} = L(m_\nu, m_\nu + 1) - L(M^{(\alpha)} - m_\nu - 1, M^{(\alpha)} - m_\nu)$$

and it is sufficient, in order that $f(m)$ be decreasing and monotonic, to prove the inequality

$$L(x, x + 1) > L(y, y + 1) \quad \text{when} \quad 0 < x < y.$$

However, this follows from the sign of the derivative

$$\frac{\partial L(x, x + 1)}{\partial x} = 1 - \frac{x}{x + 1} + \ln \frac{x}{x + 1},$$

which is always negative; we can thus write

$$f(m_\nu + 1) < f(m_\nu), \quad \text{when} \quad 2m_\nu \leqq M^{(\alpha)}.$$

If $2m_\nu > M^{(\alpha)}$, we must interchange m_ν and $M^{(\alpha)} - m_\nu$, by virtue of the symmetry of $W(u_0)$, and we find a maximum for $m_\nu = M^{(\alpha)}$; in addition, we have

$$W(M^{(\alpha)}) = W(0) = M^{(\alpha)} + 1,$$

with the result that we have always

$$W(u_0) < M^{(\alpha)} + 1 = S^{(\alpha)} - 1. \tag{59}$$

We now proceed to evaluate the ratio

$$\frac{W(u)}{W(u_0)} = \left(\frac{u}{u_0}\right)^{m_\nu} \left(\frac{1 - u}{1 - u_0}\right)^{M^{(\alpha)} - m_\nu}, \text{where } u_0 = \frac{m_\nu}{M^{(\alpha)}},$$

for values of u such that

$$x = u - u_0 > x_0 = \sqrt{\frac{u_0(1 - u_0)}{M^{(\alpha)}}}. \tag{60}$$

First of all, we show that

$$\ln \frac{W(u)}{W(u_0)} = m_\nu \ln\left(1 + \frac{x}{u_0}\right) + (M^{(\alpha)} - m_\nu) \ln\left(1 - \frac{x}{1 - u_0}\right)$$
$$< \frac{x}{x_0} \varphi(m),$$

where $\varphi(m)$ is a function to be determined which is independent of x; we have,

$$\left.\begin{aligned}\ln\left(1+\frac{x}{u_0}\right) &< \frac{x}{x_0}\ln\left(1+\frac{x_0}{u_0}\right),\\ \ln\left(1-\frac{x}{1-u_0}\right) &< \frac{x}{x_0}\ln\left(1-\frac{x_0}{1-u_0}\right).\end{aligned}\right\} \tag{61}$$

In the case where $2m_\nu < M^{(\alpha)}$, we can then put for $\varphi(m_\nu)$

$$\varphi(m_\nu) = m_\nu \ln\left(1+\sqrt{\frac{M^{(\alpha)}-m_\nu}{M^{(\alpha)}m_\nu}}\right)$$
$$+\,(M^{(\alpha)}-m_\nu)\ln\left(1-\sqrt{\frac{m_\nu}{M^{(\alpha)}(M^{(\alpha)}-m_\nu)}}\right).$$

Let us study the function $\varphi(m_\nu)$, dealing separately with the case where $m_\nu = 0$, since φ and x_0 are then zero at the same time; we shall show that $\varphi(m_\nu)$ passes through a maximum for $m_\nu = 1$, and then decreases monotonically. In fact, by putting

$$m_\nu \geqq 1, \quad \xi = \sqrt{\frac{m_\nu}{M^{(\alpha)}-1}}, \quad \frac{1}{\sqrt{M^{(\alpha)}}} < \xi < \sqrt{M^{(\alpha)}},$$

we have for the derivative $\partial\varphi/\partial\xi$

$$-\frac{1+\xi^2}{M^{(\alpha)}}\frac{\partial\varphi}{\partial\xi} = \frac{\xi}{\sqrt{M^{(\alpha)}}\,\xi+1} + \frac{1}{\sqrt{M^{(\alpha)}}-\xi}$$
$$-\frac{2\xi}{\xi^2-1}\left[\ln\left(1+\frac{1}{\sqrt{M^{(\alpha)}}\,\xi}\right) - \ln\left(1-\frac{\xi}{\sqrt{M^{(\alpha)}}}\right)\right]. \tag{62}$$

In order to show that the right-hand side of (62) is positive, we put

$$z = \frac{1+\dfrac{1}{\xi\sqrt{M^{(\alpha)}}}}{1-\dfrac{\xi}{\sqrt{M^{(\alpha)}}}} \qquad (1 < z < M^{(\alpha)});$$

after several transformations, we obtain for this right-hand side

$$z - \frac{1}{z} - 2 \ln z > 0$$

which is positive, since it vanishes for $z = 1$ and its derivative $\left(1 - \frac{1}{z}\right)^2$ is positive. In addition, we have

$$\varphi(1) \sim \ln 2 + (M^{(\alpha)} - 1) \ln\left(1 - \frac{1}{M^{(\alpha)}}\right) < \ln 2 - 1 = -K.$$

We derive from the foregoing results an overestimate of the probability that $u > u_0$; this is, according to what we have just seen, certainly less than

$$\exp\left[-K\sqrt{\frac{M^{(\alpha)}}{u_0(1-u_0)}}\,x + \ln(M^{(\alpha)} + 1)\right]; \tag{63}$$

this expression gives the probability that $u - u_0 = x > x_0$ with $m_\nu > 1$ and $K = 1 - \ln 2 \simeq 0.3$. We shall deduce from it the probability that

$$\frac{1}{g_\nu^{(\alpha)}}(u - g_\nu^{(\alpha)})^2 > a.$$

If $u - g_\nu^{(\alpha)} > \sqrt{(ag_\nu^{(\alpha)})}$, we deduce from it that

$$u - u_0 > \sqrt{ag_\nu^{(\alpha)}} - \frac{1}{S^{(\alpha)}},$$

according to the evaluation given previously of $|u_0 - g_\nu^{(\alpha)}|$.

In order that $u - u_0 > x_0$, it is necessary that a satisfies a certain condition; according to (60) we obtain

$$a > \frac{2}{S^{(\alpha)} - 2}$$

and we arrive at the same conclusion for the case $u < u_0$. If we put the relation $x = \sqrt{ag_\nu^{(\alpha)}} - \frac{1}{S^{(\alpha)}}$ in the overestimate (63) it becomes

$$W(a) < \exp\left[-K\sqrt{\frac{M^{(\alpha)}}{u_0(1-u_0)}}\left\{\sqrt{\frac{s_\nu^{(\alpha)}a}{S^{(\alpha)}}} - \frac{1}{S^{(\alpha)}}\right\} + \ln(S^{(\alpha)} - 1)\right]$$

$$< \exp\left[-K\sqrt{aS^{(\alpha)}} + K + \ln S^{(\alpha)}\right]. \tag{64}$$

There remains the case when $m_\nu = 0$; in this case,

$$\int_n^1 W(u)\,du = (1-a)^{S^{(\alpha)}-1} < e^{-(S^{(\alpha)}-1)a}, \quad \text{where} \quad u > u_0.$$

The case where $a < 1/S^{(\alpha)}$ does not occur here because $u_0 \sim 1/S^{(\alpha)}$; expression (64) thus remains valid and gives us an estimate for the distribution of $\frac{1}{g_\nu^{(\alpha)}}(D_{ii}^{(\nu)} - g_\nu^{(\alpha)})^2$. Using the theorem which has already served us, we shall evaluate the probability that $\sum_\nu \frac{1}{g_\nu^{(\alpha)}} \times (D_{ii}^{(\nu)} - g_\nu^{(\alpha)})^2 > a$, choosing the value $1/N^{(\alpha)}$ for the weights p_k; if

$$a > \frac{2N^{(\alpha)}}{S^{(\alpha)}-2},$$

we have

$$\text{Prob}\left\{\sum_\nu \frac{1}{g_\nu^{(\alpha)}}(D_{ii}^{(\nu)} - g_\nu^{(\alpha)})^2 > a\right\}$$
$$< \exp\left[-K\sqrt{\frac{S^{(\alpha)}}{N^{(\alpha)}}}\,a + K + \ln S^{(\alpha)}\right]. \tag{65}$$

Comparing this with (57), it can be seen that the probability for finding an observer who observes a deviation of the entropy from its canonical value larger than $2a$ is smaller than

$$A\exp\left[-b\sqrt{\frac{S^{(\alpha)}}{N^{(\alpha)}}}\,a + 2\ln S^{(\alpha)}\right], \tag{66}$$

with

$$a > \frac{2N^{(\alpha)}}{S^{(\alpha)}-2}, \quad A = \left(2 + \frac{9}{N^{(\alpha)}e^2}\right), \quad b = \frac{1}{e\sqrt{2}}.$$

Thus, we are led to postulate that the average number of states $S^{(\alpha)}/N^{(\alpha)}$ in a phase cell must be very large compared with $2\ln S^{(\alpha)}$, which seems physically probable.

APPENDIX III A

Ehrenfests' Model

1. *The function H(Z, t) and the "H-curve"*

We shall describe here the behaviour of the $H(Z, t)$ function by analysing the well-known model of P. and T. Ehrenfest. We note first of all that, because of the finite nature of the ω_i cells, the variation of $H(Z, t)$ considered as a function of the occupation numbers n_i is discontinuous; $H(Z, t)$ is a step function, the time derivative of which only takes the three values 0 and $\pm\infty$: a step is produced each time a particle leaves one of the cells to enter another, thus causing two of the n_i numbers to change by unity. The variation of $H(Z, t)$ is thus properly represented by Fig. 3a (compare Chapter V, § III.1) and the *H-curve* is obtained by choosing, on the $H(Z, t)$ function, a discrete set of points corresponding to times separated by equal intervals Δt; Δt must be small in comparison with experimental macroscopic times, but sufficiently large to contain a large number of collisions; in this case, we obtain curve 3b.

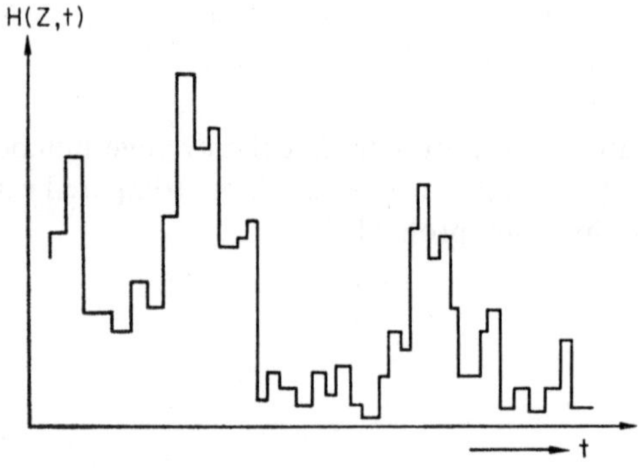

Fig. 3a.

The fundamental problem of statistical mechanics is now to reveal the irreversible tendency of a system towards an equilibrium state and to study whether this macroscopic irreversible evolution is compatible with the reversibility of the microscopic mechanical evolution. Since this problem cannot be approached by a precise study of the $H(Z, t)$ function—it would require, in fact, total integration of the equations of motion—P. and T. Ehrenfest proposed in 1907 a stochastic model which shows qualitatively the behaviour of $H(Z, t)$ and which opens the way to a solution of the Loschmidt and Zermelo paradoxes; besides its historical interest, this model (known as the "dog-flea" model) is an excellent example of the application of Markovian stochastic processes to study the problems of statistical mechanics; we have outlined this application in Chapter V (Section IV).

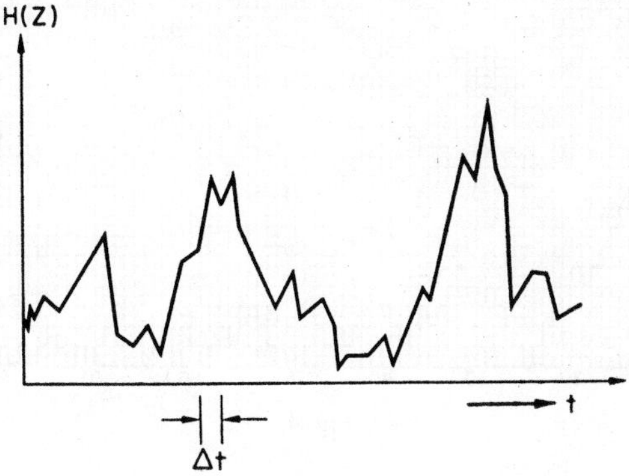

Fig. 3b.

2. *Ehrenfests' model*

Suppose that $2N$ balls, numbered from 1 to $2N$, are distributed in two urns A and B; similarly, let us suppose that a set of $2N$ cards is arranged, also numbered from 1 to $2N$. A card is drawn and the ball whose number corresponds to that of the card, is changed from one urn to the other; after replacing the card in the pack, another card is drawn and so on and so forth. We denote by $n_A(s)$ and $n_B(s)$ the respective number of balls in urns A and B after s draws; it is convenient to put

$$\left.\begin{aligned} n_A(s) + n_B(s) &= 2N \\ n_A(s) - n_B(s) &= 2k \end{aligned}\right\} \quad \text{or} \quad n_A = N + k, \quad n_B = N - k. \qquad (1)$$

Similarly, we define the absolute value Δ_s of the difference $n_A - n_B$, or:

$$\Delta_s = |n_A(s) - n_B(s)| = 2|k|, \qquad (2)$$

and, by drawing the curve of $\Delta_s = f(s)$, we obtain a curve similar to the step function $H(Z, t)$; Figure 4 shows the curve corresponding to an actual experiment carried out with 40 balls (Schaefer, 1955; Kohlrausch and Schrödinger, 1926). It is easily seen that the the excess Δ_s (which varies at each draw by 2 units) shows a marked tendency to decrease so long as it has a high value and that it fluctuates subsequently in the vicinity of $\Delta_s = 0$.

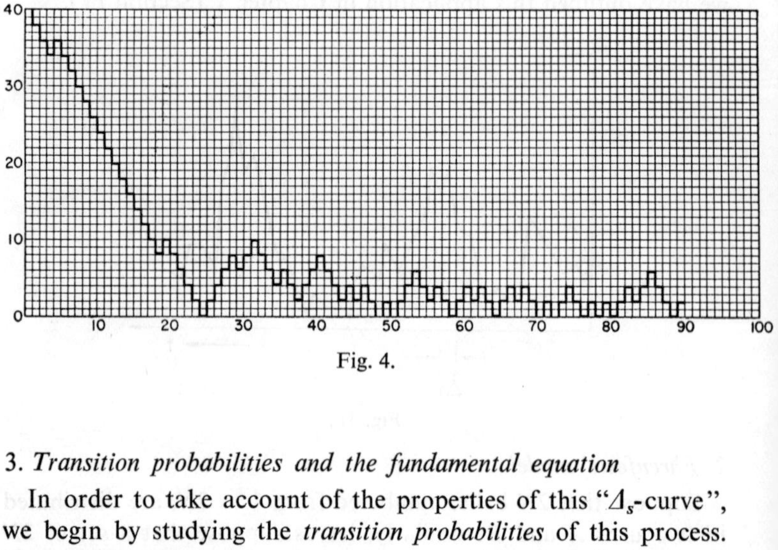

Fig. 4.

3. *Transition probabilities and the fundamental equation*

In order to take account of the properties of this "Δ_s-curve", we begin by studying the *transition probabilities* of this process. If we suppose that after s draws there are $n_A(s) = m$ balls in urn A, only the two situations

$$n_A(s + 1) = m + 1 \quad \text{or} \quad n_A(s + 1) = m - 1$$

can obtain after a new draw; in the case where $m > N$, the first possibility corresponds to an increase of Δ_s and the second to a decrease of Δ_s. According to the nature of the draws, we have immediately for the transition probabilites (since

$m = N + k$)

$$Q(m \to m+1) = \frac{2N-m}{2N} = \frac{N-k}{2N}$$

$$\left(= \frac{2N - \Delta_s}{4N} \quad \text{if} \quad k > 0, \quad \text{so that} \quad m > N\right), \tag{3}$$

$$Q(m \to m-1) = \frac{m}{2N} = \frac{N+k}{2N}$$

$$\left(= \frac{2N + \Delta_s}{4N} \quad \text{if} \quad k > 0, \quad \text{so that} \quad m > N\right). \tag{4}$$

We see from (4), that the probability of a decrease of Δ_s is the larger and, conversely, that of an increase of Δ_s is the smaller, the larger is Δ_s; this is the way in which the tendency of Δ_s to decrease is expressed and we note that this tendency is much stronger when Δ_s is larger. For example, if initially $n_A(0) = 2N$, it is clear that the first draw will involve a decrease of $\Delta_s = 2N$ of 2 units; with the second draw, however, we shall have still a probability $1 - 1/2N$ that Δ_s continues to decrease and only a probability $1/2N$ that Δ_s reassumes its initial value. If $2N \sim 10^{23}$, it can be seen that the probability of Δ_s decreasing remains very large until Δ_s has become quite small and that an irreversible decrease of Δ_s can almost certainly be expected to be observed in this case.

In addition, we can define more precisely the nature of this irreversible evolution by pointing out that P. and T. Ehrenfest's model is a special case of Markovian chain† (since the state of the system after the $(s+1)$th draw depends only on the state after the sth draw) and by writing Smoluchowski's equation satisfied by the conditional probability $P(n|m, s)$ of finding $n_A(s) = m$, with $n_A(0) = n$; with the symbols of Chapter V (Section IV) we have

$$P(n|m, s) = \sum_l P(n|l, s-1)\, Q(l|m); \tag{5}$$

according to (3) and (4), the transition probabilities $Q(l|m)$ can be written as

$$Q(l|m) = \frac{l}{2N}\delta(l-1, m) + \frac{2N-l}{2N}\delta(l+1, m). \tag{6}$$

† We point out that there is a connection between Ehrenfest's model and the Brownian movement of an elastically-bound particle.

It follows that equation (5) takes the following form (where, in order to simplify, n is omitted by putting $P(m, s) \equiv P(n|m, s)$):

$$P(m, s) = \frac{m+1}{2N} P(m+1, s-1) + \frac{2N-m+1}{2N} P(m-1, s-1), \qquad (5')$$

or again, by taking half the difference, $k(s)$, as variable,

$$P(k, s) = \frac{N+k+1}{2N} P(k+1, s-1) + \frac{N-k+1}{2N} P(k-1, s-1). \qquad (5'')$$

The solution of equations (5′) or (5″), with $P(m, 0) = \delta(n, m)$ as the initial condition, is quite tricky and we shall make only a few remarks on this subject. Nevertheless, certain interesting results can be derived from (5″) by calculating $\langle k(s) \rangle$ and $\langle k^2(s) \rangle$; we have, to begin with,

$$\langle k(s) \rangle = \sum_k kP(k, s) = \left(1 - \frac{1}{N}\right) \langle k(s-1) \rangle = n\left(1 - \frac{1}{N}\right)^s, \qquad (7)$$

with $k(0) = n$; thus, it can be seen that the average $\langle k(s) \rangle$ starts from n and approaches zero when s is sufficiently large. Subsequently, we find:

$$\langle k^2(s) \rangle = n^2 \left(1 - \frac{2}{N}\right)^s + \frac{N}{2}\left[1 - \left(1 - \frac{2}{N}\right)^s\right], \qquad (8)$$

so that if $s \to \infty$, $\langle k^2(s) \rangle \to N/2$; we shall find this value, starting from the stationary distribution for Ehrenfest's model.

4. *Stationary distribution*

An important problem, from the point of view of physical analogies, consists in studying the limit probability $\lim_{s \to \infty} P(n|m, s)$; it is expected generally that this limit is independent of the initial state n and that we could write for it in this case $W(m)$. However, Kac has shown that such was not the case in Ehrenfest's model (Kac, 1959), but that this kind of stationary distribution could be obtained provided that the value of $n = n_0$ of $n_A(0)$ is not fixed initially and that a distribution $W(n_0)$ of all possible initial values is taken (with $\sum_0^{2N} W(n_0) = 1$). We note that this amounts to replacing the system of Ehrenfest's two urns by an ensemble of such systems with initial values distributed according to $W(n_0)$. In introducing all the possible series $n_0, n_1, \ldots, n_{s-1}, m$, which progress from the initial value n_0 to the final value m, and taking account of the Markovian nature of the process, we have for the probability $\text{Prob}\,\{n_A(s) = m\}$ defined for the ensemble of systems

$$\begin{aligned}\text{Prob}\,\{n_A(s) = m\} &= \sum_{n_0} W(n_0)\, P(n_0|m, s) \\ &= \sum_{n_0, n_1, \ldots, n_{s-1}} W(n_0)\, Q(n_0|n_1) \ldots Q(n_{s-1}|m),\end{aligned} \tag{9}$$

where the $Q(n_{i-1}|n_i)$ are the transition probabilities given by (6). We then try to find a quantity $W(n_0)$ such that for every value of s

$$\text{Prob}\,\{n_A(s) = m\} = W(m);$$

for this, it is necessary that the equation

$$W(m) = \sum_{n_0=0}^{2N} W(n_0)\, Q(n_0|m) \tag{10}$$

has a solution with $W(m) \geqq 0$. We can verify easily that the only normalised solution of (10) is given by

$$W(m) = \frac{C_m^{2N}}{2^{2N}} = \left(\frac{1}{2}\right)^{2N} \frac{(2N)!}{m!(2N-m)!}; \tag{11}$$

this is the probability which would be obtained if we had assumed that the situation, with given n_A and n_B, had been obtained by playing heads or tails; heads involves placing one ball in urn A and tails involves placing one ball in B. If $2N$ is sufficiently large it can be replaced, to a good approximation, by a Gaussian law which can be written as

$$W(k) = \left(\frac{1}{\pi N}\right)^{1/2} \exp\left(-\frac{k^2}{4N}\right), \tag{11'}$$

by using the variable k; we can verify easily that we have in this case, $\langle k \rangle = 0$ and $\langle k^2 \rangle = N/2$. In view of a comparison between

Ehrenfest's Markovian process and the mechanical evolution of a system, it is interesting to note that equation (10) is the parallel of Liouville's equation and that it allows us to define the stationary distribution (11) playing the role of the invariant measure in phase space.

5. *Properties of the "Δ_s-curve"*

By considering the stationary case, it is easy to define more precisely certain properties of the Δ_s-curve, following a well-known argument of P. and T. Ehrenfest. We consider a point of the curve corresponding to $n_A(s) = m$; if m is significantly different from its average N, this point will be very probably a maximum for the Δ_s-curve. In order to see this, it is sufficient to note that there are only four possibilities for the Δ_s curve, represented by the schemes:

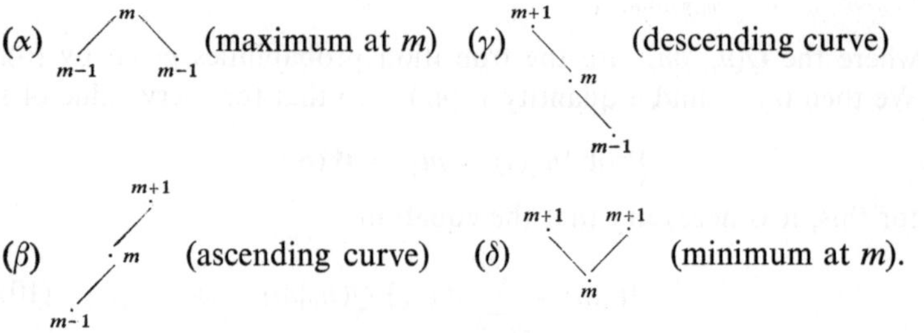

In the stationary case, the respective probabilites of these four events have, according to (3), (4) and (11), the following ratios:†

$$p_\alpha : p_\beta : p_\gamma : p_\delta = \frac{N+k}{N-k} : 1 : 1 : \frac{N-k}{N+k}. \tag{12}$$

† Let us calculate, for example, p_α; this is the conditional probability

$$\text{Prob}\{n_A(s-1) = m-1,\, n_A(s+1) = m-1 \mid n_A(s) = m\}$$

and it can be written, according to the theorem of compound probabilities, as

$$p_\alpha = \frac{\text{Prob}\{n_A(s-1) = m-1,\, n_A(s) = m,\, n_A(s+1) = m-1\}}{\text{Prob}\{n_A(s) = m\}}$$

$$= \frac{\text{Prob}\{n_A(s-1) = m-1\}}{\text{Prob}\{n_A(s) = m\}} Q(m-1 \mid m)\, Q(m \mid m-1)$$

$$= \frac{C_{m-1}^{2N}}{C_m^{2N}} \frac{2N-m+1}{2N} \frac{m}{2N} = \left(\frac{m}{2N}\right)^2.$$

Thus, it can be seen that the probability of event α is significantly greater than that of the other events, when k is not too small: this implies that if we choose in the ensemble of the series $n_0, n_1, \ldots, n_s, \ldots$, the sub-ensemble for which $n_A(s) = m$, there will be in this sub-ensemble a majority of "curves", which represents a maximum at this point.

In the same way, it can be shown that

$$\text{Prob}\{n_A(s-1) = n|n_A(s) = m\} = \text{Prob}\{n_A(s+1) = n|n_A(s) = m\},$$

that is to say, the model is *reversible*. Thus, the reversibility and the tendency of Δ_s to decrease, starting from high values, are reconciled: this is the answer to Loschmidt's paradox. Concerning Zermelo's paradox, its solution is found in calculating the time of return associated with the model; if we denote by $P(n_0|\underbrace{\bar{n}_0 \ldots \bar{n}_0}_{s-1}, n_0)$ the probability that the value n_0 is found after s draws so long as the numbers $\bar{n}_0$ corresponding to intermediate draws differ from n_0, we have, noting that we are dealing with a *strongly stationary* process,

$$\sum_{s=1}^{\infty} P(n_0|\underbrace{\bar{n}_0 \ldots \bar{n}_0}_{s-1}, n_0) = 1. \tag{13}$$

This is the recurrence theorem for a strongly stationary process, stating that every initial state n_0 must be reproduced with a probability equal to 1. The average recurrence time $\bar{T}$ can in this case be written as

$$\bar{T} = \sum_{1}^{\infty} sP(n_0|\bar{n}_0 \ldots \bar{n}_0, n_0) = \frac{1}{W(n_0)} = \frac{2^{2N}}{C_{n_0}^{2N}}; \tag{14}$$

it is generally enormous as soon as $2N$ is large and n_0 is significantly different from N.

6. *Calculation of $P(n|m, s)$*

In conclusion, we make now a few comments on the calculation of $P(n|m, s)$; because of the Markovian nature of the model we have

$$P(n|m, s) = \sum_{n_1, \ldots, n_{s-1}} Q(n|n_1)\, Q(n_1|n_2) \ldots Q(n_{s-1}|m), \tag{15}$$

so that, if we denote by Q the matrix of the transition probabilities whose (n, m) element is $Q(n|m)$, $P(n|m, s)$ is the (n, m) element of the matrix Q^s. Kac (1947b) has shown that the matrix $Q(n|m)$ can be diagonalised, although it is not symmetrical, and that its eigenvalues are the $2N + 1$ numbers $\lambda_j = j/N$ with $-N \leq j \leq N$. The probability $P(n|m, s)$ can then be expressed as a function of the eigenvectors of Q and we obtain

$$P(n|m, s) = (-1)^n \left(\frac{1}{2}\right)^{2N} \sum_{j=-N}^{+N} \left(\frac{j}{N}\right)^s D_m^{(j)} D_{N+j}^{(N-n)}, \tag{16}$$

where the $D_m^{(j)}$ are the coefficients of z^m in the expansion of the expression $(1 - z)^{N-j}(1+z)^{N+j}$. It can be seen that as $s \to \infty$, all the terms of (16) approach zero, except those corresponding to $j = N$ and $j = -N$; the term $j = N$ gives

$$(-1)^n \left(\tfrac{1}{2}\right)^{2N} D_m^{(N)} D_{2N}^{(N-n)} = \left(\tfrac{1}{2}\right)^{2N} C_m^{2N} = W(m) \tag{17}$$

which corresponds to the stationary distribution; on the other hand, the term $j = -N$ can be written as

$$(-1)^s (-1)^n \left(\tfrac{1}{2}\right)^{2N} D_m^{(-N)} D_0^{(N-n)} = (-1)^{s+n+m} W(m) \tag{18}$$

and depends therefore on n; thus, we verify that $P(n|m, s)$ does not approach a limit $W(m)$ independent of n.

Similarly, we can deduce from these formulae the results about the time of recurrence by introducing the probability $P'(n|m, s)$ such that m balls are observed in A for the *first time* after s draws when n balls were located initially in A. Then, we have the following relation:

$$P(n|m, s) = P'(n|m, s) + \sum_{k=1}^{s-1} P'(n|m, k)\, P(m|m, s-k); \tag{19}$$

if we introduce the generating functions

$$h(n|m, z) = \sum_{s=1}^{\infty} z^s P'(n|m, s), \quad g(n|m, z) = \sum_{s=1}^{\infty} z^s P(n|m, s), \tag{20}$$

equation (19) is equivalent to

$$h(n|m, z) = \frac{g(n|m, z)}{1 + g(m|m, z)}. \tag{21}$$

From this we deduce

$$\lim_{z\to 1} h(n|n, z) = h(n|n, 1)$$

$$= \sum_{s=1}^{\infty} P'(n|n, s) = 1, \quad \lim_{z\to 1} \frac{dh(n|n, z)}{dz} = \frac{1}{W(n)}, \tag{22}$$

whence the average recurrence time $\bar{T}$

$$T = \sum_{s=1}^{\infty} sP'(n|n, s) = \lim_{z\to 1} \frac{dh(n|n, z)}{dz} = \frac{1}{W(n)}. \tag{23}$$

Finally, if we wish to compare the previous stochastic process with the mechanical evolution of ensembles of systems (in Gibbs' sense) in phase space, we can liken the probability Prob $\{n_A(s) = m\} = F(m, s)$ to Ehrenfest's coarse-grained density in phase space. The function $F(m, s)$ is determined from $F(n, 0)$, which we take here to be different from $W(n)$ in order to have an evolution, and the generalised $\bar{H}$-theorem (in Gibbs' sense) would be established in this case if $\lim_{s\to\infty} F(m, s) = W(m)$. As we have seen, this is not satisfied for $F(n, 0) = \delta_{nn_0}$; we can say that the initial distribution is too fine-grained (or not sufficiently "coarse-grained"). In fact, according to (17) and (18) we have, for a very large value of s,

$$F(m, s) = \sum_{n=0}^{2N} F(n, 0)\, P(n|m, s) \simeq W(m) \tag{24}$$

$$+ (-1)^{m+s}\, W(m) \sum_{n=0}^{2N} (-1)^n F(n, 0).$$

If $F(n, 0)$ varies slowly (coarse-graining), the second term is small everywhere when $2N$ is large; thus, although $F(m, s)$ does not strictly approach $W(m)$, the second term can be neglected and this amounts to an error of the order of $1/N$.

APPENDIX III B

Notes on the Definition of Entropy

Entropy is related with the quantity H by the well-known relation $S = -kH$ (where k is Boltzmann's constant); there are two definitions of entropy by Boltzmann and Gibbs, corresponding to the two definitions of H which we analysed in Chapter V. Let us consider a system whose energy belongs to the energy shell ΔE and whose macroscopic states are defined by the stars Ω_n of Γ-space of size $W_n(\Omega_n) = \mu(\Omega_n)$: W_n represents, therefore, the measure of the microscopic states corresponding to the same macroscopic state. Having stated this, for the definition of entropy according to Boltzmann, we can put [by (V.34)]

$$S_B = k \ln W_n. \tag{25}$$

Gibbs' method consists in the introduction of a distribution function P_n' in Γ-space, where P_n' is the statistical weight of Ω_n. According to (V.27″), Gibbs' entropy is then defined by

$$S_G = -k \sum_n P_n' \ln \frac{P_n'}{W_n}. \tag{26}$$

Finally, we have also encountered in Chapter III a third definition by Fierz where we put

$$S_F = k \sum_n P_n' \ln W_n. \tag{27}$$

Here we have written only the classical form of the various definitions of entropy; the change to the quantum case can be made immediately (see Chapter VI).

If, now, we compare these definitions for the equilibrium state we note at once that the concept of equilibrium is different according to Boltzmann and Gibbs. From Boltzmann's point of view, the equilibrium state is that which makes the entropy a maximum;

if we denote by Ω_{max} the star of maximum size W_{max}, we shall have

$$(S_B)_{eq} = k \ln W_{max}. \tag{28}$$

In addition, because of the geometrical properties of N-dimensional space, it is easy to show that the definition of equilibrium entropy (28) can be replaced by $k \ln \mu(\Delta E)$ or $k \ln \mathscr{V}(E)$, as long as N is very large, where $\mu(\Delta E)$ is the measure of the shell ΔE and $\mathscr{V}(E)$ is the volume enclosed by the hypersurface $H = E$; this is the insensibility of Boltzmann's formula, pointed out by Gibbs and Lorentz (1916).

From Gibbs' point of view, the equilibrium state is defined by the distribution which makes the entropy S_G a maximum; the equilibrium state is represented, therefore, by a certain distribution. In the case that we have chosen, this equilibrium distribution is the microcanonical distribution in the shell ΔE, $P_n' = W_n/\mu(\Delta E)$, because of the extremal properties of $\bar{H}$ (and therefore of S_G: see Chapter V). In this case we have

$$(S_G)_{eq} = (S_G)_{max} = k \ln \mu(\Delta E). \tag{29}$$

Thus, we can see that the Boltzmann and Gibbs definitions are identical at equilibrium, provided that the number of degrees of freedom of the system is very large; it is easy to prove that the same is true for Fierz's entropy.

If, now, we return to Ehrenfests' model, Boltzmann's entropy for a given state m can, according to (11), be written, as

$$S_B = k \ln C_m^{2N}; \tag{30}$$

the state of maximum entropy is achieved for $m = N$ and we have

$$(S_B)_{eq} = k \ln C_N^{2N} \sim 2kN \ln 2. \tag{30'}$$

On the other hand, the function $F(m, s)$ is involved in Gibbs' entropy, which can be written as

$$S_G = -k \sum_m F(m, s) \ln \left[\frac{F(m, s)}{C_m^{2N}}\right]; \tag{31}$$

this is a maximum for $F(m, s) = W(m)$, or again

$$(S_G)_{eq} = 2kN \ln 2. \tag{31'}$$

Finally, we can show (by making use of the convexity of the expression $x \ln x$) that S_G is an increasing function of s, although $F(m, s)$ does not approach $W(m)$, in the strict sense; the monotonic behaviour of S_G must, therefore, be contrasted to the fluctuating behaviour of S_B, similar to the curve in Fig. 4.

APPENDIX IV

Note on Recent Developments in Classical Ergodic Theory

THERE have in recent years been important developments in essentially mathematical research on the ergodic properties of classical dynamic systems. This work, mainly by mathematicians from the Soviet school,† is based on the fundamental concept of an abstract dynamical system and it defines general classes of systems whose temporal evolution satisfies certain asymptotic properties (ergodicity, mixing, etc.). It employs the generalised definition of the entropy of a measure preserving transformation, in the Kolmogorov sense, and also several concepts, such as that of a *K*-system, which are used in information theory. This research has led to some remarkable results such as, for instance, ergodicity and mixing properties of geodesic flow on compact Riemannian manifolds with negative curvature or, more generally, of *K*-systems, and especially the proof of the ergodicity of the Boltzmann–Gibbs gas model.‡ These researches are therefore of major interest for the problems which we have considered in the present book; this is the reason why we have felt obliged to give in this appendix a short account of them, restricting ourselves to indicate the methods employed as well as the main theorems obtained, without insisting on rigorous proofs which would considerably go beyond the framework of our treatment.

† See, for instance, Anosov (1962, 1963), Arnold and Avez (1968; this reference contains an important bibliography of all papers by the Soviet school), Kolmogorov (1954, 1956, 1959, 1960), Kouchnirenko (1965), Rohlin (1949, 1962), Rohlin and Sinai (1961), and Sinai (1960, 1961a, b, c, 1963a, b, c).

‡ This is the model in which a gas is represented by perfectly elastic hard spheres in a parallelepiped with perfectly reflecting walls.

I. The Concept of an Abstract Dynamic System

From the analysis of the fundamental concepts of classical statistical mechanics, given in Chapter I, it follows that the underlying mathematical structure of the theory contains essentially three concepts:

Γ-space which in the case of a classical dynamic system is the same as the real Euclidean space R_n;

a measure μ defined on Γ;

a group of automorphisms with a parameter T_t which is defined on Γ and which preserves the measure μ; for a Hamiltonian system the group T_t is generated by the equations of motion for the system.

If we want to free ourselves from the particular framework of classical dynamics and forgetting about the physical meaning of these concepts, we are naturally led to consider abstract dynamical systems whose definition rests upon the following two elements:

A *measure space* represented by the triplet (X, Σ, μ) where X is a set with a sigma-algebra† Σ of measurable sub-sets of X and with a measure‡ μ defined on this σ-algebra.

A *group of automorphisms* (mod 0)†† in the measure space (X, Σ, μ) which preserves measure; according to circumstances this group may be a continuous group with a parameter t, denoted by T_t, or a discrete group generated by the automorphism T. The essential point is the measure preservation by these automorphisms, which can be expressed as $\mu(T_t A) = \mu(A)$ for all $t \in R$ and for any measurable set A; we note that the transformation T_t (or T) itself is measurable on the product space $X \times R$ and that it defines a measurable "flow" on the measure space (X, Σ, μ).

Let us straightaway remark that by considering a group of transformations we require that the transformations T (or T_t) be invertible, that is, that we can associate with each T a measure-preserving transformation S such that $ST = TS = I$. S is the

† Let us remind ourselves that a σ-algebra is a class of sets which is closed under all countable operations of the set (formation of complements and reunions, or of intersections according to the duality rule).

‡ The measure μ is an countably additive set function which is non-negative but may be infinite. We shall assume that the measure μ is σ-finite.

†† The notation (mod 0) indicates that we consider two transformations which differ merely on a set of measure zero to be equivalent.

inverse of T, which we can express as $S = T^{-1}$. Although in general the transformations considered in dynamic systems are invertible, we shall see that many results of the ergodic theory are valid for measure-preserving but non-invertible transformations, that is, for semi-groups of transformations T (or T_t).

With these definitions we call the triplet (X, μ, T) formed by the measure space X, the measure μ, and the group of measure-preserving automorphisms T (or T_t) an *abstract dynamical system*. This very general concept enables us to include systems which differ as widely as the Hamiltonian systems of classical mechanics, stochastic processes with discrete or continuous time variables, groups of geometric transformations defined on a measure space, etc. Of course, each of these cases corresponds to a particular choice of the space X and of the automorphisms T (or T_t).

If we assume that the space X has a topology, we can define on X continuous flows T_t; if, moreover, X has the structure of a differentiable manifold, we can introduce differentiable flows. To illustrate the variety of applications included in this theory, let us give some examples of such systems:

Hamiltonian systems for which the space X is the real Euclidean space R^{2n} with a Borel or Lebesgue measure; the group T_t associated with the Hamiltonian equations then defines a continuous flow on R^{2n}.

If the Hamiltonian H is infinitely differentiable (C^∞-function), the relation $H = E$ defines a C^∞-differentiable flow defined locally by the C^∞ field of the vectors tangential to the hypersurface $H = E$.

Flows on the torus where the space X for the case of a two-dimensional torus T^2 is represented by the basic square with its opposite sides (of length unity) being identical. Using as measure the usual measure $dx\, dy$ we obtain a measure space X on which we can define different automorphisms such as: translations (mod 1), linear transformations with determinant equal to 1, ...

Bernoulli stochastic schemes for which the space X is the set of bilateral sequences $x = \{x_i\}$ where the x_i are equal to one of the first n integers $0, 1, 2, \ldots, n-1$. We define a measure on X by assigning to each of these possibilities a mass p_j in such a way that

$$\sum_{j=0}^{n-1} p_j = 1.$$

The automorphism T is then a translation on X which associates with the element $x = \{x_i\}$ the element $Tx = y = \{y_i\}$ with $y_i = x_{i+1}$;

one sees easily that this transformation can be inverted and that it preserves measure.

Let us now conclude these general considerations by introducing an essential concept for the classification of abstract dynamic systems, the concept of *isomorphism of two systems*:

We say that two systems (X, μ, T) and (X', μ', T') are isomorphic if there exists a bijective application S of X on X' such that we have

$$\mu'[S(A)] = \mu(A) \quad \text{for all} \quad A \in \Sigma,$$

and

$$T' = STS^{-1},$$

where the application S clearly defines an isomorphism of the two measure spaces (X, μ) and (X', μ'); a similar definition occurs for the continuous case with a group T_t.

The importance of this concept lies in the fact that the properties of abstract dynamic systems which only involve the measure and the structure of the group T are the same for two isomorphic systems. In particular, this is true for the asymptotic properties which we shall be considering; if one proves, for instance, the ergodicity of one abstract system, this property will also hold for all other systems which are isomorphic with that system. We see thus that the ergodic theorems which we shall now state to a large extent go beyond the primitive physical framework of the theory.

II. Asymptotic Properties of Abstract Dynamic Systems

We shall now study the asymptotic properties of such systems. Let us first of all remind ourselves briefly what the notation is which we shall be using and what the essential concepts are which we have used all the time in the present book.

1. *Definitions*

(a) *Time averages.* Let (X, μ, T), or (X, μ, T_t), be a dynamic system with a discrete, or continuous, time variable, and let f be a function defined on X. The time average of f (if it exists) is by definition:

$$f^*(x) = \lim_{N\to\infty} \frac{1}{N} \sum_{n=0}^{N-1} f(T^n x), \quad \text{if} \quad x \in X, \quad n \in Z^+,$$

in the discrete case (Z^+ represents the set of positive integers), and

$$f^*(x) = \lim_{T\to\infty} \frac{1}{T} \int_0^T f(T_t x)\, dt, \quad x \in X, \quad t \in R,$$

in the continuous case.

(b) *Space averages.* By definition this is the integral of f over the space X with the condition† $\mu(X) = 1$. We have thus:

$$\bar{f} = \int_X f(X)\, d\mu,$$

(c) *Ergodic systems.* With the above definitions, we say that a system is ergodic if for any μ-summable function the time and space averages are almost everywhere equal, or,

$$f^*(x) \underset{\text{a.e.}}{=} \bar{f}.$$

We see immediately that the time average is for an ergodic system independent of the point x.

(d) *Decomposable systems.* These are systems for which X is the disjoint union of two sets X_1 and X_2 which are invariant under T (or T_t) and have positive measure. We can thus write:

$$X = X_1 \cup X_2 \quad \text{with} \quad X_1 \cap X_2 = \emptyset,$$

$$TX_1 = X_1, \quad TX_2 = X_2,$$

and

$$\mu(X_1) > 0, \quad \mu(X_2) > 0.$$

Such a system is clearly not ergodic since when one takes for $f(x)$ the characteristic function of X_1 (or of X_2), we have $f^*(x) = f(x)$ so that the time average depends on x. An indecomposable system is called metrically transitive; therefore this property is the necessary and sufficient condition for the ergodicity of a system, as we have seen already.

(e) *Mixing systems.* By analogy with Gibbs' example (where we consider a mixture of water and black ink) we say that a system (X, μ, T_t) is a mixing one if for any measurable sets A and B we have

$$\lim_{t\to\infty} \mu[T_t A \cap B] = \mu(A)\, \mu(B).$$

† We clearly assume here that the measure μ is finite and is then chosen in such a way that $\mu(X) = 1$.

This property which plays an important role in the study of the foundations of statistical mechanics is stronger than ergodicity; every mixing system is, by the way, ergodic. One can similarly introduce other concepts, such as weak mixing.

2. *Asymptotic properties*

(a) *Recurrence.* The abstract dynamic systems which we have just defined satisfy very general asymptotic properties which are similar to Poincaré's recurrence theorem for a continuous Hamiltonian flow T_t. From an abstract point of view we can define the concept of recurrence as follows:

Let $E \subset X$ be a sub-set of X; we call a point $x \in E$ recurrent (with respect to E and to T) if $T^n x \in E$ for at least one positive integer n. (To simplify matters we shall consider the discrete case.) One can the state the *Recurrence Theorem*:

If E is a measurable sub-set of a space X of finite measure and if T is a measure-preserving transformation of T onto itself, practically all points of E are infinitely recurring; that means that if $x \in E$, we have $T^n x \in E$ for an infinite number of positive values of n.

The essential elements of the proof are the following ones: X has a finite measure, $\mu(X) < \infty$, and T preserves measure. We note that the invertibility of T is not invoked so that the recurrence properties are valid for transformations which cannot be inverted.

(b) *Ergodic theorems for abstract dynamic systems.* We have often indicated that the study of the ergodic properties of the transformations T contains two well separated stages. First of all, we must prove the existence of limits of time averages associated with certain classes of functions f defined on the measure space X; this is the object of the ergodic theorems which we are about to state. Then we have to determine under what conditions these time averages reduce (almost everywhere) to constants which moreover are equal to the space averages of the functions f; this is the real ergodicity property to which we shall come back later on.

The functions f which we are considering introduce different functional spaces in which the transformations T induce transformations U. We shall essentially consider the class of summable functions, $f \in L_1(X, \mu)$ and that of the quadratically summable functions, $f \in L_2(X, \mu)$. Before stating the ergodic theorems let us briefly remind ourselves of the properties of the operators U which are induced in the two functional spaces.

(A) *Properties of the induced operators* U (or U_t)

If f is some function on X, the operator U induced by T is defined by

$$Uf(x) = f(Tx);$$

this is a covering operation acting in the functional space associated with f. U possesses two fundamental properties which are a consequence of its definition and of the fact that T conserves measures:

U is *linear*.

U is an *isometry* on $L_1(X, \mu)$ if T preserves measure.

It follows that U is an isometry on $L_2(X, \mu)$. If, moreover, the transformation T can be inverted, U is an invertible isometry; and we have thus a third property:

U is a *unitary* operator on $L_2(X, \mu)$ if T preserves measure and is invertible. This is in the discrete case the equivalent of the theorem stated for the continuous case by Koopman (1931):

Any measurable and invertible flow T_t induces on $L_2(X, \mu)$ a continuous and unitary group U_t with one parameter which is defined by $U_t f(x) = f(T_t x)$.

We can now state the ergodic theorems for the abstract dynamic systems; they are a generalisation of the theorems which we have discussed in Chapter I of the present book. As we have already noted the kind of convergence adopted depends naturally on the functional space considered; we shall thus be dealing with a convergence in quadratic mean in the case of L_2 and a point (or nearly certain) convergence in the case of L_1.

(B) *Ergodic theorem in quadratic mean*

This is a generalisation of von Neumann's theorem proved for unitary operators. It is the *Mean Ergodic Theorem*:

If U is an isometry on the complex Hilbert space $L_2(X, \mu)$ and if P is the projection on the space of all vectors which are invariant under U the series $N^{-1} \sum_{j=0}^{N-1} U^j f$ converges to P for all $f \in L_2$.

We note that this theorem, which was proved by Riesz (1945) is not limited to unitary operators, as the transformations T are not necessarily invertible.

(C) *Individual ergodic theorem*

This is a generalisation of Birkhoff's theorem and was also proved by Riesz (1945). It is the following theorem:

If T is a measure preserving transformation, which is not necessarily invertible, on the measure space (X, Σ, μ) (whose measure may be infinite) and if $f \in L_1(X, \mu)$, $N^{-1} \sum_{j=0}^{N-1} f(T^j x)$ will converge almost everywhere. The limit function f^* is integrable and invariant (we have almost everywhere $f^*(Tx) = f^*(x)$); if, moreover, $\mu(X) < \infty$, we have

$$\int f^* \, d\mu(x) = \int f \, d\mu(x).$$

The proof of this theorem is based upon an auxiliary statement due to Yosida and Kakutani (1939) which is called the maximal theorem. It states the *Maximal Ergodic Theorem*:

Let (X, Σ, μ) be a measure space and f an integrable function on X. If E is the set of points x such that at least one of the sums $f(x) + f(Tx) + \cdots + f(T^n x)$ is positive, we have

$$\int_E f(x) \, d\mu(x) \geqq 0.$$

These theorems establish the existence of "time" limits in very general circumstances. We can now return to the real ergodicity property.

(c) *Ergodicity and mixing.* We have seen that a decomposable system is non-ergodic by virtue of the definition itself of ergodicity. Conversely, if a system (X, μ, T) is non-ergodic it is decomposable; indeed, there then exists a function whose time average $f^*(x)$ is not constant almost everywhere. As f^* is invariant under T one can define two sub-sets of X which are invariant under T and which have non-zero measure. We are thus led to the statement:

The necessary and sufficient condition that an abstract dynamic system be ergodic is that it is indecomposable, that is, that any invariant measurable set has measure 0 or 1.

Following what we have seen so far we can also say that a system is ergodic if, and only if, any invariant function $f \in L_1(X, \mu)$ is almost everywhere constant.

On the other hand it is easy to see from the definition of the mixing property that an abstract dynamic mixing system is ergodic. Indeed, if A is an invariant measurable sub-set and if we form $B = X - A$, we have:

$$(T_t A) \cap B = A \cap B = \emptyset.$$

According to the mixing property we must thus have $\mu(A)\,\mu(B) = 0$, whence $\mu(A) = 0$ or 1. On the other hand, one can easily construct examples of ergodic systems which are not mixing (for instance, the ergodic translations defined on tori).

The mixing property is thus much more restrictive that that of ergodicity. One can introduce concepts intermediate between ergodicity and mixing; this is the case for weak mixing, introduced by Halmos (1956):

A dynamic system (X, μ, T_t) is weakly mixing if

$$\lim_{T\to\infty} \frac{1}{T} \int_0^T |\mu(T_t A \cap B) - \mu(A)\,\mu(B)|\, dt = 0,$$

in the continuous case (with an analogous definition for the discrete case). Note that one can also introduce mixings of a higher degree.

We have several times emphasised that the essential problem is to determine under what conditions an abstract dynamic system is effectively ergodic (or mixing). The condition relating to the metric transitivity of a system is difficult to check and it is thus good to look for other conditions which are equivalent to it. We may look for a solution of this problem by studying the spectral structure of the induced operators.

(d) *Spectral properties of the induced unitary operators.* We consider here the case of the unitary operators U induced in $L_2(X, \mu)$ by invertible transformations. We note that if two dynamic systems are isomorphic, they induce in L_2 unitary operators U and U' which are equivalent. The invariants of the induced operator correspond thus to certain invariants of the dynamic system: we call them spectral invariants; the spectrum of U also belongs to them. (On the other hand, the operators U and U' may be equivalent without the systems being isomorphic.) It now turns out that certain ergodic properties have a spectral counterpart. This is an important consequence of the following *Eigenvalue Theorem*:

An invertible transformation T which preserves measure on a finite measure space is ergodic if and only if the number 1 *is a simple eigenvalue of the induced unitary operator U.*

Thus, if T is ergodic, we have:

(a) The absolute value of any eigenfunction of U is constant a.e.

(b) Every eigenvalue is a simple one.

(c) The set of eigenvalues of U is a sub-group of the circular group.

(d) If the system is a mixing one, the only eigenvalue is 1.

In the case of a continuous transformation T_t the spectrum of U_t is the spectrum associated with the resolution of the identity E (compare Stone's theorem, p. 16); the spectral resolution of U_t can be written as

$$U_t = \int_{-\infty}^{+\infty} e^{2\pi i\lambda t}\, dE(\lambda).$$

We now see that an abstract dynamic system is ergodic if and only if $\lambda = 0$ has multiplicity 1 in the spectrum of U_t.

The mixing property can still be translated in another way in the language of induced operators. Using the definition of mixing we obtain the following theorem: an abstract dynamic system is mixing if and only if for all $f, g \in L_2(X, \mu)$ we have

$$\lim_{T\to\infty} (U_t f, g) = (f, 1)\,(1, g).$$

On the other hand, we say that an ergodic dynamic system has a truly continuous spectrum if constants are the only eigenfunctions of U_t. We can then prove that the necessary and sufficient condition that an ergodic dynamic system has a truly continuous spectrum is that it is weakly mixing.

We see thus the importance of the spectral properties of the induced operator in the study of the ergodicity of dynamic systems. For the developments which we shall consider it is useful also to consider the case of a discrete spectrum.

We say that an abstract dynamic system (X, μ, T) has a truly discrete spectrum if the eigenfunctions of the unitary induced operator U form a complete base of $L_2(X, \mu)$. With this definition one can then prove the *Theorem of the Discrete Spectrum*:

Two ergodic dynamic systems with truly discrete spectra are isomorphic if and only if their induced unitary operators are equivalent.

This theorem must in principle enable us to answer any question about ergodic transformations with a discrete spectrum; for this one uses the following result:

Theorem: A measure conserving ergodic transformation with a truly discrete spectrum is isomorphic with a rotation on a compact Abelian group.

This theorem has an important corollary:

Corollary. Any sub-group of the circular group is the spectrum of a measure-preserving ergodic transformation with a truly discrete spectrum.

The importance and generality of these results suggest that one of the problems of ergodic theory is to study how far the consequences of the theorem of the discrete spectrum remain valid for systems whose spectrum is not necessarily discrete. In this context we mention in conclusion a special case of a continuous spectrum which often occurs, namely that of systems with a Lebesgue spectrum which is defined as follows (compare Halmos, 1956; Arnold and Avez, 1968):

An abstract dynamic system (X, μ, T) with induced unitary operator U is said to have a Lebesgue spectrum L^I if there exists a complete orthonormal base of $L_2(X, \mu)$ formed of the function 1 and the functions f_{ij} such that

$$Uf_{ij} = f_{i,j+1} \quad \text{for all} \quad i \in I, \quad j \in Z$$

(where Z denotes the set of positive or negative integers).

The power of the set I is called the order of multiplicity of the Lebesgue spectrum; if I is countably infinite, it has a countably infinite Lebesgue spectrum and if I contains one single element, we are dealing with a simple Lebesgue spectrum.

With this definition we can now establish the following important theorem:

Theorem: An abstract dynamic system with a Lebesgue spectrum is mixing.

We note that one can construct automorphisms on the torus which possess this property (compare Arnold and Avez, 1968). Similarly one can show that the Bernoulli schemes have an infinite Lebesgue spectrum; they are thus all mixing and of the same spectral type.

The application of these general results to the case of classical dynamic systems goes thus through the study of the spectral properties of the unitary evolution operator U_t, as we saw in Chapter I. This programme poses very difficult problems which we can only mention in this short note. We shall conclude this brief account by indicating another approach which is based upon a geralised definition of entropy in the Kolmogorov sense (1959) and on the introduction of a particular class of abstract dynamic systems called K-systems.

III. Entropy and K-systems

We have seen that the ergodicity and mixing properties are connected with those of the spectral invariants of dynamic systems. We shall now introduce the concept of a generalised entropy which follows from information theory and which enables us to define a new invariant of dynamic systems. We have combined in the same section the study of the generalised entropy and that of the K-systems as these ideas are both based upon a consideration of the algebras of measurable decompositions.

1. *Measurable decompositions*

If (X, Σ, μ) is a measurable space, a decomposition $\alpha = \{A_i, i \in I\}$ of X is a family of measurable non-empty sub-sets such that

$$\mu(A_i \cap A_j) = 0, \text{ if } i \neq j; \quad \mu(X - \bigcap_{i \in I} A_i) = 0, \quad 0 < \mu(A_i) < \infty.$$

The decomposition α is measurable if there exists a countable family of measurable sets $\{B_j, j \in J\}$ such that:

(1) Each B_j is the reunion of sets of α;

(2) For each pair A_i, A_j of α there exists a B_k such that $A_i \subset B_k$, $A_j \not\subset B_k$, or $A_i \not\subset B_k$, $A_j \subset B_k$.

According to this definition, the decomposition α is measurable, if it is finite or countable. On the other hand, we can say that two decompositions α and β are identical if they are the same except possibly on a set of measure zero; we then write $\alpha = \beta \pmod 0$.

We can assign a partial ordering relation to the family of measurable decompositions, and we define this as follows: we write $\alpha \leqq \beta$ if each set of α is a reunion of sets of β except perhaps for a set of zero measure. We can then introduce an operation $\vee$ defined as follows:

Let $\{\alpha_l, l \in L\}$ be a family of measurable decompositions; we denote by

$$\alpha = \bigvee_{l \in L} \alpha_l$$

the smallest decomposition containing all α_l. For example, for the two decompositions $\alpha = \{A_i, i \in I\}$ and $\beta = \{B_j, j \in J\}$ we have

$$\alpha \curlyvee \beta = \{A_i \cap B_j | \mu(A_i \cap B_j) \neq 0, i \in I, j \in J\}.$$

It follows immediately from this definition that the operation $\curlyvee$ is commutative, $\alpha \curlyvee \beta = \beta \curlyvee \alpha$, and associative, $(\alpha \curlyvee \beta) \curlyvee \gamma = \alpha \curlyvee (\beta \curlyvee \gamma)$. Finally, if $\alpha \leqq \alpha'$ and $\beta \leqq \beta'$, we have

$$\alpha \curlyvee \beta \leqq \alpha' \curlyvee \beta'.$$

We conclude these considerations by introducing the concept of the algebra of a measurable decomposition. The algebra $\mathscr{A}(\alpha)$ generated by a measurable decomposition α is the set formed by the countable reunions of elements of α. With this definition one finds the following properties:

$$\mathscr{A}(\alpha) = \mathscr{B}(\beta), \quad \text{if} \quad \alpha = \beta;$$

$$\mathscr{A}(\alpha) \subset \mathscr{B}(\beta), \quad \text{if} \quad \alpha \leqq \beta;$$

and

$$\mathscr{A}\Big(\bigvee_{l \in L} \alpha_l\Big) = \bigvee_{l \in L} \mathscr{A}(\alpha_l).$$

The algebras of measurable decompositions are thus special cases of all measurable sets which we shall denote by $\mathscr{M}$. By definition, a sub-algebra of measurable sets $\mathscr{F}$ of $\mathscr{M}$ is a part of $\mathscr{M}$ closed for countable reunions and for changing to the complement. We can thus establish the following result:

For any sub-algebra $\mathscr{F}$ of $\mathscr{M}$ there exists a measurable decomposition α such that $\mathscr{F} = \mathscr{A}(\alpha)$.

2. *Entropy*

Before introducing the concept of the entropy of an automorphism we shall define the entropy of a decomposition. We define the entropy of a decomposition α as follows:

$$h(\alpha) = -\sum_{i \in I} \mu(A_i) \ln \mu(A_i).$$

We note that we come back to the usual definition when we consider, for instance, a decomposition of N elements of the same measure; indeed, we have in that case: $h(\alpha) = \ln N$. We note also that two equivalent decompositions have the same entropy; finally, if $h(\alpha) = 0$, all $\mu(A_i)$ vanish, bar one of them which is equal to 1.

By considering two decompositions $\alpha = \{A_i, i \in I\}$ and $\beta = \{B_j, j \in J\}$ we can then introduce the concept of a conditional entropy, $h(\alpha|\beta)$, defined by

$$h(\alpha|\beta) = - \sum_{i \in I, j \in J} \mu(A_i \cap B_j) \ln [\mu(A_i \cap B_j)/\mu(B_j)].$$

We can then prove the following properties:

$h(\alpha \vee \beta|\beta) = h(\alpha|\beta)$;

$h(\alpha|\beta) \geqq 0$, the equal sign holding if and only if $\alpha \leqq \beta$;

$h(\alpha \vee \beta) = h(\beta) + h(\alpha|\beta)$;

$h(\alpha) \leqq h(\alpha')$, if $\alpha \leqq \alpha'$; $h(\alpha|\beta) \leqq h(\alpha|\beta')$, if $\beta \leqq \beta'$.

We can now turn to the concept of the entropy of an automorphism with respect to a decomposition:

If (X, μ, T) is a dynamic system, the number

$$h(\alpha, T) = \lim_{n \to \infty} [h(\alpha \vee T\alpha \vee \cdots \vee T^{n-1}\alpha)]/n$$

is called the entropy of the automorphism T with respect to a, finite or countable, measurable decomposition α. Using the above-mentioned properties of $h(\alpha|\beta)$, we can show that the limit $h(\alpha, T)$ exists and is equal to

$$\lim_{n \to \infty} h(T^n\alpha|\alpha \vee T\alpha \vee \cdots \vee T^{n-1}\alpha).$$

We now get to the definition of the entropy of an automorphism (see Kolmogorov, 1960; Sinai, 1960):

The entropy $h(T)$ of the automorphism T is:

$$h(T) = \sup h(\alpha, T),$$

where α denotes the set of finite measurable decompositions. This entropy satisfies the following important theorem (which one can prove by considering two isomorphic systems):

The entropy $h(T)$ is an invariant of the dynamic system (X, μ, T).

To calculate the entropy of an automorphism we can use the Kolmogorov–Sinai theorem (Kolmogorov, 1959, 1960; Sinai, 1960):

If the decomposition α is a generator with respect to T, we have $h(T) = h(\alpha, T)$. (We say that α is a generator with respect to T if the closure of $\bigvee_{n=-\infty}^{+\infty} T^n\mathscr{A}(\alpha)$ is identical with $\mathscr{M}$.)

We can thus define the entropy of Bernoulli schemes, the entropy of ergodic automorphisms on a torus, etc. We can also prove the important *Kouchnirenko theorem* (1965):

Classical systems have a finite entropy.

It follows that abstract systems with an infinite entropy cannot be realised by classical systems; this is the case for diffusion-type Markovian processes. We shall now define a particular class of systems, the K-systems, which possess remarkable properties.

3. *K-systems*

A dynamic system (X, μ, T) is a K-system if there exists a subalgebra $\mathscr{F}$ of the algebra $\mathscr{M}$ of measurable sets (mod 0) such that

$$\mathscr{F} \subset T\mathscr{F},$$

$$\bigcap_{n=-\infty}^{+\infty} T^n\mathscr{F} = \mathscr{N},$$

where $\mathscr{N}$ represents the algebra of the sets with measure 0 or 1 and

$$\overline{\bigvee_{n=-\infty}^{+\infty} T^n\mathscr{F}} = \mathscr{M},$$

where the bar indicates the closure operation.

There is a similar definition for the continuous case, with T_t replacing T. These systems possess important properties illustrated by the following theorems:

A K-system has a countably infinite Lebesgue spectrum on the orthocomplement of constant functions.

As a consequence, the K-systems are mixing and ergodic, according to the spectral theorems of Section II.

The entropy of a K-system is positive. As a consequence we can show that there are systems with a Lebesgue spectrum which are not K-systems.

The K-systems form a very large class of abstract systems; they contain for example the Bernoulli schemes; also many classical systems such as the automorphisms of the tori and the geodesic flows on compact Riemannian manifolds with negative curvature. Using these results one can then prove that the Boltzmann–Gibbs gas model is not only ergodic but, moreover, is a K-system, a theorem which is essential from our point of view; of course, the elastic collisions between the spheres which represent the gas molecules are the origin of the ergodicity and mixing properties of this model.

The rigorous proof of this theorem is very tedious and long and we cannot dream of reproducing it here. We restrict ourselves to indicate that it rests upon the properties of a class of dynamic systems, the so-called *C*-systems, which are characterised by the existence of asymptotic orbits and of corresponding "sheets" of the phase space. More exactly, one can say that the positive character of the entropy and the properties of a *K*-system are connected with the existence of bundels of asymptotic trajectories which approach each other exponentially; this is the case for the geodesic flow on Riemannian manifolds with negative curvature. Sinai's proof (1963c) then consists in being guided by these properties in studying the Boltzmann–Gibbs model. He first of all proves that this dynamic system has a positive entropy; then he defines transverse sheets of the phase space similar to the ones one obtains in the case of the geodesic flow when one considers the families of manifolds orthogonal to the bundle of asymptotic trajectories. The ergodicity of the Boltzmann–Gibbs model then follows from the study of the σ-algebra generated by the decomposition of the phase space in transverse sheets.

We hope that this brief account is sufficient to show the importance of modern research in ergodic theory and the interest which this research has for the study of the foundations of statistical mechanics. It seems certain to us that this research will lead in the near future to many important developments.

Bibliography†

ABRIKOSOV, A. A., L. P. GOR'KOV and I. E. DZYALOSHINSKII (1965) *Quantum Field Theoretical Methods in Statistical Physics*, Pergamon Press, Oxford.

AESCHLIMANN, F. (1952) *J. Phys. Rad.* **13**, 600.

ALBERTONI, S., P. BOCCHIERI and A. LOINGER (1960) New Theorem in the Classical Ensemble Theory, *J. Math. Phys.* **1**, 244.

ANOSOV, D. V. (1962) Roughness of geodesic flows on compact riemannian manifolds of negative curvature, *Sov. Math.-Doklady* **3**, 1068.

ANOSOV, D. V. (1963) Ergodic properties of geodesic flows on closed riemannian manifolds of negative curvature, *Sov. Math.-Doklady* **4**, 1153.

ARNOLD, V. I., and A. AVEZ (1968) *Ergodic Problems of Classical Mechanics*, Benjamin, New York.

*BARTLETT, M. S., (1955) *Introduction to Stochastic Processes*, Cambridge Univ. Press.

BERGMANN, P. G., and J. L. LEBOWITZ (1955) *Phys. Rev.* **99**, 578.

*BILLNGSLEY, P. (1965) *Ergodic Theory and Information*, J. Wiley, New York.

BIRKHOFF, G. D., (1931 a) Proof of a Recurrence Theorem for Strongly Transitive Systems, *Proc. Nat. Acad. Sci. U.S.A.* **17**, 650.

BIRKHOFF, G. D. (1931 b) Proof of the Ergodic Theorem, *Proc. Nat. Acad. Sci. U.S.A.* **17**, 656.

*BIRKHOFF, G. D. (1932) Probability and Physical Systems, *Bull. Amer. Math. Soc.* p. 361.

*BIRKHOFF, G. D., and B. O. KOOPMAN (1932) Recent Contributions to the Ergodic Theory, *Proc. Nat. Acad. Sci. U.S.A.* **18**, 279.

*BLANC-LAPIERRE, A., and R. FORTET (1953) *Théorie des fonctions aléatoires*, Masson, Paris.

BLANC-LAPIERRE, A., P. CASAL and A. TORTRAT (1959) *Méthodes mathématiques de la mécanique statistique*, Masson, Paris.

*BLANC-LAPIERRE, A., and A. TORTRAT (1955) Statistical Mechanics and Probability Theory, *Proceedings of the Third Berkeley Symposium on Mathematical Statistics and Probability*, vol. III, p. 145, University of California Press.

*BLATT, J. M. (1959) *Progr. Theor. Phys.* **22**, 745.

BLOCH, C. (1960) On the Theory of Imperfect Fermi Gases, *Physica*, **26**, S 62.

† Apart from the papers mentioned in the text, this bibliography also contains many papers (indicated by asterisks) which are not quoted in the text, but which contain material relevant to the topic of the present book.

BLOCH, C., and C. DE DOMINICIS (1958) Un développement du potentiel de Gibbs d'un système quantique composé d'un grand nombre de particules, *Nucl. Phys.* **7**, 459.

BLOCH, E. (1930) *Théorie cinétique des gaz*, 3rd ed., A. Colin, Paris.

*BOCCHIERI, P. (1959) *Rend. Sem. Mat. Fis. di Milano*, **30**, 3.

BOCCHIERI, P., and A. LOINGER (1957) Quantum Recurrence Theorem, *Phys. Rev.* **107**, 337.

BOCCHIERI, P., and A. LOINGER (1958) Ergodic Theorem in Quantum Mechanics, *Phys. Rev.* **111**, 668.

BOCCHIERI, P., and A. LOINGER (1959) Ergodic Foundation of Quantum Statistical Mechanics, *Phys. Rev.* **114**, 948.

DE BOER, J. (1949) Molecular Distribution and Equation of State of Gases, *Rep. on Progr. in Phys.* **12**, 305.

BOGOLYUBOV, N. N. (1946a) *J. Phys. U.S.S.R.* **10**, 256 and 265.

BOGOLYUBOV, N. N. (1946b) *Problems of a Dynamical Theory in Statistical Physics*, Tech. Press, Moscow. (English translation in *Studies in Statistical Mechanics*, vol. I).

BOHM, D., and W. SCHÜTZER (1955) *The General Statistical Problem in Physics and the Theory of Probability*, Suppl. vol. II, p. 1004.

*BOLTZMANN, L. (1896 and 1898) *Vorlesungen über Gastheorie*, vols. 1 and 2 Barth, Leipzig (French translation by A. GALOTTI with introduction and notes by M. BRILLOUIN, 1902–5, 2 vols.. Gauthier-Villars, Paris. English translation by S. G. Brush, 1964, Univ. of California Press, Berkeley.)

*BOLTZMANN, L. (1866) Über die mechanische Bedeutung der zweiten Hauptsatzes der Wärmetheorie, *Wien. Ber.* **53**, 195.

*BOLTZMANN, L. (1871) Über das Wärmegleichgewicht zwischen mehratomigen Gasmolekülen, *Wien. Ber.* **63** (2), 397.

*BOLTZMANN, L. (1871) Einige allgemeine Sätze über Wärmegleichgewicht, *Wien. Ber.* **63** (2), 679.

*BOLTZMANN, L. (1872) Weitere Studien über Wärmegleichgewicht unter Gasmolekülen (*H*-Theorem), *Wien. Ber.* **66** (2), 275.

*BOLTZMANN, L. (1875) Über das Wärmegleichgewicht von Gasen, auf welche äußere Kräfte wirken, *Wien. Ber.* **72** (2), 427.

*BOLTZMANN, L. (1877) Bemerkungen über einige Probleme der mechanischen Wärmetheorie, *Wien. Ber.* **75**, 62.

*BOLTZMANN, L. (1877) Über die Beziehung zwischen dem zweiten Hauptsatze der mechanischen Wärmetheorie, *Wien. Ber.* **76**, 373.

*BOLTZMANN, L. (1878) Weitere Bemerkungen über einige Probleme der mechanischen Wärmetheorie, *Wien. Ber.* **78**, 7.

*BOLTZMANN, L. (1887) Über die mechanischen Analogien des zweiten Hauptsatzes der Thermodynamik, *J. für reine und angewandte Mathematik*, **100**, 201.

*BOLTZMANN, L. (1887) Neuer Beweis zweier Sätze über Wärmegleichgewicht unter mehratomigen Gasmolekülen, *Wien. Ber.* **95** (2), 153.

*BOLTZMANN, L. (1895) *Nature*, **51**, 413 and 581; **52**, 221.

*BOLTZMANN, L. (1896–7) Entgegnung auf die wärmetheoretische Betrachtung des Hrn. Zermelo, *Wied. Ann.* **57**, 773; **60**, 392.

*BOLTZMANN, L. (1897) Über einen mechanischen Satz von Poincaré, *Wien. Ber.* **106**, (2), 12.

*BOLTZMANN, L. (1898) Über die sogenannte *H*-curve, *Math. Ann.* **50**, 325.

BOREL, E. (1906) Sur les principes de la théorie cinétique des gaz, *Ann. Ec. Norm. Sup.* **23** (3), 9.

BOREL, E. (1914) *Introduction géométrique à quelques théories physiques*, Gauthier-Villars, Paris.

BOREL, E. (1925) *Mécanique statistique classique*, Gauthier-Villars, Paris.

BORN, M. (1948) Die Quantenmechanik und der zweite Hauptsatz der Thermodynamik, *Ann. Physik*, **3**, 107.

BORN, M. (1949) The Foundation of the Quantum Statistics, *Nuovo Cimento*, **6** (suppl), 161.

*BORN, M. (1949) *Natural Philosophy of Cause and Chance*, Clarendon Press, Oxford.

BORN, M., and H. S. GREEN (1946) A General Theory of Liquids: The Molecular Distribution Functions, *Proc. Roy. Soc.* A**188**, 10.

BORN, M., and H. S. GREEN (1947a) A General Theory of Liquids: Dynamical Properties, *Proc. Roy. Soc.* A**190**, 455.

*BORN, M., and H. S. GREEN (1947b) Quantum Mechanics of Fluids, *Proc. Roy. Soc.* A**191**, 168.

BORN, M. and H. S. GREEN (1948) The Kinetic Basis of Thermodynamics, *Proc. Roy. Soc.* A**192**, 166.

*BORN, M., and H. S. GREEN (1949) *A General Kinetic Theory of Liquids*, Cambridge University Press.

*BRILLOUIN, L. (1930) *Les statistiques quantiques*, Les Presses Universitaires de France, Paris.

*BRILLOUIN, L. (1953) The Negentropy Principle of Information, *J. Appl. Phys.* **24**, 1152.

*BRILLOUIN, L. (1956) *Science and Information Theory*, Academic Press, New York.

*DE BROGLIE, L. (1930) *Introduction à l'étude de la Mécanique ondulatoire*, Hermann, Paris.

*DE BROGLIE, L. (1932) *Théorie de la quantification dans la Nouvelle Mécanique*, Hermann, Paris.

*DE BROGLIE, L. (1939) *La Mécanique ondulatoire des systèmes de corpuscules*, Gauthier-Villars, Paris.

DE BROGLIE, L. (1948) La statistique des cas purs en mécanique ondulatoire et l'interférence des probabilités, *La Revue scientifique*, **86**, 259.

*DE BROGLIE, L. (1953) *Eléments de théorie des quanta et de Mécanique ondulatoire*, Gauthier-Villars, Paris.

DE BROGLIE, L. (1957) *La Théorie de la Mesure en Mécanique ondulatoire* (*Interprétation usuelle et interprétation causale*), Gauthier-Villars, Paris.

BROUT, R. (1956) Statistical Mechanics of Irreversible Processes. VII. Boltzmann Equation, *Physica*, **22**, 509.

*BROUT, R., and I. PRIGOGINE (1956) Statistical Mechanics of Irreversible Processes. VIII. General Theory of Weakly Coupled Systems, *Physica*, **22**, 621.

*BRUSH, S. G. (1965) *Kinetic Theory*, vol. 1, Pergamon Press, Oxford.

*BRUSH, S. G. (1966) *Kinetic Theory*, vol. 2, Pergamon Press, Oxford.

*BURBURY, S. H. (1899) *A Treatise on the Kinetic Theory of Gases*, Cambridge University Press.

Bibliography

*Burbury, S. H. (1894) On the Law of Distribution of Energy, *Phil. Mag.* **37**, 143.

*Burbury, S. H. (1894) The Second Law of Thermodynamics, *Phil. Mag.* **37**, 574.

*Burbury, S. H. (1894–1895) Discussion between S. H. Burbury, H. Bryan, L. Boltzmann, E. P. Culverwell, G. F. Fitzgerald, A. Schuster, and H. W. Watson, *Nature*, **51** and **52**.

*Burbury, S. H. (1903) On the Variation of Entropy as treated in W. Gibb's Statistical Mechanics, *Phil. Mag.* **6**, 251.

*Burbury, S. H. (1903) On Jeans' Theory of Gases, *Phil. Mag.* **6**, 529.

*Burbury, S. H. (1904) Theory of Diminution of Entropy, *Phil. Mag.* **8**, 43.

*Caldirola, P. (1960) Recenti sviluppi dei metodi ergodici nella meccanica statistica, *Nuovo Cimento* (suppl.), **16**, 50.

*Caldirola, P., and A. Loinger (1959) In *Max Planck-Festschrift*, Deutsche Akad. der Wissenschaften zu Berlin, sec. 3.

Carathéodory, C. (1919) Über den Wiederkehrsatz von Poincaré, *Sitzgs. Preuss. Akad. Wiss.* p. 580.

*Casimir, H. B. G., (1945) On Onsager's Principle of Microscopic Reversibility, *Rev. Mod. Phys.* **17**, 343.

Chandrasekhar, S. (1943) Stochastic Problems in Physics and Astronomy, *Rev. Mod. Phys.* **15**, 1.

Chapman, S., and T. G. Cowling (1953) *The Mathematical Theory of Non-uniform Gases*, 2nd. ed., Cambridge University Press.

Cherry, T. M. (1925) *Proc. Cambridge Phil. Soc.* **22**, 287, 325, 510.

*Clausius, R. (1868 and 1869) *Théorie mécanique de la chaleur* (French translation by F. Folie), Eugène Lacroix, Paris.

*Clausius, R. (1857) Über die Art der Bewegung, welche wir Wärme nennen, *Poggendorf Ann.* **100**, 353.

*Clausius, R. (1858) Über die mittlere Länge der Wege, welche bei der Molekularbewegung gasförmiger Körper von den einzelnen Molekülen zurückgelegt werden; nebst einigen anderen Bemerkungen über die mechanische Wärmetheorie, *Poggendorf Ann.* **105**, 239.

*Clausius, R. (1862) Über die Wärmeleitung gasförmiger Körper, *Poggendorf Ann.* **115**, 1.

*Clausius, R. (1870) Über einen auf die Wärme anwendbaren mechanischen Satz, *Poggendorf Ann.* **141**, 124.

*Clausius, R. (1871) Über die Zurückführung des zweiten Hauptsatzes der mechanischen Wärmetheorie auf allgemeine mechanische Prinzipien, *Poggendorf Ann.* **142**, 433.

Cohen, E. G. D. (1962) *Fundamental Problems in Statistical Mechanics*, North-Holland, Amsterdam.

Cohen, E. G. D., and T. H. Berlin (1960) Note on the Derivation of the Boltzmann Equation from the Liouville Equation, *Physica*, **26**, 717.

Costa de Beauregard, O. (1958) Équivalence entre les deux principes des actions retardées et de l'entropie croissante, *Cahiers de Phys.* **12**, 317.

Costa de Beauregard, O. (1960) Équivalence entre le principe de l'entropie croissante et le principe des ondes retardées, *Rev. Quest. Sc.*, p. 41.

*Costa de Beauregard, O. and B. d'Espagnat (1947) Quelques remarques sur les paradoxes de la Mécanique statistique classique, *Rev. Quest. Sc.*, pp. 351 and 527.

DARWIN, C. G., and R. H. FOWLER (1923) *Proc. Cambridge Phil. Soc.* **21**, 391 and 730.

*DAVYDOV, B. (1947) Quantum Mechanics and Thermodynamic Irreversibility, *J. Phys. U.S.S.R.* **11**, 33.

DELBRÜCK, M., and G. MOLIÈRE (1936) *Abhandl. Preuss. Akad. Wiss. Physik-math. Kl.*, No. 1.

*DESTOUCHES, J. L. (1941) *Corpuscules et systèmes de corpuscules*, Gauthier-Villars, Paris.

DIRAC, P. A. M. (1929) The Basis of Statistical Quantum Mechanics, *Proc. Camb. Phil. Soc.* **25**, 62.

*DIRAC, P. A. M. (1930–1) Quelques problèmes de Mécanique quantique, *Ann. Inst. H. Poincaré*, **1**, 357.

DIRAC, P. A. M. (1947) *The Principles of Quantum Mechanics*, 3rd ed., Clarendon Press, Oxford.

DOETSCH, G. (1943) *Laplace Transformations*, Dover, New York.

*DOOB, J. L. (1953) *Stochastic Processes*, John Wiley, New York.

*DUNFORD, N., and D. S. MILLER (1946) On the Ergodic Theorem, *Trans. Amer. Math. Soc.* **60**.

*EHRENFEST, P. (1906) Zur Planckschen Strahlungstheorie, *Physik. Z.* **7**, 528.

*EHRENFEST, P. (1911) Welche Züge der Lichtquantenhypothese spielen in der Theorie der Wärmestrahlung eine wesentliche Rolle? *Ann. Physik*, **36**, 91.

EHRENFEST, P. (1914) Zum Boltzmannschen Entropie. Wahrscheinlichkeitstheorem, *Physik. Z.* **15**, 657.

*EHRENFEST, P. (1916) Adiabatische Invarianten und Quantentheorie, *Ann. Physik*, **51**, 327.

*EHRENFEST, P. and T. (1906) Zur Theorie der Entropiezunahme in der statistischen Mechanik von Gibbs, *Wien. Ber.* **115** (2), 89.

EHRENFEST, P. and T. (1907) Über zwei bekannte Einwände gegen das Boltzmanns *H*-theorem, *Physik. Z.* **8**, 311.

EHRENFEST, P. and T. (1911) Begriffliche Grundlagen der statistischen Auffassung in der Mechanik, *Encykl. math. Wiss.* **4**, No. 32. (English translation: *The Conceptual Foundations of the Statistical Approach in Mechanics*, Cornell University, Press, Ithaca, New York, 1959).

*EINSTEIN, A. (1902) Kinetische Theorie des Wärmegleichgewichts und des Zweiten Hauptsatzes der Thermodynamik, *Ann. Physik*, **9**, 417.

*EINSTEIN, A. (1903) Eine Theorie der Grundlagen der Thermodynamik, *Ann. Physik*, **11**, 170.

*EINSTEIN, A. (1905) Über die von der molekular kinetische Theorie der Wärme geforderte Bewegung von in ruhenden Flüssigkeiten suspendierten Teilchen, *Ann. Physik*, **17**, 549.

*EINSTEIN, A. (1906) Zur Theorie der Brownschen Bewegung, *Ann. Physik*, **19**,371.

*EINSTEIN, A. (1910) Theorie der Opaleszenz von homogenen Flüssigkeiten und Flüssigkeitsgemischen in der Nähe des kritischen Zustandes, *Ann. Physik*, **33**, 1275.

EINSTEIN, A. (1956) *Investigations on the Theory of the Brownian Movement*, 2nd ed., Dover, New York.

*EKSTEIN, H. (1957) Ergodic Theorem for Interacting Systems, *Phys. Rev.* **107**,333.

ENSKOG, D. (1917) The Kinetic Theory of Phenomena in Fairly Rare Gases, Dissertation, Upsala.

Bibliography

*EPSTEIN, P. S. (1936) In *Commentary on the Scientific Writings of J. Willard Gibbs*, Yale University Press, New Haven.

FANO, U. (1957) Description of States in Quantum Mechanics by Density Matrix and Operator Techniques, *Rev. Mod. Phys.* **29**, 74.

*FARINELLI, U., and A. GAMBA (1956) Entropy in Quantum Mechanics, *Nuovo Cimento*, **3**, 1033.

*FARQUHAR, I. E. (1964) *Ergodic Theory in Statistical Mechanics*, John Wiley, New York.

FARQUHAR, I. E., and P. T. LANDSBERG (1957) On the Quantum-Statistical Ergodic and *H*-theorems, *Proc. Roy. Soc.* A**239**, 134.

*FELLER, W. (1940) The Integrodifferential Equations of Completely Discontinuous Markov Processes, *Trans. Amer. Math. Soc.* **48**, 488.

*FELLER, W. (1957) *An Introduction to Probability Theory and its Applications*, John Wiley, New York.

*FERMI, E. (1923) *Physik, Z.* **24**, 261.

*FERMI, E. (1926) Zur Quantelung des idealen einatomigen Gases, *Z. Physik*, **36**, 902.

FIERZ, M. (1955) Der Ergodensatz in der Quantenmechanik, *Helv. Phys. Acta*, **28**, 705.

FOWLER, R. H. (1936) *Statistical Mechanics*, Cambridge University Press.

FOWLER, R. H., and E. A. GUGGENHEIM (1939) *Statistical Thermodynamics*, Cambridge University Press.

*GAMBA, A. (1955) Thermodynamics and Quantum Mechanics, *Nuovo Cimento*, **1**, 358.

GIBBS, J. W. (1902) *Elementary Principles in Statistical Mechanics*, Yale University Press, New Haven.

GOLDEN, S., and H. C. LONGUET-HIGGINS (1960) *J. Chem. Phys.* **33**, 1479.

GRAD, H. (1949) On the Kinetic Theory of Rarefied Gases, *Comm. Pure and Appl. Math.* **2**, 331.

GRAD, H. (1952) Statistical Mechanics, Thermodynamics and Fluid Dynamics of Systems with an Arbitrary Number of Integrals, *Comm. Pure and Appl. Math.* **5**, 455.

GRAD, H. (1958) Principles of the Kinetic Theory of Gases, *Handbuch der Physik*, **12**, 205, Springer-Verlag, Berlin.

*GRAD, H. (1960) On Molecular Chaos and the Kirkwood Superposition Hypothesis, *J. Chem. Phys.* **33**, 1342.

GREEN, C. D., and D. TER HAAR (1955) Fluctuations in Simple Mechanical Models, *Physica*, **21**, 63.

GREEN, H. S. (1947) A General Theory of Liquids. II. Equilibrium Properties, *Proc. Roy. Soc.* A**189**, 103.

*GREEN, H. S. (1952) *The Molecular Theory of Fluids*, North-Holland, Amsterdam.

*GREEN, M. S. (1956a) Nat. Bur. of Standards, Rep. 3327.

GREEN, M. S. (1956b) Boltzmann Equation from the Statistical Mechanical Point of View, *J. Chem. Phys.* **25**, 836.

*GROENEWOLD, H. J. (1952) Information in Quantum Measurements, *Proc. Kon. Ned. Akad. v. Wet* (*Amsterdam*) B**55**, 219.

*DE GROOT, S. R. (1951) *Thermodynamics of Irreversible Processes*, North-Holland, Amsterdam.

TER HAAR, D. (1954) *Elements of Statistical Mechanics*, Rinehart, New York.
TER HAAR, D. (1955) Foundations of Statistical Mechanics, *Rev. Mod. Phys.* **27**, 289.
TER HAAR, D. (1961) *Repts. Progr. Phys.* **24**, 304.
TER HAAR, D. (1966) *Elements of Thermostatistics*, Holt, Rinehart and Winston, New York.
TER HAAR, D., and C. D. GREEN (1953) The Statistical Aspect of Boltzmann's *H-Theorem*, *Proc. Phys. Soc.* A**66**, 153.
TER HAAR, D., and C. D. GREEN (1955) The Ehrenfests' Wind-Wood Model in Two Dimensions, *Proc. Cambridge Phil. Soc.* **51**, 141.
*HADAMARD, J. (1906) La Mécanique statistique, *Amer. Math. Soc. Bull.* **2** (2), 194.
*HALMOS, P. R. (1946) An Ergodic Theorem, *Proc. Nat. Acad. Sci. U.S.A.* **32**, 156.
*HALMOS, P. R. (1950) *Measure Theory*, Van Nostrand, New York.
HALMOS, P. R. (1956) *Lectures on Ergodic Theory*, Chelsea Pub. Cy., New York.
*HARTMAN, S., E. MARCZEWSKI and C. RYLL-NARDZEWSKI (1951) Théorèmes ergodiques et leurs applications, *Coll. Math.* **2**.
HILBERT, D. (1912) Begründung der Kinetischen Gastheorie, *Math. Ann.* **72**, 562.
HIRSCHFELDER, J. O., C. F. CURTISS and R. B. BIRD (1954) *Molecular Theory of Gases and Liquids*, John Wiley, New York.
*HÖFLICH, P. (1927) Wahrscheinlichkeitstheoretische Begründung der Ergodenhypothese, *Z. Physik*, **41**, 636.
*HOPF, E. (1932a) On the Time Average Theorem in Dynamics, *Proc. Nat. Acad. Sci. U.S.A.* **18**, 93.
HOPF, E. (1932b) Complete Transitivity and the Ergodic Principle, *Proc. Nat. Acad. Sci. U.S.A.* **18**, 204.
HOPF, E. (1932c) Proof of Gibbs' Hypothesis on the Tendency toward Statistical Equilibrium, *Proc. Nat. Acad. Sci. U.S.A.* **18**, 333.
HOPF, E. (1937) *Ergodentheorie*, Verlag-Springer, Berlin.
VAN HOVE, L. (1955) Quantum Mechanical Perturbations Giving Rise to a Statistical Transport Equation, *Physica*, **21**, 517.
*VAN HOVE, L. (1957a) Statistical Mechanics; A Survey of Recent Lines of Investigation, *Rev. Mod. Phys.* **29**, 200.
VAN HOVE, L. (1957b) The Approach to Equilibrium in Quantum Statistics, *Physica*, **23**, 441.
VAN HOVE, L. (1959) The Ergodic Behaviour of Quantum Many-Body Systems, *Physica*, **25**, 268.
*HUREWICZ, W. (1944) Ergodic Theorem without Invariant Measure, *Ann. Math.* **45**, 192.
HUSIMI, K. (1940) Some Formal Properties of the Density Matrix, *Proc. Phys. Math. Soc. Japan*, **22**, 264.
*JANCEL, R. (1955a) Sur la théorie ergodique en Mécanique quantique, *C. R. Acad. Sci.* **240**, 1693.
*JANCEL, R. (1955b) Sur l'hypothèse fondamentale de la Mécanique statistique quantique, *C. R. Acad. Sci.* **240**, 1864.
JANCEL, R. (1956) Sur le théorème *H* en Mécanique quantique, *C. R. Acad. Sci.* **242**, 1268.

Bibliography

*JANCEL, R. (1957) Les fondements de la Mécanique statistique quantique, Thèse de Doctorat, Paris.

JANCEL, R. (1960a) Propriétés des observables macroscopiques et théorie ergodique quantique, *C. R. Acad. Sci.* **250**, 671.

JANCEL, R. (1960b) Comparaison entre les aspects classique et quantique de la théorie ergodique, *C. R. Acad. Sci.* **250**, 2152.

JANCEL, R., and T. KAHAN (1955) Chaînes de mesures quantiques et principe de néguentropie, *C. R. Acad. Sci.* **240**, 54.

JANCEL, R., and T. KAHAN (1963) *Electrodynamique des Plasmas*, vol. I, Dunod, Paris.

JAYNES, E. T. (1957) Information Theory and Statistical Mechanics, I and II, *Phys. Rev.* **106**, 620; **108**, 171.

*JEANS, J. (1903) Kinetic Theory of Gases Developed from a New Standpoint, *Phil. Mag.* **5**, 597.

*JEANS, J. (1925) *Dynamical Theory of Gases*, Cambridge University Press.

*JORDAN, P. (1933) *Statistische Mechanik auf quantentheoretischer Grundlage*, Vieweg, Braunschweig.

JORDAN, P., and O. KLEIN (1927) Zum Mehrkörperproblem der Quantentheorie, *Z. Physik*, **45**, 751.

JORDAN, P., and E. P. WIGNER (1928) Über das Paulische Äquivalenzverbot, *Z. Physik*, **47**, 631.

KAC, M. (1947a) On the Notion of Recurrence in Discrete Stochastic Processes, *Bull. Amer. Math. Soc.*, **53**, 1002.

KAC, M. (1947b) Random Walk and the Theory of Brownian Motion, *Amer. Math. Monthly*, **54**, 369.

*KAC, M. (1954) Foundations of Kinetic Theory, Lecture Notes, Department of Math., Cornell University, New York.

KAC, M. (1959) *Probability and Related Topics in Physical Sciences*, Interscience, New York.

VAN KAMPEN, N. G. (1954) Quantum Statistics of Irreversible Processes, *Physica*, **20**, 603.

*VAN KAMPEN, N. G. (1958) Correspondence Principle in Quantum Statistics, *Proc. Int. Symposium in Transport Processes in Statistical Mechanics* (Brussels, 1956), Interscience, New York.

*KEMBLE, E. C. (1939a) Fluctuations, Thermodynamic Equilibrium and Entropy, *Phys. Rev.* **56**, 1013.

*KEMBLE, E. C. (1939b) The Quantum Mechanical Basis of Statistical Mechanics, *Phys. Rev.* **56**, 1146.

*KENNARD, E. H. (1938) *Kinetic Theory of Gases*, McGraw-Hill, New York.

*KHINCHIN, A. I. (1932) Zu Birkhoffs Lösung des Ergodenproblems, *Math. Ann.* **107**, 485.

*KHINCHIN, A. I. (1933) The Method of Spectral Reduction in Classical Dynamics, *Proc. Nat. Acad. Sci. U.S.A.* **19**, 567.

KHINCHIN, A. I. (1949) *Mathematical Foundations of Statistical Mechanics* (translated from Russian by G. Gamow), Dover, New York.

KHINCHIN, A. I. (1950) On the Analytical Apparatus of Physical Statistics, *Trudy Steklov Mat. Inst.* **33**, 3.

KHINCHIN, A. I. (1960) *Mathematical Foundations of Quantum Statistics* (translation of the first Russian edition, 1951), Graylock Press, Albany.

KIKUCHI, S., and L. NORDHEIM (1930) Über die kinetische Fundamentalgleichung in der Quantenstatistik, *Z. Physik*, **60**, 652.

KIRKWOOD, J. G. (1946) The Statistical Mechanical Theory of Transport Processes. I. General Theory, *J. Chem. Phys.* **14**, 180; (1947) II. Transport in Gases, *J. Chem. Phys.* **15**, 72.

*KIRKWOOD, J. G. (1949) The Statistical Mechanical Theory of Irreversible Processes, *Nuovo Cimento* (Suppl.), **6**, 233.

KIRZHNITS, D. A. (1967) *Field Theoretical Methods in Many-Body Systems*, Pergamon Press, Oxford.

KLEIN, M. J. (1952) The Ergodic Theorem in Quantum Statistical Mechanics, *Phys.. Rev.* **87**, 111.

*KLEIN, M. J. (1956) Entropy and the Ehrenfest Urn Model, *Physica*, **22**, 569.

KLEIN, O. (1931) Zur quantenmechanischen Begründung des zweiten Hauptsatzes der Wärmelehre, *Z. Physik*, **72**, 767.

KOHLRAUSCH, K., and E. SCHRÖDINGER (1926) Das Ehrenfestsche Modell der *H*-Kurve, *Physik. Z.* **27**, 306.

KOLMOGOROV, A. N. (1954) La théorie générale des systémes dynamiques et la mécanique classique, Amsterdam Congress, Vol. 1, 315; *Math. Rev.* **20**, 4066 (1959).

KOLMOGOROV, A. N. (1956) *Foundations of Probability Theory*, Chelsea, New York.

KOLMOGOROV, A. N. (1959) On the entropy per unit time as a metric invariant of automorphisms, *Dokl. Akad. Nauk* **124**, 754.

KOLMOGOROV, A. N. (1960) A new metric invariant of transitive systems and automorphisms of Lebesgue spaces, *Math. Rev.* **21**, 2035a.

KOOPMAN, B. O. (1931) Hamiltonian Systems and Transformations in Hilbert Space, *Proc. Nat. Acad. Sci. U.S.A.* **17**, 315.

KOUCHNIRENKO, A. G. (1965) An estimate from above for the entropy of a classical system, *Sov. Math.-Doklady* **6**, 360.

*KRAMERS, H. A. (1938) Didaktisches zur Verwendung der grand Ensembles in der Statistik, *Proc. Kon. Ned-Akad. Wet* (*Amsterdam*), **41**, 10.

*KRAMERS, H. A. (1938) *Grundlagen der Quantentheorie*, Akad. Verlag, Leipzig (English translation published by North Holland, Amsterdam).

*KRAMERS, H. A. (1949) On the Behaviour of a Gas Near a Wall, *Nuovo Cimento* (Suppl.), **6**, 297.

KUBO, R. (1957) *J. Phys. Soc. Japan*, **12**, 570.

KURTH, R. (1958) *Z. Naturf.* **13**a, 110.

*LANDAU, L. D., and E. M. LIFSHITZ (1958) *Statistical Physics*, Pergamon Press, London.

LEBOWITZ, J. L., and P. G. BERGMANN (1957) *Ann. Phys.* **1**, 1.

LEWIS, R. M. (1960) *Arch. Rational Mech. Anal.* **5**, 355.

*LIÉNARD, A. (1903) Sur la théorie cinétique des gaz, *J. Phys. Rad.* **2** (4), 677.

*LOÈVE, M. (1951) On Almost Sure Convergence, *Proc. Sec. Berkeley Symp. on Stat. and Prob.*

*LOÈVE. M. (1960) *Probability Theory*, 2nd ed. Van Nostrand, New York.

*LOINGER, A. (1958) *Rend. Sem. Mat. Fis. di Milano*, **29**, 3.

*LONDON, F., and E. BAUER (1939) La théorie de l'observation en Mécanique quantique, *Act. Sci. Ind.* No. 775, Hermann, Paris.

LORENTZ, H. A. (1905) The Motion of Electrons in Metallic Bodies, *Proc. Amsterdam Acad.* **7**, 438, 585 and 684.
LORENTZ, H. A. (1907) *Abhandlungen über theoretische Physik*, Vol. XI, § 79, p. 132, Teubner, Leipzig.
LORENTZ, H. A. (1916) *Les théories statistiques en Thermodynamique*, Leipzig.
LOSCHMIDT, J. (1876 and 1877) Über den Zustand des Wärmegleichgewichtes eines Systems von Körpern mit Rücksicht auf die Schwerkraft, *Wien. Ber.* **73** (2), 135 and 366; **75** (2), 67.
*LUDWIG, G. (1953) Der Messprozess, *Z. Physik*, **135**, 483.
LUDWIG, G. (1954) *Die Grundlagen der Quantentheorie*, Springer-Verlag, Berlin.
LUDWIG, G. (1958a, b) Zum Ergodensatz und zum Begriff der makroskopischen Observablen, I and II, *Z. Physik*, **150**, 346; **152**, 98.
LUDWIG, G. (1961) Axiomatic Quantum Statistics of Macroscopic Systems in "Ergodic Theories" (14th Course of the Varenna Summer School), Academic Press, New York.
MASSIGNON, D. (1957) *Mécanique statistique des fluides*, Dunod, Paris.
MATHEWS, P. M., I. I. SHAPIRO and D. L. FALKOFF (1960) Stochastic Equations for Non-Equilibrium Processes, *Phys. Rev.* **120**, 1.
MATSUBARA, T. (1955) A New Approach to Quantum-Statistical Mechanics, *Prog. Theor. Phys.* **14**, 351.
MAXWELL, J. C. (1867) On the Dynamical Theory of Gases. *Phil. Trans. Roy. Soc.* **157**, 49 (*Scientific Papers* **2**, 26).
MAXWELL, J. C. (1879) On Stresses in Rarefied Gases arising from Inequalities of Temperature, *Phil. Trans. Roy. Soc.* **170**, 231 (*Scientific Papers* **2**, 681).
*MAXWELL, J. C. (1879) On Boltzmann's Theorem on the Average Distribution of Energy in a System of Material Points, *Cambridge Phil. Soc. Trans.* **12**, 547 (*Scientific Papers* **2**, 713).
MAYER, J. E. (1937) *J. Chem. Phys.* **5**, 67. (1961) *J. Chem. Phys.* **34**, 1207.
*MAYER, J. E., and M. G. (1940) *Statistical Mechanics*, John Wiley, New York.
MAZUR, P., and J. v. d. LINDEN (1963) *J. Math. Phys.* **4**, 271.
*VON MISES, R. (1920) Ausschaltung der Ergodenhypothese in der physikalischen Statistik, *Physik. Z.* **21**, 225 and 256.
*VON MISES, R. (1936) Les lois de probabilité pour les fonctions statistiques, *Ann. Inst. H. Poincaré*, **6**, 185.
*VON MISES, R. (1951) *Wahrscheinlichkeit, Statistik und Wahrheit*, 3rd ed., Springer, Vienna.
MONTROLL, E. W. (1960) *Boulder Lectures in Theoretical Physics*, **3**, 221.
MONTROLL, E. W. (1962) In *Fundamental Problems in Statistical Mechanics* (E.G.D. Cohen editor), North Holland, Amsterdam, p. 230.
MORI, H., and S. ONO (1952) The Quantum-Statistical Theory of Transport Phenomena, I., *Prog. Theor. Phys.* **8**, 327.
MOYAL, J. E. (1949) Stochastic Processes and Statistical Physics, *J. Roy. Stat. Soc.*. (Symposium on Stochastic Processes) **11**, 150.
MÜNSTER, A. (1959) Prinzipien der statistischen Mechanik, *Handbuch der Physik*, **3** (2), Springer-Verlag, Berlin.
NAKAJIMA, S. (1958) *Progr. Theor. Phys.* **20**, 948.
VON NEUMANN, J. (1927) *Nachr. Akad. Wiss. Göttingen, Math. Physik. Kl.*, pp. 245 and 271.

VON NEUMANN, J. (1929) Beweis des Ergodensatzes und des *H*-Theorems in der Neuen Mechanik, *Z. Physik*, **57**, 30.

VON NEUMANN, J. (1932a) Proof of the Quasi-ergodic Hypothesis, *Proc. Nat. Acad. Sci. U.S.A.* **18**, 70.

VON NEUMANN, J. (1932b) Physical Applications of the Ergodic Hypothesis, *Proc. Nat. Acad. Sci. U.S.A.* **18**, 263.

VON NEUMANN, J. (1932c) Über einen Satz von Herrn M. Stone, *Ann. Math.* **33**, 567.

*VON NEUMANN, J. (1932d) Zur Operatorenmethode in der klassischen Mechanik, *Ann. Math.* **33**, 587 and 789.

VON NEUMANN, J. (1932e) *Mathematische Grundlagen der Quantenmechanik*, Springer, Berlin. [English translation published by Princeton University Press].

NORDHEIM, L. (1928a) On the Kinetic Method in the New Statistics and its Application in the Electron Theory of Conductivity, *Proc. Roy. Soc.* A **119**, 689.

NORDHEIM, L. (1928b) *Proc. Roy. Soc.* A **121**, 626.

ONO, S. *Mem. Fac. Eng. Kyushu Univ.* **11**, 125.

*ONO, S. (1958) The Boltzmann Equation in Quantum-Statistical Mechanics, *Proc. Int. Symposium in Transport Processes in Stat. Mechanics* (Brussels, 1956), Interscience, New York.

*ONSAGER, L. (1931) Reciprocal Relations in Irreversible Processes, *Phys. Rev.* **37**, 405 and **38**, 2265.

*OXTOBY, J. C. (1948) On the Ergodic Theorem of Hurewicz, *Ann. Math.* **49**, 872.

*OXTOBY, J. C., and S. M. ULAM (1939) On the Existence of a Measure Invariant under a Transformation, *Ann. Math.* **40**, 560.

*OXTOBY, J. C., and S. M. ULAM (1941) Measure Preserving Homeomorphism, *Ann. Math.* **42**, 874.

PAULI, W. (1928) Über das *H*-theorem vom Anwachsen der Entropie vom Standpunkt der neuen Quantenmechanik. In *Probleme der Modernen Physik*, Sommerfeld-Festschrift (Hirzel, Leipzig), p. 30.

*PAULI, W. (1933) *Handbuch der Physik*, 2nd edition, **24** (1), 149, Springer, Berlin.

PAULI, W. (1949) Note on the Conference of Prof. M. Born, *Nuovo Cimento*, (Suppl.) **6**, 169.

PAULI, W., and M. FIERZ (1937) Über das *H*-theorem in der Quantenmechanik, *Z. Physik*, **106**, 572.

PERCIVAL, I. C. (1961) *J. Math. Phys.* **2**, 235.

*PERRIN, F. (1939) *Mécanique statistique quantique*, Gauthier-Villars, Paris.

*PLANCHEREL, M. (1913) Beweis der Unmöglichkeit ergodischer mechanischer Systeme, *Ann. Physik*, **42**, 1061.

*PLANCK, M. (1904) Über die mechanische Bedeutung der Temperatur und der Entropie (*Festschrift L. Boltzmann*), J. A. Barth, Leipzig, p. 113.

*PLANCK, M. (1913) *Leçons de Thermodynamique* (translation of the third German edition), Hermann, Paris.

*PLANCK, M. (1930) *Einführung in die Theorie der Wärme*, Leipzig.

POINCARÉ, H. (1890) Sur le problème des trois corps et les équations de la dynamique, *Acta Math.* **13**, 1.

POINCARÉ, H. (1892) *Méthodes Nouvelles de la Mécanique Celeste*, Gauthier-Villars, Paris.

Bibliography

*Poincaré, H. (1906) Réflexions sur la théorie cinétique des gaz, *J. Phys. Rad.* **5** (4), 369.

Prigogine, I. (1962) *Non-Equilibrium Statistical Mechanics*, John Wiley, New York.

Prigogine, I., and R. Balescu (1959) Irreversible Processes in Gases; I. The Diagram Technique, *Physica*, **25**, 281; II. The Equations of Evolution, *Physica*, **25**, 302.

Prigogine, I., and R. Balescu (1960) *Physica*, **26**, 145.

Prigogine, I., and P. Résibois (1961) *Physica*, **27**, 629.

Prosperi, G. M., and A. Scotti (1959) Ergodicity Theorem in Quantum Mechanics; Evaluation of the Probability of an Exceptional Initial Condition, *Nuovo Cimento*, **13**, 1007.

Prosperi, G. M., and A. Scotti (1960) Ergodicity Conditions in Quantum Mechanics, *J. Math. Phys.* **1**, 218.

Résibois, P. (1963) *Physica*, **29**, 721.

Riesz, F. (1945) Sur la théorie ergodique, *Comm. Math. Helv.* **17**, 221.

*Riesz, F., and B. Sz. Nagy (1952) *Leçons d'analyse fonctionelle*, Acad. Sc. de Hongrie.

Rohlin, V. A. (1949) In general a measure preserving transformation is not mixing, *Dokl. Akad. Nauk* **13**, 329.

Rohlin, V. A. (1962) On the fundamental ideas of measure theory, *Am. Math. Soc. Transl.* (1) **10**, 1.

Rohlin, V. A., and Ya. Sinai (1961) Construction and properties of invariant measurable partitions, *Sov. Math.-Doklady* **2**, 1611.

Rosenfeld, L. (1952) Statistical Mechanics (Lecture Notes of the Les Houches Summer School in Theoretical Physics).

Rosenfeld, L. (1955) On the Foundations of Statistical Thermodynamics, *Acta Phys. Polon.* **14**, 3.

*Rosenthal, A. (1913) Beweis der Unmöglichkeit ergodischer Gassysteme, *Ann. Physik*, **42**, 796.

*Rosenthal, A. (1914) Aufbau der Gastheorie mit Hilfe der Quasiergodenhypothese, *Ann. Physik*, **43**, 894.

*Rushbrooke, G. S. (1949) *Introduction to Statistical Mechanics*, Oxford, Clarendon Press.

Schaefer, C. (1955) *Einführung in die Theoretische Physik*, vol. 2, W. de Gruyter-Berlin.

*Schrödinger, E. (1948) *Statistical Thermodynamics*, Cambridge University Press.

Siegert, A. J. F. (1949) On the Approach to Statistical Equilibrium, *Phys. Rev.* **76**, 1708.

Sinai, Ya. (1960) On the concept of entropy of a dynamical system, *Math. Rev.* **21**, 2036a.

Sinai, Ya. (1961a) Dynamical systems with countably multiple Lebesgue spectra, *Am. Math. Soc. Transl.* (2) **39**, 83.

Sinai, Ya. (1961b) The central theorem for geodesic flows on manifolds of constant negative curvature, *Sov. Math.-Doklady* **1**, 983.

Sinai, Ya. (1961c) Geodesic flows on compact surfaces of negative curvature, *Sov. Math.-Doklady* **2**, 106.

Sinai, Ya. (1963a) Some remarks on the spectral properties of ergodic dynamical systems, *Russian Math. Surveys* **18**, No 5, p. 37.

SINAI, YA. (1963b) Properties of spectra of ergodic dynamical systems, *Sov. Math.-Doklady* **4**, 875.

SINAI, YA. (1963c) On the foundations of the ergodic hypothesis for a dynamical system of statistical mechanics, *Sov. Math.-Doklady* **4**, 1818.

*VON SMOLUCHOWSKI, M. (1904) Über Unregelmässigkeiten in der Verteilung von Gasmolekülen (*Festschrift für L. Boltzmann*), Leipzig, p. 626.

*VON SMOLUCHOWSKI, M. (1906) Sur le chemin moyen parcouru par les molécules, *Cracovie*, *Bull. de l'Acad.*, p. 202.

*VON SMOLUCHOWSKI, M. (1906) Kinetische Theorie der Brownschen Bewegung, *Ann. Physik*, **21**, 756.

*VON SMOLUCHOWSKI, M. (1908) Molekular-kinetische Theorie der Opaleszenz im kritischen Zustande, sowie einiger verwandter Erscheinungen, *Ann. Physik*, **25**, 205.

*VON SMOLUCHOWSKI, M. (1912) Experimentell nachweisbare, der üblichen Thermodynamik widersprechende Molekular-Phänomene, *Physik Z.* **13**, 1069.

*VON SMOLUCHOWSKI, M. (1916a, b) *Physik. Z.* **17**, 557 and 587.

*SOMMERFELD, A. (1919) *Atombau und Spektrallinien*, Vieweg, Braunschweig.

*SOMMERFELD, A. (1956) Thermodynamics and Statistical Mechanics, *Lectures on Theoretical Physics*, vol. 5, Academic Press, New York.

STECKI, J., and M. S. TAYLOR (1965) *Rev. Mod. Phys.* **37**, 762.

STONE, M. H. (1930) Linear Transformations in Hilbert Space. III., *Proc. Nat. Acad.. Sci. U.S.A.* **16**, 172.

*STUECKELBERG, E. C. G. (1952) Théorème *H* et unitarité de *S*, *Helv. Phys. Acta*, **25**, 577.

*SZILARD, L. (1929) Über die Entropieverminderung in einem thermodynamischen System bei Eingriffen intelligenter Wesen, *Z. Physik*, **53**, 840.

TERLETSKY, J. P. (1949) *Les lois dynamiques et statistiques de la Physique*, Moscow University Publication.

*THOMSON, W. (Lord Kelvin) (1882) *Mathematical and Physical Papers*, Cambridge University Press.

*THOMSON, W. (1893) *Constitution de la matière* (transl. Lugol), with a Preface by M. Brillouin, Gauthier-Villars, Paris.

TOLMAN, R. C. (1938) *The Principles of Statistical Mechanics*, Oxford University Press.

TRUESDELL, C. (1961) In *Ergodic Theories* (14th Course of the Varenna Summer School), Academic Press, New York.

TRUESDELL, C., and D. MORGENSTERN (1958) Neuere Entwicklungen in der klassischen statistischen Mechanik und in der kinetischen Gastheorie, *Ergebnisse exact. Naturw.* **13**, 286.

UEHLING, E. A., and G. E. UHLENBECK (1933) Transport Phenomena in Einstein–Bose and Fermi–Dirac Gases, *Phys. Rev.* **43**, 552.

*UHLENBECK G. E. (1955) The Statistical Mechanics of Non-Equilibrium Phenomena (Lecture Notes of the Les Houches Summer School).

*UHLENBECK, G. E., and L. S. ORNSTEIN (1930) On the Theory of the Brownian Motion, *Phys. Rev.* **36**, 823.

UHLHORN, U. (1960) *Arkiv. Fys.* **17**, 193.

UHLHORN, U. (1961) In *Ergodic Theories* (14th Course of the Varenna Summer School), Academic Press, New York.

URSELL, H. D. (1927) *Proc. Camb. Phil. Soc.* **23**, 685.

Bibliography

*VONSOVSKY, S. (1946) Derivation of Fundamental Kinetic Equations in Quantum Mechanics, *J. Phys. U.S.S.R.* **10**, 367.

WANG, M. C., and G. E. UHLENBECK (1945) On the Theory of the Brownian Motion, *Rev. Mod. Phys.* **17**, 323.

*WANNIER, G. H. (1965) Quantum-mechanical proof of the second law, *Am. J. Phys.* **33**, 222.

WATANABE, S. (1935) Le deuxieme théorème de la Thermodynamique et la Mécanique ondulatoire, *Act. Sci. Ind.* No. 308, Hermann, Paris.

WATANABE, S. (1951) Reversibility of Quantum Electrodynamics, *Phys. Rev.* **84**, 1008.

WATANABE, S. (1955a, b, c) Symmetry of Physical Laws, *Rev. Mod. Phys.* **27**, 26, 40 and 179.

*WERGELAND, H. (1958) Simple proof of the ergodic theorem, *Acta Chem. Scand.* **12**, 1117.

WHITTAKER, E. (1943) *Proc. Phys. Soc.* **55**, 459.

*WIENER, N. (1948) *Cybernetics*, John Wiley, New York.

WIGNER, E. P. (1932) On the Quantum Correction for Thermodynamic Equilibrium, *Phys. Rev.* **40**, 749.

WINTNER, A. (1941) *The Analytical Foundations of Celestial Mechanics*, Princeton Univ. Press.

YOSIDA, K., and S. KAKUTANI (1939) Birkhoff's Ergodic Theorem and the Maximal Ergodic Theorem, *Proc. Imp. Acad. Tokyo*, **15**, 165.

*YOSIDA, K., and S. KAKUTANI (1941) Operator-Theoretical Treatment of Markov's Process and Mean Ergodic Theorem, *Ann. Math.* **42**, 188.

YVON, J. (1935) Statistique des fluides et l'équation d'état, *Act. Sci. Ind.* No. 203, Hermann, Paris.

YVON, J. (1937a) Recherches sur la théorie cinétique des liquides, I. Fluctuations en densité, *Act. Sci. Ind.* No. 542, Hermann, Paris.

YVON, J. (1937b) Recherches sur la théorie cinétique des liquides, II. La propagation et la diffusion de la lumière, *Act. Sci. Ind.* No. 543, Hermann, Paris.

*YVON, J. (1958) Mise en équilibre d'un petit système couplé à un thermostat, *Proc. Int. Symposium in Transport Processes in Statistical Mechanics* (Brussels, 1956), Interscience, New York.

*YVON, J. (1969) *Correlations and Entropy in Classical Statistical Mechanics*, Pergamon Press, Oxford.

ZERMELO, E. (1896a) Über einen Satz der Dynamik und die mechanische Wärmetheorie, *Wied. Ann.* **57**, 485.

ZERMELO, E. (1896b) Über mechanische Erklärungen irreversibler Vorgänge. Eine Antwort auf Hrn. Boltzmanns "Entgegnung", *Wied. Ann.* **59**, 793.

ZWANZIG, R. (1960) *Boulder Lectures in Theoretical Physics*, **3**, 106.

ZWANZIG, R. (1964) *Physica*, **30**, 1109.

Index

Index

Index

Index

Index